T0310220

ELECTROMAGNETIC COMPATIBILITY

ELECTROMAGNETIC COMPATIBILITY

Analysis and Case Studies in Transportation

DONALD G. BAKER

WILEY

Published by John Wiley & Sons, Inc., Hoboken, New Jersey
Published simultaneously in Canada

For general information on our other products and services or for technical support, please contact our
Customer Care Department within the United States at (800) 762-2974, outside the United States at
(317) 572-3993 or fax (317) 572-4002.

Wiley also publishes its books in a variety of electronic formats. Some content that appears in print may
not be available in electronic formats. For more information about Wiley products, visit our web site at
www.wiley.com.

Library of Congress Cataloging-in-Publication Data

Baker, Donald G., 1935– author.
  Electromagnetic compatibility : analysis and case studies in transportation / Donald G. Baker.
    pages cm
  Includes bibliographical references and index.
  ISBN 978-1-118-98539-7 (cloth)
1. Electromagnetic compatibility. 2. Transportation–Case studies. I. Title.
  TK7867.2.B35 2015
  629.04'6015376–dc23

                                                                    2015021088

Cover image courtesy of miluxian/Getty.

Set in 10/12pt Times by SPi Global, Pondicherry, India

Printed in the United States of America

10 9 8 7 6 5 4 3 2 1

1   2016

Dedicated to:
My wife Barbara
Daughters Tricia and Stephanie
Grandchildren Aidan, Addie, Evan and Marlaina

CONTENTS

Preface   xi

About the Author   xiii

About the Companion Website   xv

1 Introduction   1

    1.1   Introduction, 1
    1.2   Definitions of Commonly Used Terms, 2
    1.3   Book Sections and Content Overview, 8
    1.4   Regulations, 10
    1.5   Background, 16
    1.6   EMC Testing Methods for FCC Part 15 Radiation
          Measurements, 17
    1.7   Canadian Regulations, 24
    1.8   European Union Regulations, 24
    1.9   Review Problems, 57
    1.10 Answers to Review Problems, 57

2 Fundamentals of Coupling Culprit to Victim   59

    2.1   Radiation Effects on Equipment and Devices, 59
    2.2   Various Types of Emission Coupling, 61
    2.3   Intermodulation, 64
    2.4   Common Mode Rejection Ratio, 67
    2.5   Susceptibility and Immunity, 69
    2.6   Filters for EMC, 79
    2.7   Lightning Stroke Analysis, 81

2.8   Skin Effect in Wire, 83
2.9   Conclusion, 86
2.10   Review Problems, 86
2.11   Answers to Review Problems, 88

3 Introduction to Electromagnetic Fields   **91**

3.1   An Introduction to Electromagnetic Fields, 91
3.2   Wave Equation Solutions for Cylindrical
       Coordinate Systems, 98
3.3   Wave Equation Solutions for Spherical
       Coordinate Systems, 102
3.4   Review Problems, 113
3.5   Answers to Review Problems, 114

4 Case Studies and Analysis in Transportation Systems   **115**

4.1   Background Information for Subway Systems, 115
4.2   Case Studies, 118
4.3   Tunnel Radiation from a Temporary Antenna
       Installed on the Catwalk in a Tunnel, 142
4.4   Simulcast Interference at the End of the Cut and Cover
       Subway Tunnel, 145
4.5   Tracks Survey, 165
4.6   Leaky Radiating Coaxial Cable Analysis, 177
4.7   Effect of Rail on 26 Pair Cable Buried
       Along Right of Way, 187
4.8   Radiation Leakage from Way Side Communication
       Houses and Cabinets, 190
4.9   Lightning Rod Ground EMC Installation, 192

**5 Case Studies and Analysis of LRT Vehicle and Bus Top Antenna
    Farm Emissions and Other Radio Related Case Studies**   **199**

5.1   Introduction, 199
5.2   Circulation Currents in the Ground Plane, 201
5.3   Antenna Installation on a Radio Mast Case Study, 203
5.4   Unique Testing Technique for EMI and Police Vehicles, 210
5.5   Antenna Close to the Edge of the Ground Plane, 217
5.6   Case Study: Possible Fade Problem due to Antenna
       Reflections on the Rooftop of a Locomotive, 219
5.7   Case Study: Antenna Reflection and Diffraction
       at the Edge of the Ground Plane, 229
5.8   Antenna Application with Reflection also at
       the Edge of the Ground Plane, 234
5.9   Antenna Application with Reflection between
       Antennas in a Rooftop Antenna Farm, 239

5.10   Antenna Farm Application with Patch Antennas, 247
5.11   Review Problems, 253
5.12   Answers to Review Problems, 255

**6 Case Studies and Analysis of Communications
   Equipment and Cable Shielding and Grounding
   for Bus and Ferry Operations**   **263**

6.1   Introduction, 263
6.2   Communication System Overview, 264
6.3   Reflections (Ferry and Bus), 272
6.4   Review Problems, 279
6.5   Answers to Review Problems, 279

**7 Health and Safety Issues with Exposure
   Limits for Maintenance Workers and the Public**   **281**

7.1   Electromagnetic Emission Safety Limits, 281
7.2   EMI Prevention and Control, 290
7.3   Analysis of Rails as a Shock Hazard, 292
7.4   Lightning and Transient Protection, 293
7.5   Power Line Safety Calculations, 294
7.6   FCC Regulations, 297
7.7   Review Problems, 301
7.8   Answers to Review Problems, 302

**8 Miscellaneous Information Test Plans and
   Other Information Useful for Analysis**   **305**

8.1   Introduction, 305
8.2   EMC Plan, 306
8.3   EMC/EMI Performance Evaluation of
      Communications Equipment, 308
8.4   EMC/EMI Design Procedures, 317
8.5   Fresnel Zone Clearance, 333
8.6   Diffraction Losses, 335
8.7   Review Problems, 337
8.8   Answers to Review Problems, 338

9 Track Circuits and Signals   **341**

9.1   Introduction, 341
9.2   AF Track Circuits, 344
9.3   Loop Calculations, 352
9.4   Circuit Theory in Loop Calculations, 354
9.5   Review Problems, 359
9.6   Answers to Review Problems, 359

10   Useful Examples   **361**

   10.1   Introduction, 361
   10.2   Examples, 361

References   **379**

Index   **381**

PREFACE

A contributor to this book both directly and indirectly is my friend and colleague Dr. Kent Chamberlin of the University of New Hampshire (UNH). He was my professor while taking graduate school courses toward a PhD degree (did not finish because of health problems). He was instrumental in teaching me the finer points in vector analysis applied to EMC issues. Previously I was working and analyzing EMC problems that could be reduced to a Cartesian form of equations. These were much easier to work with than the cylindrical and spherical differential equations. Many times the wave equation was unnecessary to do an analysis. Under his tutelage I could read and study more complex books on the subject, such as the ones written by Dr. Balanis (located in the References).

This book is written in several chapters, the first being the regulations for electromagnetic emissions for electric and magnetic fields. The second chapter is an introduction to electromagnetic compatibility (EMC). This has some simple examples, as shown by illustration in equations that are necessary for a PE without previous training or a person wishing to delve into this field. The third chapter of this book catalogues the solutions to the wave equation and Maxwell's equations in Cartesian, cylindrical and spherical coordinate systems and also has several examples for the use of these systems.

The next three chapters are devoted to communication issues in transportation requiring EMC analysis. These include analysis of communication houses, signals bungalows/houses and the effects of magnetic and electric fields on the equipment inside, external radiation from licensed radios, cell phones, spread-spectrum devices, power lines, power supplies and other types of emissions that are induced on communication lines and PC boards. These chapters have many examples that can be used as a guide for the engineer in deciding how to analyze a particular anomaly caused by electric and magnetic fields. As emphasized previously, never try to overextend an analysis of frequency airspace without knowing the limitations of the equations. One must always keep vigilant when understanding that the equations are only a tool and would be equivalent to a mechanic using a hammer to remove his spark plugs.

The seventh chapter of this book is related to health and safety issues and catalogs many of the safety issues that must be observed due to electromagnetic emissions, with examples. In each of the chapters of this book, problems are provided at the end of the chapter to reinforce the knowledge gained by studying the chapter. Answers are provided at the end and in many cases the answers are provided with equations with the numbers shown so as to guide the engineer reading the chapter to a result and in some cases the engineer can use the equations by just changing the numbers slightly.

The eighth chapter of this book has miscellaneous documents and functions that may be useful in generating or answering a requirements document in transportation with a report, as is often required. More often than not, test results are required for the EMC analysis until the integration phase is complete. Then, only if an EMC issue occurs after commissioning, all test results are released and generally can be found in DOT documents. During the 1980s when working in research and not in systems, test results were usually required for EMC analysis for military-type projects. But most commercial and consumer products require test results that must be provided to the FCC, usually through test laboratories, such as Underwriter Laboratories, if the company producing a product does not have facilities for testing. Since all the products installed in the system must be FCC approved, with care no emissions will be present due to the system. Often the analysis is only a guide used by the system engineer to prevent anomalies from occurring.

The ninth chapter deals with signals and tracks and the effects of electromagnetic emission signals, both from track and signals. For each of these entities examples of signals equipment functions and how these affect communications is the object of this study. Signals equipment operates using both copper and fiber optic networking and rails function similar to transmission lines and these are low-frequency communications on the rails themselves. However many new spread-spectrum devices are used in signals for conveying information from the rails to the operational control center (OCC). Examples of signals are provided at the end of the chapter, as mentioned previously, to reinforce the knowledge of the person studying EMC effects.

The last chapter of this book provides useful examples that may be used in EMC analysis. These consist of both equations and situations where these anomalies may be examined. These not only apply to communications and transportation but can be generally used for other analyses as required. The audience will find that some of the information in this book is used for other EMC analyses outside the realm of transportation, such as emissions within the home that may be causing EMC issues, the design of cabinets and enclosures that require strict EMC shielding from emissions both internal and external, the automotive industry where harsh environments with radiation emission is present from electric car drives, ignition systems, GPS, emissions from cell phones, wireless games, shielded buildings with security issues, navigational aids emissions, airports and many others that are outside the realm of how this book may be used. There is a course that was originally a one-week seminar in PowerPoint for PEs that is now available at www.wiley.com/go/electromagneticcompatibility. This PowerPoint presentation is meant for the layman and is not heavily involved with vector analysis.

DONALD G. BAKER PE

ABOUT THE AUTHOR

Donald G. Baker began his experience in 1965 at the Motorola Corporation after graduation from the Illinois Institute of Technology with a BSEE in electronics. Motorola required that each engineer with less than one year's experience attend their plant school. The first design project was a 70 MHz phase lock loop for a Tract 92 Tropo-Scatter Radio System. At this time a transistor design at 70 MHz was an advanced project. The next design with some patents was a military grade audio signal generator for Holt Instrument with the patent for the feedback circuit in 1968.

The next series of designs was for the Magnaflux Corporation from 1968 until 1972: during which time the following equipment test equipment was designed: (i) a conductivity meter requiring a patent for the bridge circuit (one of these meters is used as a federal standard for calibration of conductivity meters), (ii) an ultrasonic crack detector with a oscilloscope type readout designed for detecting cracks at one-10 000th inch below the material surface, (iii) a meter type and (iv) an ultrasonic crack detection unit for large cracks below the surface that did not require the accuracy of the initial crack detector.

The next series of designs were for the Sundstrand Corporation (machine tool division). The author obtained a MSEE from IIT night school in 1972 and worked from 1972 to 1978 on the following design projects:

1. The control system for the Clinch River Nuclear Breeder Reactor for refueling. This design required using an analog computer design of differential equations that were sampled and converted to digital format for the government of the refueling system as a safety precaution. The plug drives for alignment to the fueling grapple were all controlled by a digital computer composed by the Digital Equipment Corporation (DEC).

2. The design of a six phase motor to be used for a spindle drive at 75 000 rpm to be used on milling machines.

3. Transistorized H drives to be used for the digital control of milling machine positioning.
4. An ultrasonic method for correcting spring-back in milling machines to increase accuracy.

The author was employed by EXTEL Corporation designing audio modems from 1978 to 1979. While employed by Microtek, 1980 to 1982, he designed a telephone caller ID system for analog phones using spread-spectrum technology. During employment by MIT Research (MITRE) from 1982 to 1990, he worked on several projects that are classified as secret and cannot be divulged at this time. Even their titles are secret; however most of the designs were used for fiber-optic networking. Employment from 1990 to 1991 with the Deleuw Cather involved EMC analysis and reliability work.

Employment from 1991 to 2013 was for the SESCO, Harmon and GE Corporations. This was all at the same workplace as the various companies changed hands but the work remain the same. The author's tasks were as follows: (i) analysis of all EMC issues for transportation communications and sensor systems, (ii) reliability studies, (iii) maintainability studies and (iv) communication computer timing issues. The last work while retired is writing this book from March 2013 to March 2015 for the first iteration of the manuscript. Miscellaneous work from 1966 to 1968 was teaching elementary courses in electronics and instrumentation at high schools and from 1972 to 1990 teaching as an adjunct professor at several junior colleges and graduate schools.

Several of the corporations where the author was employed were involved in mergers or went out of business completely; but some of the information about the author's work can be found in the author's books on fiber optics written in 1985, 1986 and 1987.

ABOUT THE COMPANION WEBSITE

This book is accompanied by a companion website:

www.wiley.com/go/electromagneticcompatibility

The website includes:
- PowerPoint slides for PEs based on an Electromagnetic Compatibility EMC Seminar
- Appendix A

1

INTRODUCTION

1.1 INTRODUCTION

This book presents a vast number of areas of industry beside transportation. Transportation is one of the harshest environments for communications. Electromagnetic compatibility (EMC) is in most of the industrialized world today. As computer and other electronic components get smaller, the need for EMC analysis and testing becomes more acute. Systems are generally designed and built with components that meet or exceed requirements for emissions. However, a piece of equipment may pick up extraneous noise from emissions through a host of poor practices in grounding and wiring.

The engineer designing system components must be vigilant during the design phase to check for emissions during prototyping, production and final design phases. The closer to final product the component gets, the more expensive becomes the correction in design. As an example, a circuit board design with a poor layout can be very costly in the final stage of design. While doing consulting work, the author was asked to help a particular manufacturer get a production board into production. The board had so many defects that the FCC sent a notice the equipment could not be connected to telephone lines. The solution was not very simple. The designer did not have the correct isolation transformer and the output and input lines were not separated sufficiently to maintain the isolation. There were many other problems with the design but the point is the printed circuit (PC) board had to be redesigned and several optical isolators added to complete the design.

The case studies are the result of several analyses required to satisfy the various State Authority Requirements. More often, the testing is part of the overhaul testing of the final systems during commissioning of a transportation system. The analysis brings to light some of the EMC issues that may arise. Often the specification sheets

Electromagnetic Compatibility: Analysis and Case Studies in Transportation, First Edition.
Donald G. Baker.
© 2016 John Wiley & Sons, Inc. Published 2016 by John Wiley & Sons, Inc.
Companion website: www.wiley.com/go/electromagneticcompatibility

for system components such as amplifiers, radios signals equipment and so on will have certain minimum Immunity Requirements that the system component must operate under with no effect in performance.

1.2 DEFINITIONS OF COMMONLY USED TERMS

Electromagnetic Compatibility (EMC)   This is the ability of equipment, systems or devices to operate without deficiencies in performance in an electromagnetic environment. The system, equipment or device must also be non-polluting to the electromagnetic environment, that is it must not have emissions (both radiated and conducted) that affect other systems, equipment or devices. The electromagnetic environment is composed of both radiated and conducted emissions.

Susceptibility   This is the ability of a system, equipment or device to respond to electromagnetic emissions interference. The emissions may be either radiated, conducted or both. Susceptibility is noise that affects the performance of system, equipment or device.

Immunity   The ability of equipment to operate with the required performance in the presence of electromagnetic interference noise.

Electromagnetic Interference (EMI)   Electromagnetic Interference (EMI) is noise due to electromagnetic energy through emissions, either radiated, conducted or both. This does not include distortion due to non-linearities in the system, equipment or device.

Radio Frequency Interference (RFI)   This is radiation due to intentional and unintentional radiators. The limits are shown in the tables presented in the sections on standards.

Culprit   This is the source of the emissions that result in a reduction in performance of the victim equipment, device, circuit or system. The culprit can be manmade or extraneous signals from galactic noise.

Victim   This is the device, equipment, circuit or system that is affected by the culprit. It depends on the coupling from the culprit. Coupling can be due to electric fields, magnetic fields, poor grounds, lack of proper supply filtering or combinations of these.

Supervisory Control and Data Acquisition (SCADA) System   This system monitors and controls complex equipment. It automates the complex system with control and monitor functions at an operation central control (OCC) room. A simplified version of the control room is shown in Figure 1.1. The project configuration is a large display the size of a wall in the OCC, that is 9 × 14 feet. The display shown in

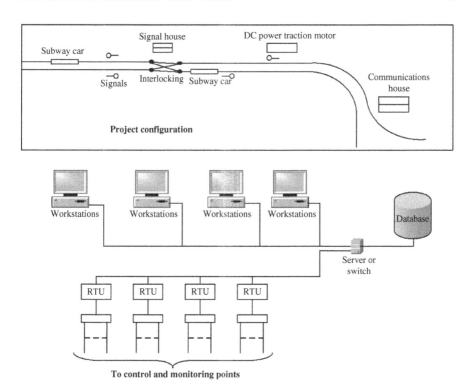

FIGURE 1.1 Operation control center simplified wall project display and workstation layout

the figure is only an example of what might be shown on the actual display. It may have as many as 15 or 20 interlockingss and signal houses and 10–12 communication houses in many miles of track, all displayed on this one board in symbolic form. It is the whole subway or bus system that is displayed. It may also have highway crossings shown with crossing gates and warning lights. The signals shown are positioned along the rails showing the direction of the traffic flow. The DC power supply houses for traction motors are shown. There may be many of these also depending on the project size. In the actual display the subway cars are shown moving in various directions and the display may indicate flow against the signals traffic. This is controlled from the OCC.

The workstations are arranged connected to a central server. Generally two servers are connected in tandem (one is a backup for the other). The primary runs the functions and the secondary shadows the primary. In the event of a failure an automatic switch over to the secondary occurs so that the service is restored; this provides a failsafe operation. Each of the workstations generally has the same software but some are dedicated to maintenance personnel, others are traffic control and one is dedicated for managerial functions. They all have logon passwords and the managerial station may have a lock to prevent tampering, with further identification functions so that only personnel with the correct credentials can use the workstation.

A large database holds information in the archives that are used later for statistical purposes and record the maintenance functions that have been performed on the equipment in the field. The network connects the workstations to the server and this is all done with fiber optics. The connections between the server and/or switch and the remote terminal units (RTUs) have a fiber optic self-healing ring with a SONET unit that connects several RTUs to a single node on the network. As can be observed, Figure 1.1 is a very simplified version of the communications between the control and monitor of devices. More details on the communications network are shown in Chapter 2 under the heading communications.

All OCCs have a backup control room, not in the same building. In the event of a catastrophe these control rooms are smaller and will not have all the functionality of the major control room. They have enough functionality to keep the subway or bus system functional if the main control room is damaged or destroyed. The backup control room will have a limited number of workstations, usually about half the number of the main control room. It will have an alternate site server/switch with the backup function of the main control room. As can be observed, signals carry the signal house data via RS 232 or RS 422 fiber optic connections to the communication house to be transported to the OCC for updating the project configuration screen. Occasionally in large systems a heartbeat is required from each RTU to determine if data is there and needs to be transported to the OCC. The heartbeat is a polling method for the RTUs. Some systems have interrupts instead of the heartbeat; this is all embedded in the software at the server/switch. The reason for designating a server/switch is some systems are small and only require servers; others are very large and require a switch and server.

Remote Terminal Unit (RTU)   These units interface to objects and equipment that either monitor or control pieces of equipment such as radio systems, PA systems on platforms, visual displays on platforms, ticket collection, pumps, ventilating fans in tunnels, fire and intrusion alarm systems, power for communications and traction power supplies. This unit is also equipped with a programmable logic controller (PLC).

Programmable Logic Controller (PLC)   These controllers are used for signals. They monitor and control interlockings and signage along the right of way, monitor headway between subway trains switch and control block information and other functions that are necessary for signaling.

The Communications Network   The simplified workstations shown in Figure 1.1 have more than one display, usually from three to four depending on the size of the project. The reason being that dispatchers can magnify a part of the network shown on the display board for use on his/her part of the rail system. The dispatcher also has a two-way radio to be used to communicate directly with the motorman and con-ductor on the subway. In the event of a complete failure of the network, the dispatcher can keep in touch with the motorman and conductor via the radio system. Sometimes both radio and network are used simultaneously, depending on the traffic on the system, that is during rush hours or emergencies.

Synchronous Optical Networking (SONET) The SONET network is composed of two counter-rotating rings, as shown in Figure 1.2. The rings carry data in both directions simultaneously. This particular network has a total of 25 nodes. The head end nodes are connected to the primary OCC and backup OCC or as shown in the diagram. If a break should occur in the cable or if a node is damaged it may be removed from service and a single ring will exist that supports the other 24 nodes. Automatic switching occurs within SONET nodes that allows self-healing of the ring. If two nodes are damaged and taken out of service the ring will form two islands, that is two separate single ring nodes. Most of the newer installations have high-speed rings, for example OC-768 or the equivalent STM-256 have a transmission rate of 38.5 Gbits/s. The base rates of OC-1 and STM-1 are 51.84 and 155.52 Mbits/s respectively. All of the others are integer numbers of these base rates. All of the data from the various houses and cabinets are transported by the SONET nodes to the OCC and backup OCC. The nodes arrange all data in a digital form and arrange it in frames to perform a seamless transmission network. The bandwidth used by most authorities is much greater than necessary in most cases; they plan for large expansions that occasionally never come. Occasionally another ring is added in a gateway or a switch is used to produce a much larger ring topology.

The RTU connections are all bidirectional. They may either have fiber optic modems or be wired with copper cabling. A listing of the network functions is shown in Table 1.1.

As can be observed in Table 1.1, the communications system is not only for voice and data. It is used for command and control of the entire subway system. As a

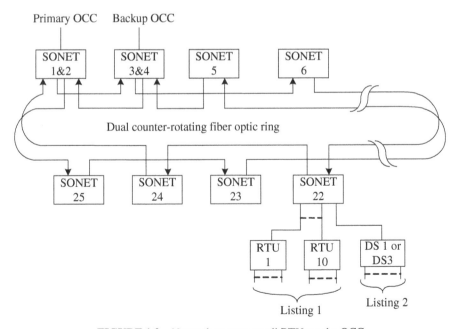

FIGURE 1.2 Network to connect all RTUs to the OCC

TABLE 1.1   Functions of Equipment in Houses, Bungalows, Cabinets and Stations Along the Right of Way

Equipment	List 1	List 2
Station platform RTUs	PLC and SCADA equipment	Telephone services
	Station communications	Emergency telephone
	Security alarms	Police Department emergency line
	Fire alarms	Fire Department direct line
	PA system control	Fare collection
	Camera monitoring	
	Parking lots camera monitoring	
	Camera control	
	Communication outage	
Communication house RTUs	AC power to the house	Telephone and extension service
	Fire alarm and suppression such as Halon gas	Fire Department direct line
	Security alarm.	
	Communication outage	
	Battery monitor for Uninterrupted Power Supply (UPS)	
	UPS monitor and control	
	Emergency panel switch over	
	Smoke alarms	
	Heat alarms	
	Backup portable generator for longtime outage	
	Ventilator fan control	
Data from signal houses	Data is from vital and non-vital logic	Signal house telephone and extension
Tunnel data	Monitor and control of several vent fans in tunnels	
	Damper monitor and control in tunnels	
	Sump pump and water levels and controls in tunnels	
	Carbon monoxide level monitors in tunnels	
	Fire and smoke alarms in tunnels	

(Continued)

TABLE 1.1 (Continued)

Equipment	List 1	List 2
	Stations in tunnels of monitor and control similar to his above ground stations	
	More attention is communications houses built into tunnels i.e. extra security measures	
Data from traction power houses	Data points for control and monitoring are regulated by the DC power contractor.	Traction power house telephone and extension
Cabinets on the right of way	Signals equipment monitoring and control	Telephone service
	Audio equipment monitor and control	
	Parking lot displays control	
	SCADA equipment to monitor and control all cabinet equipment may/may not have a local RTU in the cabinet	

backup for the dispatcher, all of the subsystems have a backup by some means so that failure is always circumvented by a backup of some means. Most AC power is provided by multiple substations in the event of AC power failure. EMC is a very important aspect of the communication systems because all data is stored at the OCC and backup OCC. This is required to analyze failures using all this data to determine cause, affect and the necessary maintenance to prevent future failures. Since most of the internal wiring of a station's electronic enclosures are wired in copper, EMI emissions are always present but at a very low level if precautions are taken during the design and installation of the various pieces of equipment.

The SCADA system is similar to the nervous system of a human body. It monitors the health of a particular subsystem such as the PA system and it makes corrections or circumvents a failure. The data is all sent to the OCC that functions similar to the brain of the system. The SCADA system has analog inputs that monitor such items as radio signal strength with a set-point that will result in an alarm if the signal strength drops below a certain level or the noise level is excessive. It also monitors temperature but is also an analog function with set-points that will send an alarm to the OCC if the temperature cannot be controlled, such as air conditioning or heating system failure.

RTUs, vital logic and non-vital logic for signals all have the latter logic embedded when programmed. They use relay contacts similar to the old-style relay logic used in older installations. This relay logic is not physical; it is all done in the software and the design engineer uses the logic as it would be employed for physical relays. The software in this equipment pre-processes the data points and a series of digital coded words are sent to the OCC indicating the health of the particular area being monitored, such as station platforms, communication cabinets along the right of way, signal house data, traction power supply and mechanical maintenance data from tunnels.

The OCC and backup OCC each has two SONET terminals. This allows them each to monitor the network in both directions. A complete failure of either of these SONET nodes will allow control by the surviving node. The RTUs implement a host of other monitor and control functions, such as grounds maintenance around the buildings and in shop areas where maintenance is being performed on subway cars. Building functions such as such as fire alarms, security systems, card swipe units, interlocking doors, cameras and camera controls are all monitored and controlled in the central control room.

Security is very strict in these OCC areas. Everyone must wear a swipe edge badge with a picture. For obvious reasons it is an ideal place for vandals or terrorists to do damage and bring down the subway system. Even some office space is highly restricted, such as where the database computers or the telephone system are kept. In some office spaces some canned messages are produced to send out both with audio and video information to PA systems and display units and stations. These of course cannot be compromised to show unauthorized messages that may even cause panic at stations; this is especially true in tunnels.

Elevators at the OCC are monitored by cameras and the data is sent back to a security workstation which is equipped with several displays, including those for stations. The operator at any time can control the camera to look at a particular event and report his findings, for example vandalism on the platform or assaults. The same holds true in tunnels where accidents may be observed and recorded and saved in the event of litigation. In this section the author has provided the reader with a good overview of how the communication system functions. This allows the reader to understand where all the EMC issues may occur.

1.3   BOOK SECTIONS AND CONTENT OVERVIEW

This book is divided into five sections. The first is introduction and standards; this includes FCC, CSA and European Union standards. Some of the testing techniques are also presented to introduce the reader to facilities that will be necessary to conduct the testing. The techniques for testing are not cast in stone. Standards are living documents that must be checked before designing systems, equipment or devices. Usually the changes are minor but these subtle differences may result in costly fixes later.

The second section is devoted to the coupling between victim and culprit circuits or equipment in general. These fundamentals are used throughout the book. It may be a refresher to some readers; however, the presentation makes the book easier to read.

The third section of the book is a discussion of Maxwell's equations and the wave equations and solutions that will be used throughout the case studies. The derivation of the solutions will not be shown in detail. References are provided to assist the reader who desires to observe how the solutions were derived. In some of the case studies a derivation will be provided, but this will be on a case by case basis.

The fourth section of the book is the largest. It involves past experiences of the author; there about 20 case studies in all. These have all been in the transportation industry. They generally deal with communications in a harsh environment. The case studies are rather diverse and can be applied in other industries as well. One such case is: shielding of a communication house due to the rebar embedded in the concrete. This same technique can be used to shield a building. Some structures that may require security can use these techniques for shielding with a little modification required for the windows and vent. The tunnel case study has applications where confined spaces have RF devices used such as cell phones and Bluetooth devices, such as the automotive industry. Radio engineers analyzing antenna farms on buildings can use some of the case studies as a guide. Subway car case studies have wideband transient analysis that can be applied to the steel industry with its overhead cranes, rolling mills, electric furnaces, shears, arc welders and other equipment where arcing may be present.

The aircraft industry engineers may use some of the information provided in this book. Present-day aluminum skin aircraft provide a good ground plane for most electronics. The newer composite aircraft may require more shielding and filtering of electronic equipment. Radiation coupling between suites of equipment can result in a degradation in performance. The transportation, rail and bus systems have similar grounding problems.

Medical facilities with electronic instrumentation engineers can use some of the case studies, such as tunnel applications. Tunnels have leaky coaxial cable to extend communications underground. The same leaky cable can be used in hospitals with microcellular phones.

The fifth section is radiation exposure safety issues for maintenance and public exposure. This will be discussed briefly and tables with exposure limits provided. A particular case is provided in the case studies. This is a case study in a tunnel.

Table 1.2 is a short list of emission sources that can lead to device, equipment and system failure or performance degradation. The first five are radiation sources due to radio transmissions; several are analyzed in the case studies. The primary part is how these sources affect the performance of various equipment and systems. Most of the devices used in transportation will have a fairly good immunity, but equipment and systems have a means of implementing either cabling or wireless communication.

The sixth and 15th sources are due to computer emissions. These are in some cases very difficult to analyze, due to a combination of radiated and conducted emissions. In a particular situation, there were two open racks (one with fiber optic communication equipment, the other with a VHF police radio). The clock for the communication equipment happened to be near the VHF radio band and, when the radio was keyed, the communication link began to drop bits. As computer device clock speeds get higher they also increase their emissions for radio and other wireless communications.

TABLE 1.2 Conduction and Radiation Emission Sources

Item	Description of source	Remarks
1	Radio transmitter broadcasts	AM and FM band radios
2	Communication narrow band radio	Two-way handheld and base station
3	Cell phones	Phones and towers
4	Wireless devices	Low level radiation, mW region
5	Radars	Weather, military and handheld
6	Receiver local oscillators from computers	High frequencies, 1–10 GHz
7	Motors	Brushes and commutation
8	Switches	Arcing
9	Fluorescent lights	Harmonics
10	Light dimmers	Track switches, phase control
11	Diathermy	Track switches, phase control
12	Dielectric heaters	Track switches, phase control
13	Welders	Arcs
14	Subway centenary and hot rail	Arcing
15	Engine ignition	Radiation from spark plug wires
16	Computer peripherals	1–10 GHz switching
17	Lightning	High rise time pulse and energy
18	Galactic noise electrostatic discharge	Space communications
19	Electromagnetic pulse (EMP; nuclear blast)	Not discussed

Sources 8–14 are due to transients. These of course are in some cases very difficult to analyze due to intermittent behavior. A transient will generally need to be analyzed with a storage device unless it is cyclic. However, the random case is usually the type that most often occurs. The case studies have some transient analysis included as part of the analysis.

1.4 REGULATIONS

Regulations are living documents, that is they are continually changing. The designer that is working on a system, equipment or device must look up the regulations to ensure compliance at the time of the design. If there are regulation changes after the design is completed, there is generally a time span before the change takes effect.

1.4.1 United States FCC Regulations

FCC Part 15 radiation regulations are represented in Tables 1.3–1.5. These are radiated emission limits for systems, equipment and devices. These limits are in terms of electric fields (E). The full range of the radiation measurement is from 9 kHz to 3 GHz.

The tables can be found or generated using CFR 47 Regulations Part 15, which has a wealth of information for EMC design. The regulations provide guidance in several areas for obtaining FCC certification.

Some of the tables in this section of the book may not appear to be very useful. However, the various emission tables will provide the EMC practitioner with a lead on where to look for possible culprit sources. For example wireless devices are now very widespread and Table 1.8 provides the frequency range for these devices. When investigating a particular problem that appears to be related to radio, FCC subparts C, D, F and H can be examined if the problem appears to be related to radio communications. Tables are not provided in these sections to prevent the book from being outdated at the time of printing. The frequency spectrum allocation is continually modified by various users.

The remarks column FM 88–108 MHz radio can cause interference for very sensitivity devices. The station power is under very strict licensing; shielding or filtering in most cases will be required.

CB radios have limits but they are sometimes violated when the culprit transmitter has higher gain ("linear amplifiers called foot warmers") than allowed by FCC rules, or very high gain antennas. Generally the FCC will impose fines for these types of installations.

TABLE 1.3   FCC Emission Limit Regulations Measured at 3 m

Frequency (MHz)	Class B magnitude	Class A magnitude	Remarks
30–88	(28.9 µV/m) 29.5 dBµV/m	(100 µV/m) 40 dBµV/m	FM, CB band
88–216	(44.7 µV/m) 33.0 dBµV/m	(150 µV/m) 43.5 dBµV/m	FM, two way radio
216–960	(59.6 µV/m) 35.5 dBµV/m	(200 µV/m) 46.0 dBµV/m	Two way radio, CellP
>960	(298 µV/m) 49.5 dBµV/m	(500 µV/m) 54.0 dBµV/m	Wireless dev

TABLE 1.4   FCC Emission Limit Regulations 9 KHz To 30 MHz

Frequency	Electric field	Measurement distance (m)	Remarks
9–490 KHz	2400/f µV/m	300	AM L band
490–1705 KHz	2400/f µV/m	30	AM U band
1705 KHz to 30 MHz	30 µV/m	3	Multiple bands

TABLE 1.5   FCC Emission Limit Regulations Measured at 10 m

Frequency (MHz)	Class B magnitude	Class A magnitude	Remarks
30–88	(28.2 μV/m) 29 dBμV/m	(90 μV/m) 39 dBμV/m	FM, CB band
88–216	(47.3 μV/m) 33.5 dBμV/m	(150 μV/m) 43.5 dBμV/m	FM, two way radio
216–960	(59.6 μV/m) 35.5 dBμV/m	(210 μV/m) 46.5 dBμV/m	Two way radio, CellP
>960	(149.6 μV/m) 43.5 dBμV/m	(300 μV/m) 49.5 dBμV/m	Wireless dev

Two-way radios can cause interference due to their mobility. VHF and UHF radios and cell phones are always a source of noise. They are all licensed radiators and must be used prudently to prevent interference problems. As an example, a measurement was taken in a Paris subway station at rush hour. The noise produced by cell phone transmissions produced sufficient noise to disrupt train communications.

The emission limits apply to unintentional radiators the remarks column in Table 1.3 is to remind the audience that intentional radiators can be present. These must be measured, if present, and not be part of the measured emissions. Open air test systems (OATS) are discussed in Section 1.6 *EMC Testing Methods*. All radio signals can enter though grounds, power supplies, radiation (when a product has an extraneous projection acting as an antenna) and induction (where magnetic fields can couple into circuits). Coupling techniques are discussed in Chapter 2 with examples.

Two different distances are provided in the tables, for tests to making comparisons between Classes A and B. The distance of 3 m can easily be extrapolated to 10 m using Equation 1.1. The reason this is possible is the 3 m dipole used to measure 30 MHz is outside the near field limit $2*(\lambda/2)^2/\lambda$ or 2.5 m.

$$\Delta_{10} = 20*(1 - \log 3) = 10.46\,\text{dB} \qquad (1.1)$$

FCC part 15 is divided into subparts as follows:

1. A General information
2. B Unintentional radiators
3. C Intentional radiators
4. D Unlicensed personal communication devices
5. E Unlicensed national information infrastructure devices
6. F Ultra wideband operation
7. G Access broadband over power line wideband operation
8. H TV band devices.

Equation 1.1 is the result of the electric field reduction as a function of 1/r where r is in meters. Class B for digital or analog and digital devices are more stringent than Class A. This classification of limits is for residential environments where emissions are more likely to cause interference. Class A is for commercial, industrial or business environments where anomalies that result in interference and the culprit causing noise can be identified and eliminated. Class B limits for residential areas may not solve the problems (Tables 1.6–1.9). Interference with TV or home entertainment equipment from radiated emissions that represent culprit noise sources are the responsibility of the user of the device to solve, even if the offender meets the radiation standards (see References [4] and [5]). It is also useful to adhere to Class B limits to give the designer of equipment a better chance of passing Class A emission tests; a 10 dB margin for error is apparent. The pre-compliance testing encourages the designer to have a margin for errors.

1.4.2   United States Department of Transportation

Excerpts from rapid transit documents   In the course of the study, excerpts will be taken when necessary to do the case study. The total regulations for Department of Transportation (DOT) will not be presented. An entire text would be required

TABLE 1.6   Measurement Range Devices for Tuned and Untuned Circuits

Highest frequency generated in device (MHz)	Upper frequency of measurement range
<1.705	30 MHZ
1.705–108	1 GHz
108–500	2 GHz
500–1000	5 GHz
>1000	Fifth harmonic of the highest frequency or 40 GHz, whichever is lower

TABLE 1.7   Maximum Electric Field Strength for Spread Spectrum Radios

Unlicensed international band device frequency range (MHz)	E field (mV/m)
902–928	500
2435–2465	500
5785–5815	500

TABLE 1.8   FCC/CISPR Conducted Limit Regulations, Class A. CISPR = Comité Internionale Special Des Perturbations, Special international Committee on Radio interference, Founded 1934

Frequency	Quasi-peak magnitude	Average magnitude	Remarks
150–500 KHz	(8.9 mV) 79 dBμV	(2 mV) 66 dBμV	AM radio bands
0.5–30 MHz	(4.47 mV) 73 dBμV	(1 mV) 60 dBμV	AM, FM, CB radio bands

TABLE 1.9 FCC/CISPR Conducted Limit Regulations, Class B

Frequency	Quasi-peak magnitude	Average magnitude	Remarks
150–500 KHz[a]	(2–0.63 mV) 66–56 dBμV	(0.63–0.2 mV) 56-46 dBμV	AM radio bands
0.5–5 MHz	(4.47 mV) 56 dBμV	(0.2 mV) 46 dBμV	AM, FM
5–30 MHz	(1.0 mV) 60 dBμV	(0.32 mV) 50 dBμV	CB radio

[a]Linear slope =35 KHz/dBμV

TABLE 1.10 DOT Radiation Emission Measurements

Frequency	BW	Field	Measurement	Remarks
14 KHz to 30 MHz 140 KHz to 30 MHz	1 KHz 10 KHz	H (field)	Loop antenna H (field) sensor	Vertical and parallel, 15 m from centerline of track
14–140 KHz		E (field)	Rod vertical E (field) sensor	Vertical, 15 m from centerline of track
30–400 MHz 400 MHz to 1 GHz	100 KHz 1 MHz	E (field)	Log-periodic horizontal polarized	Horizontal, 15 m from centerline of track

to completely describe all of the regulations and tests. Most of the equipment requiring EMC in transit systems is tested to various standards. They are not all listed in the book. Table 1.10 does not have test standards. The subway car manufacturer will have a set of test results that they must pass for DOT compliance. Since all of the scenario case studies are on equipment installed after manufacture, these devices or equipment must not create noise in communications houses, signal houses, subway or freight train radio systems or wayside wireless equipment. This installed equipment or device must have a measured immunity from radiated emissions of the subway cars. Table 1.10 indicates the type of antenna used in the measurements and the bandwidth (BW) of the instrumentation (see Figure 1.3 for the antennas). The measurements are taken at 15 m from the center of the track but measurements at 30 m is also common when possible. The 15 m measurements are used in case studies when equipment placement is hampered by a lack of access to the rails. Most of the equipment is within 15 m of the center of the right of way (track).

1.4.3 Canadian Regulations

The Canadian standards are very similar to the United States FCC standards, with cooperation between the two countries on the use of standards. The two countries accept each other's standards. The Canadians have a standard for immunity to electromagnetic radiation. The grades of immunity are as follows:

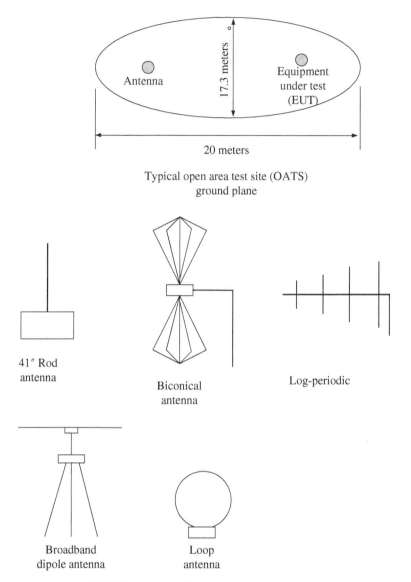

FIGURE 1.3 OATS ground plane and antennas

Canadian immunity grades

Grade 1 meets a 1 V/m test; the equipment is most likely to have performance deficiencies.

Grade 2 meets a 3 V/m test; the equipment is not likely to have an adequate performance

Grade 3 meets a 10 V/m test; the equipment is not likely to fail only under very harsh conditions.

1.5   BACKGROUND

This is a discussion of the author's background in the transportation industry. Overviews, questions at the end of each chapter and case studies are used to peak the interest of the audience. Other industries are also mentioned, where the electromagnetic compatibility (EMC) case studies may apply. DOT regulations are discussed that are in some cases more stringent than FCC regulation. For example, the DOT regulations have immunity regulations that must be met that are not required in all FCC regulations.

The emission of all system components must meet as a minimum FCC CFR 47 part 15 before they can be sold in the United States. The susceptibility of such components, devices and equipment is a de facto standard of 3 V/m for commercial and an actual standard of 10 V/m for medical. The equipment manufacturers will be asked to produce documentation confirming compliance with this standard when the study is conducted. The light rail transit (LRT) vehicle and signals emission data and susceptibility is unnecessary for CTS, PA/VMB, fire, intrusion, telephone and UPS equipment; however, an EMI/EMC analysis is required. Electromagnetic interference (EMI) from external sources for both intentional and unintentional radiators, TPSS, light rail and substations are investigated in the case studies when deemed necessary.

EMC/EMI will be coordinated with the cable and wire installation contractor to insure emission and susceptibility integrity. Wiring and coupling through ground are investigated to insure no extraneous EMI signals are coupled into the system components supplied for the case studies. Radio communications systems are some of the major case studies. Subway vehicles have several radio systems, not only for voice communications but also for monitoring the progress of vehicular travel using various sensors. The data is usually relayed to a large screen showing the route where each vehicle is bound and its progress as the vehicle travels along the right of way. The dispatcher will control the trains from the control room by observing the screens. Two-way radios serve as a backup in the event the main system fails. Wayside wireless sensors are often used for reading the bill of lading for freight train traffic. SCADA equipment is the nervous system of the subway or freight rail systems. It provides the operators in the control room with data about the health of all other systems. Some examples are radio system maintenance problems, unauthorized persons entering a communication, signal or DC power houses, fire alarm, fire suppression, public address and video monitoring camera failures, building equipment failures such as air conditioning and heating system, signals equipment and any other subsystems failures.

Induced, conducted and radiated emissions will be suppressed by various methods to provide a signal to noise ratio (SNR) consistent with good communications. In all cases, safety and reliable operation are the primary concern.

The most popular OATS sites are the 3 m emissions measurement type for obvious reasons; the sites take the least amount of space. The larger 10 m emission sites are more difficult to implement. The problems with operating in the far field are eliminated, for example measurements at 30 MHz in the near field is 20 m. Even measurements at 10 m require corrections for the near field but operating at 3 m requires even more extensive corrections. Therefore, measuring emissions about 300 MHz should not be a problem, but below this frequency the near field may begin to require corrections for machine measurements.

Radiated emissions for military equipment are taken at 1 m. These require more extensive measurements which can be found in MIL-STD-461. Most of the equipment used in transportation has had many of these measurements taken and they are not required; however, occasionally panels with components are fabricated in the shop and may need emissions testing. A case in point is a panel which was constructed with relays, controllers and audio amplifiers and was to be installed in a communication house. The panel required testing for emissions, mainly due to the wiring of the various components on the panel.

Fire and intrusion alarms are also fabricated on panels; however these are tested by the manufacturer in various configurations and require no emissions testing provided the assemblies are configured as recommended by the manufacturer. Control room panel configurations are often done using several manufacturers of the various components emission testing, as a total system is usually required and is often neglected by many of the authorities. However, when panels are constructed with shielded wire or coaxial cable and terminations are completed so that no standing waves or traveling standing waves occur, the radiation from the various areas in the panel are minimized.

Generally when testing racks of equipment for performance, poor grounding or wiring that radiates emissions is usually the culprit. During commissioning testing of subway systems some of these anomalies are found, which of course may be only require a simple shielding, changing the orientation of the particular culprit or providing simple filtering such as a line filter installed in the power line for a particular piece of equipment.

One particularly difficult instance of electromagnetic compatibility occurred when several radios were connected to a leaky radiating cable in a tunnel. The combiner network used to sum all the radios onto the leaky radiating cable worked fine in the shop when installed in a rack but, once the racks were placed in the control room in the tunnel and connected to the leaky radiating cable, intermodulation products produced many problems for the engineers installing the equipment. The racks had to be removed and the combiner network retuned in the shop. After many iterations of retuning the combiner network, the 17 channels on the leaky radiating cable finally performed satisfactorily. The cost to perform the repeated removal and retuning of the network was more than the bid for the radios and their installation. A mockup of the installation should have been done prior to installing the racks in the tunnel; this may have alleviated several problems.

1.6 EMC TESTING METHODS FOR FCC PART 15 RADIATION MEASUREMENTS

The FCC Part 15 radiated emissions open area test site (OATS) is conducted on a ground plane, as shown in Figure 1.3. The ground plane is constructed of copper mesh. The dimensions are for an elliptical minimum area but a rectangular area is generally installed at 20 m long or more and at least 18 m wide. The ground plane is sized for 10 m radiated emission measurements. For 3 m radiated emission measurements the ellipse is 6 m major axis and 5.2 m minor axis as the minimum area.

The rectangular is generally over 6 m long and at least 5.2 m wide. The ground plane is for a rotating turntable for the equipment under test (EUT).

Circular ground planes will be necessary where the antennas are moved and the EUT is on a stationary platform. The ground plane radius is 15 and 4.5 m respectively for 10 and 3 m emissions measurements. The dimensions for a square plane are 30 and 9 m square respectively for 10 and 3 m emissions. The rotating turntable allows the smallest ground plane footprint. For ground planes installed on a wooden flooring no metallic framing or fasteners can be used for construction, only glue. For concrete structures, a mesh of fiberglass for reinforcing can be used; no metallic rebar can be used. It is advisable to make a small test area of a 3 m square with concrete and reinforcing material and test its properties for construction of the main site. The testing should be done with and without the copper ground plane. Concrete is poured and should be sealed to prevent water vapor from entering it.

The ground plane area must be free on any obstructions that will result in reflections, such as trees, posts, metallic or non-metallic objects. It is best to level the area before installing the copper mesh. The weather can result in errors, if measurements are taken when the soil is damp when copper mesh is used instead of solid copper. The test site should be covered when not in use. Walls, trees, posts, building or other reflective objects at the periphery of the test area can produce errors. The periphery area should be cleared at least three times distance from the edge of the periphery of the major axis of antenna and the EUT. The antenna and focal point of the ellipse is at the antenna and EUT. For a 3 m test the cleared distance is 1.5 m to the edge of the ellipse along the major axis plus three times that distance from the edge or 6 m from the antenna and EUT. For the 10 m emission test the measurement is three times the focal distance (5 m) plus three times that from the edge of the major axis of the ellipse or 20 m from antenna and EUT.

To eliminate ground reflections while making measurements, the antenna height is moved up and down during a set of spectral measurements. Measurements for radiated emission are taken at least at 45° increments for either turntable or moving the antenna. A set of measurements are taken at various antenna heights. If deep nulls or peaks are noted during a measurement, ground reflections are occurring. The antenna height should be adjusted for each set of measurements; this will be based on the emission frequency. For heights that produce reflections, the reflected wave and the line of site wave will result in deep nulls or peaks that can result in large errors. As a first pass at measurements, the antenna should be raised from ground level to 3 m above the EUT on the turntable and reduced in increments of 1 m. The peaks and nulls are noted as the test progresses and refinements can be made in the tests to allow for reflections.

1.6.1 Measurement Errors

The analyzer and attenuator (if required) can be setup poorly; this can contribute errors of 1–2 dB depending on the skill of the operator. Reflection errors can be as high as ± 10 dB that will be either a null or peak above the RMS value of the emission.

The cable to EUT and attenuator or spectrum analyzer can give a mismatch of –2 dB. The far field at 30 MHz is approximately 2.5 m for a 0.5 wavelength antenna. A 3 m measurement distance can result in a small error if the near field does not have a sharp transition to the far field. If the correct antenna is installed that is made for EMC testing, this should not be a problem. Most of the antennas have a matching network built in; these usually have a 2–3 dB loss and generally the manufacturer will provide an insertion loss value. When measurements must be taken for vertical and horizontal polarization there is approximately 20 dB difference between them. For turntable use, the cable between the spectrum analyzer or attenuation and the antenna needs to be extra long to allow for coiling of the cable as the EUT is rotated. The coil cannot be tight or this will introduce a large loss in the cable. The cable should be calibrated before the test to provide a correction table that can be stored and used during the tests. The calibration should be plotted with sufficient accuracy to allow the operator to interpolate losses from the curve or curves. Generally a skilled operator can produce a reading within ± 1 dB.

1.6.2 Antenna Selection

The antennas implemented for radiation testing are shown in Figure 1.3. When setting up an OATS facility, the facilitator should always check for newer antennas that may be an improvement over those depicted in Figure 1.3. For frequencies below 30 MHz a rod antenna will be used to measure frequencies down to 9 kHz. This antenna will require a counterpoise which is generally available from the supplier. There are some that have automatic switching of the matching networks and conversion of dBμV to dBμV/m (antenna factor). The antenna must be calibrated at least yearly, due to component aging. These antennas have a preamplifier and switching built into the unit. The biconical antenna is used for bands between 30 and 216 MHz and the conversion of dBμV to dBμV/m (antenna factor) must be stated or automatically corrected for distance. The antenna requires calibration at least on a yearly basis. The log periodic broadband antenna is designed for frequency measurements between 30 MHz and 15 GHz; it can be used for the entire band of frequencies above 30 MHz. This antenna has high directivity and a narrow beam-width. However, it does not require changing antennas and that can be a source of errors. The di-pole broadband antenna in Figure 1.3 from 20 to 200 MHz has many of the attributes of the others; it is shown as an alternative to the biconical and it will be somewhat lower in cost. The final antenna shown is for measuring magnetic fields at frequencies below 30 MHz. The construction, that is loop diameter and turns, is determined by the frequency, sensitivity and accuracy to be measured.

To set up an actual site, check the FCC title CFR 47 regulations. As stated previously, these change and the book representation has excerpts that typify an OATS facility (Figure 1.4). The dimensions are accurate for the construction of a 3 m emission test at the time the book is written. The copper is shown as very thick; this is not the case. The minimum thickness depends or the mechanical strength and the skin affects the depth of the emissions. Since the lowest radiation frequency is 30 MHz, the copper thickness is calculated using Equations 2.18 and 2.19

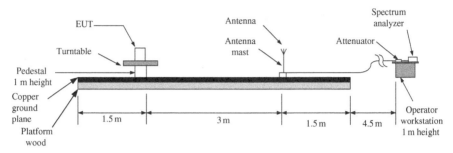

FIGURE 1.4   Typical 3 m emission test facility. (The components are not to scale)

in Chapter 2; $1/\alpha \approx 21.5$ µm skin depth. This is 63% of the thickness of the copper mat, it must be three times the thickness or 64.5 µm; however for mechanical strength 1/8 inch or more is a better choice. The pedestal is where connections to the EUT turntable drive mechanism are installed. The antenna mast can be an electronic telescoping or manual type to change the antenna height during testing. An attenuator (optional) and spectrum analyzer is required with sufficient bandwidth and sensitivity to complete measurements. The selection should be determined based on oscillators, clocks or expected transients of the EUT. As an example, a fiber optic component such as an audio modem will have a bit stream of about 1.5 Mbits/s.

OATS measurements and testing is an expensive undertaking. For a test facility to be certified by the FCC is rather costly (US $ 30 000–40 000. Manufacturing of equipment or devices depending on size, production runs and skilled personnel will determine whether the outlay of the cost is beneficial or not. Private FCC certified laboratories are the best alternative to designing and getting FCC certification. Underwriters Laboratory (UL) has a large facility that can test large or small devices or equipment, even aircraft. There are many others that may be better suited for small-scale tests. To do a low-cost solution, pre-compliance testing can be done and later, when the tests provide a level of confidence the equipment or device will pass the emission test, it can be done at a FCC certified laboratory. Pre-compliance for microwave ovens, radios and other physically small-size equipment can be bench tested. Once compliance is achieved, the manufacturing methods are established and only a spot check of equipment or device may be required.

1.6.3   Pre-compliance Radiation Measurement and Testing

Most equipment will fail EMC FCC radiation testing unless some tests are done prior to the final FCC certification tests. Pre-compliance testing will assist in achieving the final FCC certification where perhaps only minor adjustments to the equipment or device are needed to pass. The EMC radiation emission tests must be at least 5 or 10 dB better than the standards in the tables in Section 1.4 to provide a margin for measurement errors alluded to in OATS testing. When a piece of equipment or device has passed, a sample should be kept if possible to use as a sort of an in-house standard to be used when testing random units. It can be used to set up the test facility for pre-compliance testing at a later date. This may not be possible if the device or equipment is quite expensive, such as test equipment. Keeping accurate records of the test setup and equipment that passed including all environment factors such as temperature, humidity and all equipment in the test area. References [4] and [5] have an excellent treatment of pre-compliance issues.

1.6.4   Conducted Emission

The test conducted in Figure 1.5a is for single phase input power to the EUT. The line impedance stabilization network (LISN) is necessary on the power input and ground return. The return allows the harmonics and noise from switched power supplies to be measured. For an isolated and very clean 120 V AC input the LISN may not be necessary. For any of the LISN not used for a measurement, it must have a 50 Ω termination. The instruments connected to measurement port must have a 50 Ω output impedance to prevent mismatch reflections. When dealing with power lines, it is necessary to place an attenuator between the measurement device and the LISN instrument port. Transients on power lines in particular transportation applications can be destructive to the instrument. The harsh environment from arcing of catenaries and the third hot rail can also be very destructive to other circuits. The tests are conducted on a ground plane with LISN bonded to it. The power lines are connected to the EUT with feed-through capacitors connected to a ground connection through copper plate at a 90° angle to the ground plane, as shown in Figure 1.6. The AC mains should have line filters or isolations transformers and not inject noise into the measurements. LISN can be constructed or purchased; however, if constructed poorly they can be a hindrance to measurement accuracy. The LISN circuit design is provided in Figure 1.7.

Figure 1.5b is the block diagram for a three phase conducted emission test. A LISN is installed on the neutral due to several types of switching DC power supplies that may be connected to the 120 V AC mains and filtering or isolation transformers does not take out all the noise. The switching DC power supplies can be at various frequencies and mixing may occur, thus providing a host of harmonics. Unlicensed radiators such as 902–928 MHz equipment may be operating on one phase and a 857 MHz radio bay station may be operating on another leg of the power mains with a potential for 5–31 MHz for mixing if the grounds have poor connections due to corrosion. Therefore the operator must be vigilant when making measurements not inject these frequencies into the results. Any LISN not connected to a measurement device must have a termination. This diagram depicts a four channel spectrum analyzer. If it is not available each phase may be tested separately.

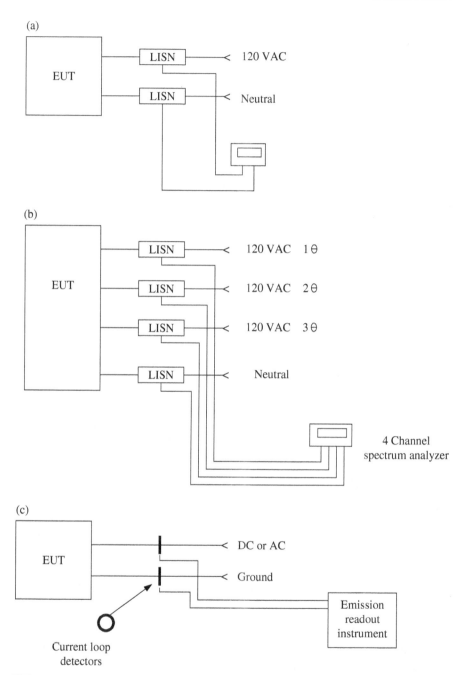

FIGURE 1.5 Typical test apparatus for conducting emission tests. (a) Single phase. (b) Three phases. (c) DC or AC emission current measurement

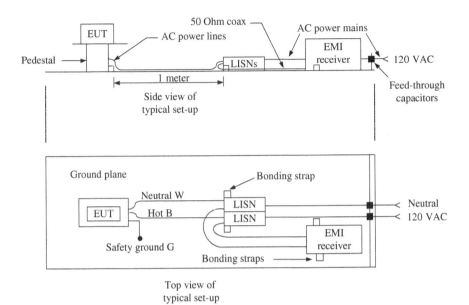

FIGURE 1.6   Typical test set-up for conducting voltage emission measurements

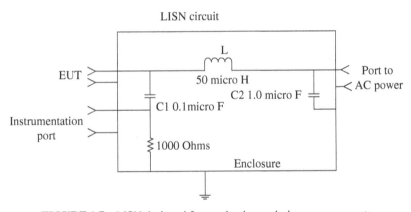

FIGURE 1.7   LISN designed for conducting emission measurements

Conducted emission current is measured using the set-up shown in Figure 1.5c. There is a large variety of loop detectors available. It is best to use an online search for these types of detectors. The technology is changing rapidly and particular detectors that are available today may be either replaced or obsolete by the time of printing. The instrument for measurement should have an attenuator between the LISN instrumentation port and the measurement device. It should have a storage readout to allow for transients and other anomalies. Keying radios, AM/FM radio and transportation have other issues such as steel rails with signals with 10 KHz carriers, catenary and third rail arcing and inrush currents of traction motors. Occasionally the 13.5 kV power lines from substations are routed above ground to the DC power station for the subway. As can be expected the rail right of way authority may lease space for other utilities.

A typical measurement installation is shown in Figure 1.6. This is a description of how a single phase AC voltage measurement is made. The 120 V AC power mains will have an isolation transformer upstream from the LISNs and a copper plate fitted with feed-through capacitors in the 90° upright; this can be part of the copper ground plane bent to a 90° angle. The ground plane can be solid copper or copper mesh. The LISNs are both bonded to the ground plane. The hot side black wire and white neutral can be measured separately. Terminate the instrument port not being used for the measurements. The safety ground or green wire from the EUT is affixed to the grounded plane. In some cases, only two wires are present for double insulated devices. An EMI receiver is the best instrument to use; it saves time and is generally made to measure quasi-peak and average dBVμ measurements directly. Other instrumentation may require calculations. Not shown in Figure 1.5 is the plug to convert a three-wire output to two wires; it is located in the pedestal.

1.7   CANADIAN REGULATIONS

1.7.1   Canadian Immunity Grades

Grade 1 meets a 1 V/m test, the equipment is most likely to have performance deficiencies.

Grade 2 meets a 3 V/m test, the equipment is not likely to have an adequate performance

Grade 3 meets a 10/m test, the equipment is not likely to fail only under very harsh conditions.

1.8   EUROPEAN UNION REGULATIONS

The European Union electromagnetic compatibility (EMC) list of standards is provided in Table 1.11. These cover a host of products and systems. The reader interested in a particular standard will need to use the information provided below. The rationale for providing only the listings is that the standards are continually changing and being updated; and one who is working on a particular project that needs a standard should look up the latest version. The European Union is composed of a number of countries as compared to the United States that is one homogeneous country. Each of the partners of the European Union contributes to the generation of standards.

These standards cover a variety of products and systems. Several in particular are necessary for the designer or analyst dealing with transportation issues and these are provided in Table 1.11. The summary below only covers some of the most important standards. However there are others that are relevant, for example standards dealing with communication house and signal house construction. This summary does not cover all aspects of a particular standard; it provides a lead for the audience of where to pursue more details about the standard.

1. The EN 50121-5:2006/AC:2008 covers railway equipment including subways emissions and power supplies.
2. Electromagnetic compatibility – Road traffic signal systems – Product standard EN 50293:2000 for an example at grade crossings.
3. EN 50130-4:1995 Alarm systems, part 4: Electromagnetic compatibility – Product family standard. Immunity requirements for components of fire, intruder and social alarm systems and communication houses and signals bungalows.
4. EN 50529-2:2010 EMC network standard, part 2: Wire-line telecommunications networks using coaxial cables is necessary for communications.
5. EN 60947 Low-voltage switchgear and control-gear for SCADA systems.
6. EN 61000-3-2:2006 Electromagnetic compatibility (EMC), part 3-2: Limits for harmonic current emissions (equipment input current $\leq$ 16 A per phase). IEC 61000-3-2:2005.
7. EN 61000-6-2:2005 Electromagnetic compatibility (EMC), part 6-2: Generic standards – Immunity for industrial environments that lack a standard. IEC 61000-6-2:2005.
8. EN 61131-2:2007 Programmable controllers, part 2: Equipment requirements and tests. IEC 61131-2:2007.
9. EN 61547:2009 Equipment for general lighting purposes. EMC immunity requirements. IEC 61547:2009 + IS1:2013.
10. EN 300 386 V1.4.1 Electromagnetic compatibility and radio spectrum matters (ERM); Telecommunication network equipment; Electromagnetic Compatibility (EMC) requirements.
11. EN 300 386 V1.6.1 Electromagnetic compatibility and radio spectrum matters (ERM); Telecommunication network equipment; Electromagnetic compatibility (EMC) requirements.

For information about the content and availability of European standards, please contact the European standardization organizations (ESOs; e.g. CEN, Brussels, http://www.cen.eu; Cenelec, Brussels, http://www.cenelec.eu; ETSI, Sophia Antipolis, http://www.etsi.eu).

TABLE 1.11 List of European EMC Standards

ESO[a]	Reference and title of the harmonized standard (and reference document)	Reference of superseded standard	Date of cessation of presumption of conformity of superseded standard[b]
CEN	EN 617:2001+A1:2010 Continuous handling equipment and systems – Safety and EMC requirements for the equipment for the storage of bulk materials in silos, bunkers, bins and hoppers	EN 617:2001[c]	Date expired (30/06/2011)
CEN	EN 618:2002+A1:2010 Continuous handling equipment and systems – Safety and EMC requirements for equipment for mechanical handling of bulk materials except fixed belt conveyors	EN 618:2002[c]	Date expired (30/06/2011)
CEN	EN 619:2002+A1:2010 Continuous handling equipment and systems – Safety and EMC requirements for equipment for mechanical handling of unit loads	EN 619:2002[c]	Date expired (30/04/2011)
CEN	EN 620:2002+A1:2010 Continuous handling equipment and systems – Safety and EMC requirements for fixed belt conveyors for bulk materials	EN 620:2002[c]	Date expired (30/06/2011)
CEN	EN 1155:1997 Building hardware – Electrically powered hold-open devices for swing doors – Requirements and test methods		
CEN	EN 12015:2004 Electromagnetic compatibility – Product family standard for lifts, escalators and moving walks – Emission	EN 12015:1998[c]	Date expired (30/06/2006)

CEN	EN 12016:2013 (new) Electromagnetic compatibility – Product family standard for lifts, escalators and moving walks – Immunity	EN 12016:2004 + A1:2008^c	28/02/2014
CEN	EN 12895:2000 Industrial trucks – Electromagnetic compatibility		
N	EN 13241-1:2003+A1:2011 Industrial, commercial and garage doors and gates – Product standard – Part 1: Products without fire resistance or smoke control characteristics	EN 13241-1:2003^c	Date expired (31/10/2011)
CEN	EN 13309:2010 Construction machinery – Electromagnetic compatibility of machines with internal power supply	EN 13309:2000^c	Date expired (31/01/2011)
CEN	EN 14010:2003+A1:2009 Safety of machinery – Equipment for power driven parking of motor vehicles – Safety and EMC requirements for design, manufacturing, erection and commissioning stages	EN 14010:2003^c	Date expired (31/01/2010)
CEN	EN ISO 14982:2009 Agricultural and forestry machinery – Electromagnetic compatibility – Test methods and acceptance criteria (ISO 14982:1998)	EN ISO 14982:1998^c	Date expired (28/12/2009)
CEN	EN 16361:2013 (new) Power operated pedestrian doors – Product standard, performance characteristics – Pedestrian door sets, other than swing type, initially designed for installation with power operation without resistance to fire and smoke leakage characteristics		

(Continued)

TABLE 1.11 (Continued)

ESO[a]	Reference and title of the harmonized standard (and reference document)	Reference of superseded standard	Date of cessation of presumption of conformity of superseded standard[b]
Cenelec	EN 50065-1:2001 Signaling on low-voltage electrical installations in the frequency range 3 kHz to 148,5 kHz – Part 1: General requirements, frequency bands and electromagnetic disturbances	EN 50065-1:1991 + A1:1992 + A2:1995 + A3:1996[c]	Date expired (01/04/2003)
	EN 50065-1:2001/A1:2010	[d]	Date expired (01/10/2012)
Cenelec	EN 50065-1:2011 Signaling on low-voltage electrical installations in the frequency range 3 kHz to 148,5 kHz – Part 1: General requirements, frequency bands and electromagnetic disturbances	EN 50065-1:2001 and its amendment[c]	
Iec	EN 50065-2-1:2003 Signaling on low-voltage electrical installations in the frequency range 3 kHz to 148,5 kHz – Part 2-1: Immunity requirements for mains communications equipment and systems operating in the range of frequencies 95 kHz to 148,5 kHz and intended for use in residential, commercial and light industrial environments	Relevant generic standard(s)[e]	Date expired (01/10/2004)
	EN 50065-2-1:2003/A1:2005 EN 50065-2-1:2003/AC:2003	[d]	Date expired (01/07/2008)
Cenelec	EN 50065-2-2:2003 Signaling on low-voltage electrical installations in the frequency range 3 kHz to 148,5 kHz – Part 2-2: Immunity requirements for mains communications equipment and systems operating in the range of frequencies 95 kHz to 148,5 kHz and intended for use in industrial environments	Relevant generic standard(s)[e]	Date expired (01/10/2004)

	EN 50065-2-2:2003/A1:2005	^{d}	Date expired (01/07/2008)
	EN 50065-2-2:2003/A1:2005/AC:2006		
	EN 50065-2-2:2003/AC:2003		
Cenelec	EN 50065-2-3:2003 Signaling on low-voltage electrical installations in the frequency range 3 kHz to 148,5 kHz – Part 2-3: Immunity requirements for mains communications equipment and systems operating in the range of frequencies 3 kHz to 95 kHz and intended for use by electricity suppliers and distributors	Relevant generic standard(s)^c	Date expired (01/08/2004)
	EN 50065-2-3:2003/A1:2005	^{d}	Date expired (01/07/2008)
	EN 50065-2-3:2003/AC:2003		
Cenelec	EN 50083-2:2012 Cable networks for television signals, sound signals and interactive services – Part 2: Electromagnetic compatibility for equipment	EN 50083-2:2006^c	Date expired (21/06/2013)
Cenelec	EN 50121-1:2006 Railway applications – Electromagnetic compatibility – Part 1: General	Relevant generic standard(s)^c	Date expired (01/07/2009)
	EN 50121-1:2006/AC:2008		
Cenelec	EN 50121-2:2006 Railway applications – Electromagnetic compatibility – Part 2: Emission of the whole railway system to the outside world	Relevant generic standard(s)^c	Date expired (01/07/2009)
	EN 50121-2:2006/AC:2008		
Cenelec	EN 50121-3-1:2006 Railway applications – Electromagnetic compatibility – Part 3-1: Rolling stock – Train and complete vehicle	Relevant generic standard(s)^e	Date expired (01/07/2009)

(Continued)

TABLE 1.11 (Continued)

ESO[a]	Reference and title of the harmonized standard (and reference document)	Reference of superseded standard	Date of cessation of presumption of conformity of superseded standard[b]
	EN 50121-3-1:2006/AC:2008		
Cenelec	EN 50121-3-2:2006 Railway applications – Electromagnetic compatibility – Part 3-2: Rolling stock – Apparatus	Relevant generic standard(s)[e]	Date expired (01/07/2009)
	EN 50121-3-2:2006/AC:2008		
Cenelec	EN 50121-4:2006 Railway applications – Electromagnetic compatibility – Part 4: Emission and immunity of the signaling and telecommunications apparatus	Relevant generic standard(s)[e]	Date expired (01/07/2009)
	EN 50121-4:2006/AC:2008		
Cenelec	EN 50121-5:2006 Railway applications – Electromagnetic compatibility – Part 5: Emission and immunity of fixed power supply installations and apparatus	Relevant generic standard(s)[e]	Date expired (01/07/2009)
	EN 50121-5:2006/AC:2008		
Cenelec	EN 50130-4:1995 Alarm systems – Part 4: Electromagnetic compatibility – Product family standard: Immunity requirements for components of fire, intruder and social alarm systems	Relevant generic standard(s)[e]	Date expired (01/01/2001)
	EN 50130-4:1995/A1:1998	[d]	Date expired (01/01/2001)
	EN 50130-4:1995/A2:2003	[d]	Date expired (01/09/2007)
	EN 50130-4:1995/A2:2003/AC:2003		

Cenelec	EN 50130-4:2011 Alarm systems – Part 4: Electromagnetic compatibility – Product family standard: Immunity requirements for components of fire, intruder, hold up, CCTV, access control and social alarm systems	EN 50130-4:1995 and its amendments[c]	13/06/2014
Cenelec	EN 50148:1995 Electronic taximeters BT(IT/NOT)12	Relevant generic standard(s)[e]	Date expired (15/12/1995)
Cenelec	EN 50270:2006 Electromagnetic compatibility – Electrical apparatus for the detection and measurement of combustible gases, toxic gases or oxygen	EN 50270:1999[e]	Date expired (01/06/2009)
Cenelec	EN 50293:2000 Electromagnetic compatibility – Road traffic signal systems – Product standard	Relevant generic standard(s)[e]	Date expired (01/04/2003)
Cenelec	EN 50293:2012 (new) Road traffic signal systems – Electromagnetic compatibility	EN 50293:2000[c]	11/05/2015
Cenelec	EN 50295:1999 Low-voltage switchgear and control gear – Controller and device interface systems – Actuator Sensor interface (AS-i)	Relevant generic standard(s)[e]	Date expired (01/12/1999)
Cenelec	EN 50370-1:2005 Electromagnetic compatibility (EMC) – Product family standard for machine tools – Part 1: Emission	Relevant generic standard(s)[e]	Date expired (01/02/2008)
Cenelec	EN 50370-2:2003 Electromagnetic compatibility (EMC) – Product family standard for machine tools – Part 2: Immunity	Relevant generic standard(s)[e]	Date expired (01/11/2005)
Cenelec	EN 50412-2-1:2005 Power line communication apparatus and systems used in low-voltage installations in the frequency range 1,6 MHz to 30 MHz – Part 2-1: Residential, commercial and industrial environment - Immunity requirements	Relevant generic standard(s)[e]	Date expired (01/04/2008)

(*Continued*)

TABLE 1.11 (Continued)

ESO[a]	Reference and title of the harmonized standard (and reference document)	Reference of superseded standard	Date of cessation of presumption of conformity of superseded standard[b]
	EN 50412-2-1:2005/AC:2009		
Cenelec	EN 50428:2005 Switches for household and similar fixed electrical installations – Collateral standard – Switches and related accessories for use in home and building electronic systems (HBES)	Relevant generic standard(s)[e]	Date expired (01/01/2008)
	EN 50428:2005/A1:2007	[d]	Date expired (01/10/2010)
	EN 50428:2005/A2:2009	[d]	Date expired (01/06/2012)
Cenelec	EN 50470-1:2006 Electricity metering equipment (AC.) – Part 1: General requirements, tests and test conditions - Metering equipment (class indexes A, B and C)	Relevant generic standard(s)[e]	Date expired (01/05/2009)
Cenelec	EN 50490:2008 Electrical installations for lighting and beaconing of aerodromes – Technical requirements for aeronautical ground lighting control and monitoring systems – Units for selective switching and monitoring of individual lamps	Relevant generic standard(s)[e]	Date expired (01/04/2011)
Cenelec	EN 50491-5-1:2010 General requirements for Home and Building Electronic Systems (HBES) and Building Automation and Control Systems (BACS) – Part 5-1: EMC requirements, conditions and test set-up	EN 50090-2-2: 1996 + A2:2007[c]	Date expired (01/04/2013)
Cenelec	EN 50491-5-2:2010 General requirements for Home and Building Electronic Systems (HBES) and Building Automation and Control Systems (BACS) – Part 5-2: EMC requirements for HBES/BACS used in residential, commercial and light industry environment	EN 50090-2-2: 1996 + A2:2007[c]	Date expired (01/04/2013)

Cenelec	EN 50491-5-3:2010 General requirements for Home and Building Electronic Systems (HBES) and Building Automation and Control Systems (BACS) – Part 5-3: EMC requirements for HBES/BACS used in industry environment	EN 50090-2-2: 1996 + A2:2007[c]	Date expired (01/04/2013)
Cenelec	EN 50498:2010 Electromagnetic compatibility (EMC) – Product family standard for aftermarket electronic equipment in vehicles	Relevant generic standard(s)[c]	Date expired (01/07/2013)
Cenelec	EN 50512:2009 Electrical installations for lighting and beaconing of aerodromes – Advanced Visual Docking Guidance Systems (A-VDGS)	Relevant generic standard(s)[c]	
Cenelec	EN 50529-1:2010 EMC Network Standard – Part 1: Wire-line telecommunications networks using telephone wires		
Cenelec	EN 50529-2:2010 EMC Network Standard – Part 2: Wire-line telecommunications networks using coaxial cables		
Cenelec	EN 50550:2011 Power frequency overvoltage protective device for household and similar applications (POP)		
	EN 50550:2011/AC:2012		
Cenelec	EN 50557:2011 (new) Requirements for automatic reclosing devices (ARDs) for circuit breakers-RCBOs-RCCBs for household and similar uses		

(Continued)

TABLE 1.11 (Continued)

ESO[a]	Reference and title of the harmonized standard (and reference document)	Reference of superseded standard	Date of cessation of presumption of conformity of superseded standard[b]
Cenelec	EN 50561-1:2013 (new) Power line communication apparatus used in low-voltage installations – Radio disturbance characteristics – Limits and methods of measurement – Part 1: Apparatus for in-home use	EN 55022:2010 + EN 55032:2012[c]	
Cenelec	EN 55011:2009 Industrial, scientific and medical equipment – Radio-frequency disturbance characteristics – Limits and methods of measurement CISPR 11:2009 (Modified)	EN 55011:2007 + A2:2007[c]	Date expired (01/09/2012)
	EN 55011:2009/A1:2010 CISPR 11:2009/A1:2010	[d]	Date expired (01/07/2013)
Cenelec	EN 55012:2007 Vehicles, boats and internal combustion engines – Radio disturbance characteristics – Limits and methods of measurement for the protection of off-board receivers CISPR 12:2007	EN 55012:2002 + A1:2005[c]	Date expired (01/09/2010)
	EN 55012:2007/A1:2009 CISPR 12:2007/A1:2009	[d]	Date expired (01/07/2012)
	EN 55012 is applicable for giving presumption of conformity under directive 2004/108/EC for those vehicles, boats and internal combustion engine-driven devices that are not within the scope of directives 95/54/EC, 97/24/EC, 2000/2/EC or 2004/104/EC.		
Cenelec	EN 55013:2001 Sound and television broadcast receivers and associated equipment – Radio disturbance characteristics – Limits and methods of measurement CISPR 13:2001 (Modified)	EN 55013:1990 + A12:1994 + A13:1996 + A14:1999[c]	Date expired (01/09/2004)
	EN 55013:2001/A1:2003 CISPR 13:2001/A1:2003	[d]	Date expired (01/04/2006)
	EN 55013:2001/A2:2006 CISPR 13:2001/A2:2006	[d]	Date expired (01/03/2009)

Cenelec	EN 55013:2013 (new) Sound and television broadcast receivers and associated equipment – Radio disturbance characteristics – Limits and methods of measurement CISPR 13:2009 (Modified)	EN 55013:2001 and its amendments[c]	22/04/2016
Cenelec	EN 55014-1:2006 Electromagnetic compatibility – Requirements for household appliances, electric tools and similar apparatus – Part 1: Emission CISPR 14-1:2005	EN 55014-1:2000 + A1:2001 + A2:2002[c]	Date expired (01/09/2009)
	EN 55014-1:2006/A1:2009 CISPR 14-1:2005/A1:2008	[d]	Date expired (01/05/2012)
	EN 55014-1:2006/A2:2011 CISPR 14-1:2005/A2:2011	[d]	16/08/2014
Cenelec	EN 55014-2:1997 Electromagnetic compatibility – Requirements for household appliances, electric tools and similar apparatus – Part 2: Immunity – Product family standard CISPR 14-2:1997	EN 55104:1995[c]	Date expired (01/01/2001)
	EN 55014-2:1997/A1:2001 CISPR 14-2:1997/A1:2001	[d]	Date expired (01/12/2004)
	EN 55014-2:1997/A2:2008 CISPR 14-2:1997/A2:2008	[d]	Date expired (01/09/2011)
	EN 55014-2:1997/AC:1997		
Cenelec	EN 55015:2006 Limits and methods of measurement of radio disturbance characteristics of electrical lighting and similar equipment CISPR 15:2005	EN 55015:2000 + A1:2001 + A2:2002[c]	Date expired (01/09/2009)
	EN 55015:2006/A1:2007 CISPR 15:2005/A1:2006	[d]	Date expired (01/05/2010)
	EN 55015:2006/A2:2009 CISPR 15:2005/A2:2008	[d]	Date expired (01/03/2012)
Cenelec	EN 55015:2013 (new) Limits and methods of measurement of radio disturbance characteristics of electrical lighting and similar equipment CISPR 15:2013 + IS1:2013 + IS2:2013	EN 55015:2006 and its amendments[c]	12/06/2016

(Continued)

TABLE 1.11 (Continued)

ESO[a]	Reference and title of the harmonized standard (and reference document)	Reference of superseded standard	Date of cessation of presumption of conformity of superseded standard[b]
Cenelec	EN 55020:2007 Sound and television broadcast receivers and associated equipment – Immunity characteristics – Limits and methods of measurement CISPR 20:2006	EN 55020:2002 + A1:2003 + A2:2005[c]	Date expired (01/12/2009)
	EN 55020:2007/A11:2011	[d]	Date expired (01/01/2013)
Cenelec	EN 55022:2010 Information technology equipment – Radio disturbance characteristics – Limits and methods of measurement CISPR 22:2008 (Modified)	EN 55022:2006 + A1:2007	Date expired (01/12/2013)
	EN 55022:2010/AC:2011		
Cenelec	EN 55024:2010 Information technology equipment – Immunity characteristics – Limits and methods of measurement CISPR 24:2010	EN 55024:1998 + A1:2001 + A2:2003	Date expired (01/12/2013)
Cenelec	EN 55032:2012 (new) Electromagnetic compatibility of multimedia equipment – Emission requirements CISPR 32:2012	EN 55013:2013 + EN 55022:2010 + EN 55103-1:2009 and its amendment[c]	05/03/2017
	EN 55032:2012/AC:2013 (new)		
Cenelec	EN 55103-1:2009 Electromagnetic compatibility – Product family standard for audio, video, audio-visual and entertainment lighting control apparatus for professional use – Part 1: Emissions	EN 55103-1:1996[c]	Date expired (01/07/2012)
	EN 55103-1:2009/A1:2012 (new)	[d]	05/11/2015

Cenelec	EN 55103-2:2009 Electromagnetic compatibility – Product family standard for audio, video, audio-visual and entertainment lighting control apparatus for professional use – Part 2: Immunity	EN 55103-2:1996[c]	Date expired (01/07/2012)
Cenelec	EN 60034-1:2010 Rotating electrical machines – Part 1: Rating and performance IEC 60034-1:2010 (Modified) EN 60034-1:2010/AC:2010	Relevant generic standard(s)[c]	Date expired (01/10/2013)
Cenelec	EN 60204-31:1998 Safety of machinery – Electrical equipment of machines – Part 31: Particular safety and EMC requirements for sewing machines, units and systems IEC 60204-31:1996 (Modified) EN 60204-31:1998/AC:2000	Relevant generic standard(s)[c]	Date expired (01/06/2002)
Cenelec	EN 60204-31:2013 (new) Safety of machinery – Electrical equipment of machines – Part 31: Particular safety and EMC requirements for sewing machines, units and systems IEC 60204-31:2013	EN 60204-31:1998[c]	28/05/2016
Cenelec	EN 60255-26:2013 (new) Measuring relays and protection equipment – Part 26: Electromagnetic compatibility requirements IEC 60255-26:2013 EN 60255-26:2013/AC:2013 (new)		
Cenelec	EN 60439-1:1999 Low-voltage switchgear and control gear assemblies – Part 1: Type-tested and partially type-tested assemblies IEC 60439-1:1999	EN 60439-1:1994 + A11:1996[c]	Date expired (01/08/2002)

(Continued)

TABLE 1.11 (Continued)

ESO[a]	Reference and title of the harmonized standard (and reference document)	Reference of superseded standard	Date of cessation of presumption of conformity of superseded standard[b]
Cenelec	EN 60669-2-1:2004 Switches for household and similar fixed electrical installations – Part 2-1: Particular requirements – Electronic switches IEC 60669-2-1:2002 (Modified) + IS1:2011 + IS2:2012	EN 60669-2-1:2000 + A2:2001[c]	Date expired (01/07/2009)
	EN 60669-2-1:2004/A1:2009 IEC 60669-2-1:2002/A1:2008 (Modified)	[d]	Date expired (01/04/2012)
	EN 60669-2-1:2004/A12:2010	[d]	Date expired (01/06/2013)
	EN 60669-2-1:2004/AC:2007		
Cenelec	EN 60730-1:1995 Automatic electrical controls for household and similar use – Part 1: General requirements IEC 60730-1:1993 (Modified)		
	EN 60730-1:1995/A11:1996	[d]	Date expired (01/01/1998)
	EN 60730-1:1995/A17:2000	[d]	Date expired (01/10/2002)
	EN 60730-1:1995/AC:2007		
Cenelec	EN 60730-1:2000 Automatic electrical controls for household and similar use – Part 1: General requirements IEC 60730-1:1999 (Modified)	EN 60730-1:1995 and its amendments[c]	
	EN 60730-1:2000/A1:2004 IEC 60730-1:1999/A1:2003 (Modified)	[d]	
	EN 60730-1:2000/A16:2007	[d]	Date expired (01/06/2010)
	EN 60730-1:2000/A2:2008 IEC 60730-1:1999/A2:2007 (Modified)	[d]	Date expired (01/06/2011)
	EN 60730-1:2000/AC:2007		
	EN 60730-1:2000/A16:2007/AC:2010		

Cenelec	EN 60730-1:2011 Automatic electrical controls for household and similar use – Part 1: General requirements IEC 60730-1:2010 (Modified)	EN 60730-1:2000 and its amendments[c]	Date expired (01/10/2013)
Cenelec	EN 60730-2-5:2002 Automatic electrical controls for household and similar use – Part 2-5: Particular requirements for automatic electrical burner control systems IEC 60730-2-5:2000 (Modified)		
	EN 60730-2-5:2002/A1:2004 IEC 60730-2-5:2000/A1:2004 (Modified)	[d]	Date expired (01/12/2008)
	EN 60730-2-5:2002/A11:2005	[d]	Date expired (01/12/2008)
	EN 60730-2-5:2002/A2:2010 IEC 60730-2-5:2000/A2:2008 (Modified)	[d]	Date expired (01/03/2013)
Cenelec	EN 60730-2-6:2008 Automatic electrical controls for household and similar use – Part 2-6: Particular requirements for automatic electrical pressure sensing controls including mechanical requirements IEC 60730-2-6:2007 (Modified)	EN 60730-2-6:1995 + A1:1997	Date expired (01/07/2011)
Cenelec	EN 60730-2-7:2010 Automatic electrical controls for household and similar use – Part 2-7: Particular requirements for timers and time switches IEC 60730-2-7:2008 (Modified)	EN 60730-2-7:1991 + A1:1997[c]	Date expired (01/10/2013)
	EN 60730-2-7:2010/AC:2011		
Cenelec	EN 60730-2-8:2002 Automatic electrical controls for household and similar use – Part 2-8: Particular requirements for electrically operated water valves, including mechanical requirements IEC 60730-2-8:2000 (Modified)	EN 60730-2-8:1995 + A1:1997 + A2:1997[c]	Date expired (01/12/2008)
	EN 60730-2-8:2002/A1:2003 IEC 60730-2-8:2000/A1:2002 (Modified)	[d]	Date expired (01/12/2008)

(Continued)

TABLE 1.11 (Continued)

ESO[a]	Reference and title of the harmonized standard (and reference document)	Reference of superseded standard	Date of cessation of presumption of conformity of superseded standard[b]
Cenelec	EN 60730-2-9:2010 Automatic electrical controls for household and similar use – Part 2-9: Particular requirements for temperature sensing controls IEC 60730-2-9:2008 (Modified)	EN 60730-2-9:2002 + A1:2003 + A2:2005	Date expired (01/11/2013)
Cenelec	EN 60730-2-14:1997 Automatic electrical controls for household and similar use – Part 2-14: Particular requirements for electric actuators IEC 60730-2-14:1995 (Modified)	EN 60730-1:1995 and its amendments[e]	Date expired (01/06/2004)
	EN 60730-2-14:1997/A1:2001 IEC 60730-2-14:1995/A1:2001	[d]	Date expired (01/07/2008)
Cenelec	EN 60730-2-15:2010 Automatic electrical controls for household and similar use – Part 2-15: Particular requirements for automatic electrical air flow, water flow and water level sensing controls IEC 60730-2-15:2008 (Modified)	EN 60730-2-18:1999[c]	Date expired (01/03/2013)
Cenelec	EN 60870-2-1:1996 Tele-control equipment and systems – Part 2: Operating conditions – Section 1: Power supply and electromagnetic compatibility IEC 60870-2-1:1995	Relevant generic standard(s)[e]	Date expired (01/09/1996)
Cenelec	EN 60945:2002 Maritime navigation and radio communication equipment and systems – General requirements – Methods of testing and required test results IEC 60945:2002	EN 60945:1997[c]	Date expired (01/10/2005)

Cenelec	EN 60947-1:2007 Low-voltage switchgear and control gear – Part 1: General rules IEC 60947-1:2007	EN 60947-1:2004[c]	Date expired (01/07/2010)
	EN 60947-1:2007/A1:2011 IEC 60947-1:2007/A1:2010	[d]	Date expired (01/01/2014)
Cenelec	EN 60947-2:2006 Low-voltage switchgear and control gear – Part 2: Circuit-breakers IEC 60947-2:2006	EN 60947-2:2003[c]	Date expired (01/07/2009)
	EN 60947-2:2006/A1:2009 IEC 60947-2:2006/A1:2009	[d]	Date expired (01/07/2012)
	EN 60947-2:2006/A2:2013 IEC 60947-2:2006/A2:2013 (new)	[d]	07/03/2016
Cenelec	EN 60947-3:2009 Low-voltage switchgear and control gear – Part 3: Switches, disconnectors, switch-disconnectors and fuse-combination units IEC 60947-3:2008	EN 60947-3:1999 + A1:2001[c]	Date expired (01/05/2012)
	EN 60947-3:2009/A1:2012 IEC 60947-3:2008/A1:2012	[d]	21/03/2015
Cenelec	EN 60947-4-1:2010 Low-voltage switchgear and control gear – Part 4-1: Contactors and motor-starters – Electromechanical contactors and motor-starters IEC 60947-4-1:2009	EN 60947-4-1:2001 + A1:2002 + A2:2005[c]	Date expired (01/04/2013)
	EN 60947-4-1:2010/A1:2012 IEC 60947-4-1:2009/A1:2012 (new)	[d]	24/08/2015
Cenelec	EN 60947-4-2:2000 Low-voltage switchgear and control gear – Part 4-2: Contactors and motor-starters – AC semiconductor motor controllers and starters IEC 60947-4-2:1999	EN 60947-4-2:1996 + A2:1998[c]	Date expired (01/12/2002)
	EN 60947-4-2:2000/A1:2002 IEC 60947-4-2:1999/A1:2001	[d]	Date expired (01/03/2005)
	EN 60947-4-2:2000/A2:2006 IEC 60947-4-2:1999/A2:2006	[d]	Date expired (01/12/2009)

(Continued)

TABLE 1.11 (Continued)

ESO[a]	Reference and title of the harmonized standard (and reference document)	Reference of superseded standard	Date of cessation of presumption of conformity of superseded standard[b]
Cenelec	EN 60947-4-2:2012 (new) Low-voltage switchgear and control gear – Part 4-2: Contactors and motor-starters – AC semiconductor motor controllers and starters IEC 60947-4-2:2011	EN 60947-4-2:2000 and its amendments[c]	22/06/2014
Cenelec	EN 60947-4-3:2000 Low-voltage switchgear and control gear – Part 4-3: Contactors and motor-starters – AC semiconductor controllers and contactors for non-motor loads IEC 60947-4-3:1999	Relevant generic standard(s)[e]	Date expired (01/12/2002)
	EN 60947-4-3:2000/A1:2006 IEC 60947-4-3:1999/A1:2006	[d]	Date expired (01/11/2009)
	EN 60947-4-3:2000/A2:2011 IEC 60947-4-3:1999/A2:2011	[d]	18/04/2014
Cenelec	EN 60947-5-1:2004 Low-voltage switchgear and control gear – Part 5-1: Control circuit devices and switching elements – Electromechanical control circuit devices IEC 60947-5-1:2003	EN 60947-5-1:1997 + A12:1999[c]	Date expired (01/05/2007)
	EN 60947-5-1:2004/A1:2009 IEC 60947-5-1:2003/A1:2009 EN 60947-5-1:2004/AC:2005	[d]	Date expired (01/05/2012)
Cenelec	EN 60947-5-2:2007 Low-voltage switchgear and control gear – Part 5-2: Control circuit devices and switching elements – Proximity switches IEC 60947-5-2:2007	EN 60947-5-2:1998 + A2:2004[c]	Date expired (01/11/2010)
	EN 60947-5-2:2007/A1:2012 IEC 60947-5-2:2007/A1:2012 (new)	[d]	01/11/2015

		Relevant generic standard(s)[c]	Date expired (01/05/2002)
Cenelec	EN 60947-5-3:1999 Low-voltage switchgear and control gear – Part 5-3: Control circuit devices and switching elements – Requirements for proximity devices with defined behavior under fault conditions (PDF) IEC 60947-5-3:1999		
	EN 60947-5-3:1999/A1:2005 IEC 60947-5-3:1999/A1:2005	[d]	Date expired (01/03/2008)
Cenelec	EN 60947-5-6:2000 Low-voltage switchgear and control gear – Part 5-6: Control circuit devices and switching elements – DC interface for proximity sensors and switching amplifiers (NAMUR) IEC 60947-5-6:1999	EN 50227:1997[c]	Date expired (01/01/2003)
Cenelec	EN 60947-5-7:2003 Low-voltage switchgear and control gear – Part 5-7: Control circuit devices and switching elements – Requirements for proximity devices with analogue output IEC 60947-5-7:2003	Relevant generic standard(s)[c]	Date expired (01/09/2006)
Cenelec	EN 60947-5-9:2007 Low-voltage switchgear and control gear – Part 5-9: Control circuit devices and switching elements – Flow rate switches IEC 60947-5-9:2006		
Cenelec	EN 60947-6-1:2005 Low-voltage switchgear and control gear – Part 6-1: Multiple function equipment – Transfer switching equipment IEC 60947-6-1:2005	EN 60947-6-1:1991 + A2:1997[c]	Date expired (01/10/2008)
Cenelec	EN 60947-6-2:2003 Low-voltage switchgear and control gear – Part 6-2: Multiple function equipment – Control and protective switching devices (or equipment) (CPS) IEC 60947-6-2:2002	EN 60947-6-2:1993 + A1:1997[c]	Date expired (01/09/2005)
	EN 60947-6-2:2003/A1:2007 IEC 60947-6-2:2002/A1:2007	[d]	Date expired (01/03/2010)

(Continued)

TABLE 1.11 (Continued)

ESO[a]	Reference and title of the harmonized standard (and reference document)	Reference of superseded standard	Date of cessation of presumption of conformity of superseded standard[b]
Cenelec	EN 60947-8:2003 Low-voltage switchgear and control gear – Part 8: Control units for built-in thermal protection (PTC) for rotating electrical machines IEC 60947-8:2003	Relevant generic standard(s)[e]	Date expired (01/07/2006)
	EN 60947-8:2003/A1:2006 IEC 60947-8:2003/A1:2006	[d]	Date expired (01/10/2009)
	EN 60947-8:2003/A2:2012 IEC 60947-8:2003/A2:2011 (new)	[d]	22/06/2014
Cenelec	EN 60974-10:2007 Arc welding equipment – Part 10: Electromagnetic compatibility (EMC) requirements IEC 60974-10:2007	EN 60974-10:2003[c]	Date expired (01/12/2010)
Cenelec	EN 61000-3-2:2006 Electromagnetic compatibility (EMC) – Part 3-2: Limits – Limits for harmonic current emissions (equipment input current ≤ 16 A per phase) IEC 61000-3-2:2005	EN 61000-3-2:2000 + A2:2005[c]	Date expired (01/02/2009)
	EN 61000-3-2:2006/A1:2009 IEC 61000-3-2:2005/A1:2008	[d]	Date expired (01/07/2012)
	EN 61000-3-2:2006/A2:2009 IEC 61000-3-2:2005/A2:2009	[d]	Date expired (01/07/2012)
Cenelec	EN 61000-3-3:2008 Electromagnetic compatibility (EMC) – Part 3-3: Limits – Limitation of voltage changes, voltage fluctuations and flicker in public low-voltage supply systems, for equipment with rated current ≤ 16 A per phase and not subject to conditional connection IEC 61000-3-3:2008	EN 61000-3-3:1995 + A1:2001 + A2:2005	Date expired (01/09/2011)
Cenelec	EN 61000-3-3:2013 (new) Electromagnetic compatibility (EMC) – Part 3-3: Limits – Limitation of voltage changes, voltage fluctuations and flicker in public low-voltage supply systems, for equipment with rated current ≤ 16 A per phase and not subject to conditional connection IEC 61000-3-3:2013	EN 61000-3-3:2008[c]	18/06/2016

Cenelec	EN 61000-3-11:2000 Electromagnetic compatibility (EMC) – Part 3-11: Limits – Limitation of voltage changes, voltage fluctuations and flicker in public low-voltage supply systems – Equipment with rated current ≤ 75 A and subject to conditional connection IEC 61000-3-11:2000	Relevant generic standard(s)[e]	Date expired (01/11/2003)
Cenelec	EN 61000-3-12:2005 Electromagnetic compatibility (EMC) – Part 3-12: Limits – Limits for harmonic currents produced by equipment connected to public low-voltage systems with input current > 16 A and ≤ 75 A per phase IEC 61000-3-12:2004	Relevant generic standard(s)[e]	Date expired (01/02/2008)
Cenelec	EN 61000-3-12:2011 Electromagnetic compatibility (EMC) – Part 3-12: Limits – Limits for harmonic currents produced by equipment connected to public low-voltage systems with input current > 16 A and ≤ 75 A per phase IEC 61000-3-12:2011 + IS1:2012	EN 61000-3-12:2005[c]	16/06/2014
Cenelec	EN 61000-6-1:2007 Electromagnetic compatibility (EMC) – Part 6-1: Generic standards – Immunity for residential, commercial and light-industrial environments IEC 61000-6-1:2005	EN 61000-6-1:2001[c]	Date expired (01/12/2009)
Cenelec	EN 61000-6-2:2005 Electromagnetic compatibility (EMC) – Part 6-2: Generic standards – Immunity for industrial environments IEC 61000-6-2:2005 EN 61000-6-2:2005/AC:2005	EN 61000-6-2:2001[c]	Date expired (01/06/2008)

(Continued)

TABLE 1.11 (Continued)

ESO[a]	Reference and title of the harmonized standard (and reference document)	Reference of superseded standard	Date of cessation of presumption of conformity of superseded standard[b]
Cenelec	EN 61000-6-3:2007 Electromagnetic compatibility (EMC) – Part 6-3: Generic standards – Emission standard for residential, commercial and light-industrial environments IEC 61000-6-3:2006	EN 61000-6-3:2001 + A11:2004[c]	Date expired (01/12/2009)
	EN 61000-6-3:2007/A1:2011 IEC 61000-6-3:2006/A1:2010 EN 61000-6-3:2007/A1:2011/AC:2012 (new)	[d]	Date expired (12/01/2014)
Cenelec	EN 61000-6-4:2007 Electromagnetic compatibility (EMC) – Part 6-4: Generic standards – Emission standard for industrial environments IEC 61000-6-4:2006	EN 61000-6-4:2001[c]	Date expired (01/12/2009)
	EN 61000-6-4:2007/A1:2011 IEC 61000-6-4:2006/A1:2010	[d]	Date expired (12/01/2014)
Cenelec	EN 61008-1:2004 Residual current operated circuit-breakers without integral overcurrent protection for household and similar uses (RCCB's) – Part 1: General rules IEC 61008-1:1996 (Modified) + A1:2002 (Modified)	EN 61008-1:1994 + A2:1995 + A14:1998[c]	Date expired (01/04/2009)
	EN 61008-1:2004/A12:2009	[d]	Date expired (01/12/2011)
Cenelec	EN 61008-1:2012 (new) Residual current operated circuit-breakers without integral overcurrent protection for household and similar uses (RCCBs) – Part 1: General rules IEC 61008-1:2010 (Modified)	EN 61008-1:2004 and its amendment[c]	18/06/2017
Cenelec	EN 61009-1:2004 Residual current operated circuit-breakers with integral overcurrent protection for household and similar uses (RCBOs) – Part 1: General rules IEC 61009-1:1996 (Modified) + A1:2002 (Modified)	EN 61009-1:1994 + A1:1995 + A14:1998[c]	Date expired (01/04/2009)

	EN 61009-1:2004/A13:2009	d	Date expired (01/12/2011)
	EN 61009-1:2004/A12:2009	d	Date expired (01/12/2011)
	EN 61009-1:2004/AC:2006		
Cenelec	EN 61009-1:2012 (new) Residual current operated circuit-breakers with integral overcurrent protection for household and similar uses (RCBOs) – Part 1: General rules IEC 61009-1:2010 (Modified)	EN 61009-1:2004 and its amendments[c]	18/06/2017
Cenelec	EN 61131-2:2007 Programmable controllers – Part 2: Equipment requirements and tests IEC 61131-2:2007	EN 61131-2:2003[c]	Date expired (01/08/2010)
Cenelec	EN 61204-3:2000 Low voltage power supplies, DC output – Part 3: Electromagnetic compatibility (EMC) IEC 61204-3:2000	Relevant generic standard(s)[e]	Date expired (01/11/2003)
Cenelec	EN 61326-1:2006 Electrical equipment for measurement, control and laboratory use – EMC requirements – Part 1: General requirements IEC 61326-1:2005	EN 61326:1997 + A1:1998 + A2:2001 + A3:2003	Date expired (01/02/2009)
Cenelec	EN 61326-1:2013 (new) Electrical equipment for measurement, control and laboratory use – EMC requirements – Part 1: General requirements IEC 61326-1:2012	EN 61326-1:2006[c]	14/08/2015
Cenelec	EN 61326-2-1:2006 Electrical equipment for measurement, control and laboratory use – EMC requirements – Part 2-1: Particular requirements – Test configurations, operational conditions and performance criteria for sensitive test and measurement equipment for EMC unprotected applications IEC 61326-2-1:2005	EN 61326:1997 + A1:1998 + A2:2001 + A3:2003	Date expired (01/02/2009)

(Continued)

TABLE 1.11 (Continued)

ESO[a]	Reference and title of the harmonized standard (and reference document)	Reference of superseded standard	Date of cessation of presumption of conformity of superseded standard[b]
Cenelec	EN 61326-2-1:2013 (new) Electrical equipment for measurement, control and laboratory use – EMC requirements – Part 2-1: Particular requirements – Test configurations, operational conditions and performance criteria for sensitive test and measurement equipment for EMC unprotected applications IEC 61326-2-1:2012	EN 61326-2-1:2006[c]	06/11/2015
Cenelec	EN 61326-2-2:2006 Electrical equipment for measurement, control and laboratory use – EMC requirements – Part 2-2: Particular requirements – Test configurations, operational conditions and performance criteria for portable test, measuring and monitoring equipment used in low-voltage distribution systems IEC 61326-2-2:2005	EN 61326:1997 + A1:1998 + A2:2001 + A3:2003	Date expired (01/02/2009)
Cenelec	EN 61326-2-2:2013 (new) Electrical equipment for measurement, control and laboratory use – EMC requirements – Part 2-2: Particular requirements – Test configurations, operational conditions and performance criteria for portable test, measuring and monitoring equipment used in low-voltage distribution systems IEC 61326-2-2:2012	EN 61326-2-2:2006[c]	06/11/2015
Cenelec	EN 61326-2-3:2006 Electrical equipment for measurement, control and laboratory use – EMC requirements – Part 2-3: Particular requirements – Test configuration, operational conditions and performance criteria for transducers with integrated or remote signal conditioning IEC 61326-2-3:2006	EN 61326:1997 + A1:1998 + A2:2001 + A3:2003	Date expired (01/08/2009)

Cenelec	EN 61326-2-3:2013 (new) Electrical equipment for measurement, control and laboratory use – EMC requirements – Part 2-3: Particular requirements – Test configuration, operational conditions and performance criteria for transducers with integrated or remote signal conditioning IEC 61326-2-3:2012	EN 61326-2-3:2006^c	14/08/2015
Cenelec	EN 61326-2-4:2006 Electrical equipment for measurement, control and laboratory use – EMC requirements – Part 2-4: Particular requirements – Test configurations, operational conditions and performance criteria for insulation monitoring devices according to IEC 61557-8 and for equipment for insulation fault location according to IEC 61557-9 IEC 61326-2-4:2006	EN 61326:1997 + A1:1998 + A2:2001 + A3:2003	Date expired (01/11/2009)
Cenelec	EN 61326-2-4:2013 (new) Electrical equipment for measurement, control and laboratory use – EMC requirements – Part 2-4: Particular requirements – Test configurations, operational conditions and performance criteria for insulation monitoring devices according to IEC 61557-8 and for equipment for insulation fault location according to IEC 61557-9 IEC 61326-2-4:2012	EN 61326-2-4:2006^c	14/08/2015
Cenelec	EN 61326-2-5:2006 Electrical equipment for measurement, control and laboratory use – EMC requirements – Part 2-5: Particular requirements – Test configurations, operational conditions and performance criteria for field devices with interfaces according to IEC 61784-1, CP 3/2 IEC 61326-2-5:2006	EN 61326:1997 + A1:1998 + A2:2001 + A3:2003	Date expired (01/09/2009)

(Continued)

TABLE 1.11 (Continued)

ESO[a]	Reference and title of the harmonized standard (and reference document)	Reference of superseded standard	Date of cessation of presumption of conformity of superseded standard[b]
Cenelec	EN 61326-2-5:2013 (new) Electrical equipment for measurement, control and laboratory use – EMC requirements – Part 2-5: Particular requirements – Test configurations, operational conditions and performance criteria for devices with field bus interfaces according to IEC 61784-1 IEC 61326-2-5:2012	EN 61326-2-5:2006[c]	06/11/2015
Cenelec	EN 61439-1:2009 Low-voltage switchgear and control gear assemblies – Part 1: General rules IEC 61439-1:2009 (Modified) EN 61439-1:2009/AC:2013 (new)	EN 60439-1:1999[c]	01/11/2014
EN 61439-1:2009 does not give presumption of conformity without another part of the standard.			
Cenelec	EN 61439-1:2011 Low-voltage switchgear and control gear assemblies – Part 1: General rules IEC 61439-1:2011	EN 61439-1:2009[c]	23/09/2014
Cenelec	EN 61439-2:2009 Low-voltage switchgear and control gear assemblies – Part 2: Power switchgear and control gear assemblies IEC 61439-2:2009		
Cenelec	EN 61439-2:2011 Low-voltage switchgear and control gear assemblies – Part 2: Power switchgear and control gear assemblies IEC 61439-2:2011	EN 61439-2:2009[c]	23/09/2014
Cenelec	EN 61439-3:2012 Low-voltage switchgear and control gear assemblies – Part 3: Distribution boards intended to be operated by ordinary persons (DBO) IEC 61439-3:2012		

Cenelec	EN 61439-4:2013 **(new)** Low-voltage switchgear and control gear assemblies – Part 4: Particular requirements for assemblies for construction sites (ACS) IEC 61439-4:2012	
Cenelec	EN 61439-5:2011 Low-voltage switchgear and control gear assemblies – Part 5: Assemblies for power distribution in public networks IEC 61439-5:2010	
Cenelec	EN 61439-6:2012 **(new)** Low-voltage switchgear and control gear assemblies – Part 6: Busbar trunking systems (busways) IEC 61439-6:2012	
Cenelec	EN 61543:1995 Residual current-operated protective devices (RCDs) for household and similar use – Electromagnetic compatibility IEC 61543:1995	Relevant generic standard(s)[e] Date expired (04/07/1998)
	EN 61543:1995/A11:2003	[d] Date expired (01/03/2007)
	EN 61543:1995/A12:2005	[d] Date expired (01/03/2008)
	EN 61543:1995/A2:2006 IEC 61543:1995/A2:2005	[d] Date expired (01/12/2008)
	EN 61543:1995/AC:1997	
	EN 61543:1995/A11:2003/AC:2004	
Cenelec	EN 61547:2009 Equipment for general lighting purposes – EMC immunity requirements IEC 61547:2009 + IS1:2013	EN 61547:1995 + A1:2000 Date expired (01/07/2012)
Cenelec	EN 61557-12:2008 Electrical safety in low voltage distribution systems up to 1000 V AC and 1500 V DC. – Equipment for testing, measuring or monitoring of protective measures – Part 12: Performance measuring and monitoring devices (PMD) IEC 61557-12:2007	

(Continued)

TABLE 1.11 (Continued)

ESO[a]	Reference and title of the harmonized standard (and reference document)	Reference of superseded standard	Date of cessation of presumption of conformity of superseded standard[b]
Cenelec	EN 61800-3:2004 Adjustable speed electrical power drive systems – Part 3: EMC requirements and specific test methods IEC 61800-3:2004	EN 61800-3:1996 + A11:2000[c]	Date expired (01/10/2007)
	EN 61800-3:2004/A1:2012 IEC 61800-3:2004/A1:2011	[d]	19/12/2014
Cenelec	EN 61812-1:1996 Specified time relays for industrial use – Part 1: Requirements and tests IEC 61812-1:1996		
	EN 61812-1:1996/A11:1999	[d]	Date expired (01/01/2002)
	EN 61812-1:1996/AC:1999		
Cenelec	EN 61812-1:2011 Time relays for industrial and residential use – Part 1: Requirements and tests IEC 61812-1:2011	EN 61812-1:1996 and its amendment[c]	29/06/2014
Cenelec	EN 62020:1998 Electrical accessories – Residual current monitors for household and similar uses (RCMs) IEC 62020:1998		
	EN 62020:1998/A1:2005 IEC 62020:1998/A1:2003 (Modified)	[d]	Date expired (01/03/2008)
Cenelec	EN 62026-1:2007 Low-voltage switchgear and control gear – Controller-device interfaces (CDIs) – Part 1: General rules IEC 62026-1:2007	Relevant generic standard(s)[c]	Date expired (01/09/2010)
	EN 62026-1:2007 does not give presumption of conformity without another part of the standard.		
Cenelec	EN 62026-2:2013 **(new)** Low-voltage switchgear and control gear – Controller-device interfaces (CDIs) – Part 2: Actuator sensor interface (AS-i) IEC 62026-2:2008 (Modified)	EN 50295:1999[c]	03/12/2015

Cenelec	EN 62026-3:2009 Low-voltage switchgear and control gear – Controller-device interfaces (CDIs) – Part 3: Device Net IEC 62026-3:2008	Relevant generic standard(s)[c]	
Cenelec	EN 62026-7:2013 **(new)** Low-voltage switchgear and control gear – Controller-device interfaces (CDIs) – Part 7: CompoNet IEC 62026-7:2010 (Modified)		
Cenelec	EN 62040-2:2006 Uninterruptible power systems (UPS) – Part 2: Electromagnetic compatibility (EMC) requirements IEC 62040-2:2005	EN 50091-2:1995[c]	Date expired (01/10/2008)
	EN 62040-2:2006/AC:2006		
Cenelec	EN 62052-11:2003 Electricity metering equipment (AC) – General requirements, tests and test conditions – Part 11: Metering equipment IEC 62052-11:2003	Relevant generic standard(s)[e]	Date expired (01/03/2006)
EN 62052-11:2003 does not give presumption of conformity without a part of the EN 62053 series.			
Cenelec	EN 62052-21:2004 Electricity metering equipment (AC.) – General requirements, tests and test conditions – Part 21: Tariff and load control equipment IEC 62052-21:2004	EN 61037:1992 + A1:1996 + A2:1998 + EN 61038:1992 + A1:1996 + A2:1998[c]	Date expired (01/07/2007)
EN 62052-21:2004 does not give presumption of conformity without a part of the EN 62054 series.			
Cenelec	EN 62053-11:2003 Electricity metering equipment (AC.) – Particular requirements – Part 11: Electromechanical meters for active energy (classes 0,5, 1 and 2) IEC 62053-11:2003	EN 60521:1995[c]	Date expired (01/03/2006)

(Continued)

TABLE 1.11 (Continued)

ESO[a]	Reference and title of the harmonized standard (and reference document)	Reference of superseded standard	Date of cessation of presumption of conformity of superseded standard[b]
Cenelec	EN 62053-21:2003 Electricity metering equipment (AC.) – Particular requirements – Part 21: Static meters for active energy (classes 1 and 2) IEC 62053-21:2003	EN 61036:1996 + A1:2000[c]	Date expired (01/03/2006)
Cenelec	EN 62053-22:2003 Electricity metering equipment (AC.) – Particular requirements – Part 22: Static meters for active energy (classes 0.2 S and 0.5 S) IEC 62053-22:2003	EN 60687:1992[c]	Date expired (01/03/2006)
Cenelec	EN 62053-23:2003 Electricity metering equipment AC.) – Particular requirements – Part 23: Static meters for reactive energy (classes 2 and 3) IEC 62053-23:2003	EN 61268:1996[c]	Date expired (01/03/2006)
Cenelec	EN 62054-11:2004 Electricity metering (AC.) – Tariff and load control – Part 11: Particular requirements for electronic ripple control receivers IEC 62054-11:2004	EN 61037:1992 + A1:1996 + A2:1998[c]	Date expired (01/07/2007)
Cenelec	EN 62054-21:2004 Electricity metering (AC.) – Tariff and load control – Part 21: Particular requirements for time switches IEC 62054-21:2004	EN 61038:1992 + A1:1996 + A2:1998[c]	Date expired (01/07/2007)
Cenelec	EN 62135-2:2008 Resistance welding equipment – Part 2: Electromagnetic compatibility (EMC) requirements IEC 62135-2:2007	EN 50240:2004[c]	Date expired (01/02/2011)
Cenelec	EN 62310-2:2007 Static transfer systems (STS) – Part 2: Electromagnetic compatibility (EMC) requirements IEC 62310-2:2006 (Modified)	Relevant generic standard(s)[c]	Date expired (01/09/2009)

	Reference and title of the standard	Reference of superseded standard	Date
Cenelec	EN 62423:2009 Type B residual current operated circuit-breakers with and without integral overcurrent protection for household and similar uses (Type B RCCBs and Type B RCBOs) IEC 62423:2007 (Modified)		
Cenelec	EN 62423:2012 (**new**) Type F and type B residual current operated circuit-breakers with and without integral overcurrent protection for household and similar uses IEC 62423:2009 (modified)	EN 62423:2009^c	19/06/2017
Cenelec	EN 62606:2013 (**new**) General requirements for arc fault detection devices IEC 62606:2013 (modified)		
ETSI	EN 300 386 V1.4.1 Electromagnetic compatibility and radio spectrum matters (ERM); Telecommunication network equipment; Electromagnetic compatibility (EMC) requirements	EN 300 386 V1.3.3^c	Date expired (31/07/2011)
ETSI	EN 300 386 V1.5.1 Electromagnetic compatibility and radio spectrum matters (ERM); Telecommunication network equipment; Electromagnetic compatibility (EMC) requirements	EN 300 386 V1.4.1^c	Date expired (31/01/2014)
ETSI	EN 300 386 V1.6.1 Electromagnetic compatibility and radio spectrum matters (ERM); Telecommunication network equipment; Electromagnetic compatibility (EMC) requirements	EN 300 386 V1.5.1^c	30/11/2015
ETSI	EN 301 489-1 V1.9.2 Electromagnetic compatibility and radio spectrum matters (ERM); Electromagnetic compatibility (EMC) standard for radio equipment and services; Part 1: Common technical requirements	EN 301 489-1 V1.8.1^c	Date expired (30/06/2013)

(Continued)

TABLE 1.11 (Continued)

ESO[a]	Reference and title of the harmonized standard (and reference document)	Reference of superseded standard	Date of cessation of presumption of conformity of superseded standard[b]
ETSI	EN 301 489-34 V1.1.1 Electromagnetic compatibility and radio spectrum matters (ERM); Electromagnetic compatibility (EMC) standard for radio equipment and services; Part 34: Specific conditions for external power supply (EPS) for mobile phones		
ETSI	EN 301 489-34 V1.3.1 Electromagnetic compatibility and radio spectrum matters (ERM); Electromagnetic compatibility (EMC) standard for radio equipment and services; Part 34: Specific conditions for external power supply (EPS) for mobile phones	EN 301 489-34 V1.1.1[c]	28/02/2014
ETSI	EN 301 489-34 V1.4.1 (new) Electromagnetic compatibility and radio spectrum matters (ERM); Electromagnetic compatibility (EMC) standard for radio equipment and services; Part 34: Specific conditions for external power supply (EPS) for mobile phones	EN 301 489-34 V1.3.1[c]	28/02/2015

[a] ESO: European standardization organization: CEN, http://www.cen.eu; CENELEC, http://www.cenelec.eu; ETSI, http://www.etsi.eu.

[b] Generally the date of cessation of presumption of conformity will be the date of withdrawal ("dow"), set by the European standardization organization, but attention of users of these standards is drawn to the fact that in certain exceptional cases this can be otherwise.

[c] The new (or amended) standard has the same scope as the superseded standard. On the date stated, the superseded standard ceases to give presumption of conformity with the essential or other requirements of the relevant Union legislation.

[d] In the case of amendments, the referenced standard is EN CCCCC:YYYY and its previous amendments, if any, and the new, quoted amendment. The superseded standard therefore consists of EN CCCCC:YYYY and its previous amendments, if any, but without the new quoted amendment. On the date stated, the superseded standard ceases to give presumption of conformity with the essential or other requirements of the relevant Union legislation.

[e] The new standard has a narrower scope than the superseded standard. On the date stated the (partially) superseded standard ceases to give a presumption of conformity with the essential or other requirements of the relevant European Union legislation for those products or services that fall within the scope of the new standard. Presumption of conformity with the essential or other requirements of the relevant European Union legislation for products or services that still fall within the scope of the (partially) superseded standard, but that do not fall within the scope of the new standard, is unaffected.

1.9 REVIEW PROBLEMS

Problem 1.1
Immunity to radiated emissions has a de facto standard and medical standard of _____ and _____ V/m respectively.

Problem 1.2
Immunity allow a system or equipment to function with no reduction in performance in the presence of EMI noise:
a) True
b) False.

Problem 1.3
FCC emission is limited to _____ at 3 m at 35.5 MHz.

Problem 1.4
The European Union has no limits on immunity to EMI noise:
a) True
b) False.

Problem 1.5
Construction of the communication house or signal bungalow may require the use of generic standards if the railway standards do not comply with the mission requirements:
a) True
b) False.

Problem 1.6
If you are checking emissions from the signaling system which European standard would you use?

Problem 1.7
To check whether an FM station is affecting your emission, using an OATS measurement at 3 m for signal system components, what is the maximum emission _____ V/m or _____ dBµV/m?

1.10 ANSWERS TO REVIEW PROBLEMS

Problem 1.1
3 and 10 V/m respectively.

Problem 1.2
The ability of equipment or devices to function as required in the presence of EMI noise without a degradation in performance.

Problem 1.3
40 dBµ/m.

Problem 1.4
False.

Problem 1.5
True.

Problem 1.6
EN 50121-4:2006.

Problem 1.7
FCC emission limit regulations measured at 3 m, class B 44.7 μV/m or 33.0 dBμV/m.

2

FUNDAMENTALS OF COUPLING CULPRIT TO VICTIM

2.1 RADIATION EFFECTS ON EQUIPMENT AND DEVICES

The electric field in volts/meter can be calculated using a simple formula, see Equation 2.1.

$$E = \left(30 * \text{ERP}\right)^{1/2} / R \tag{2.1}$$

Where ERP = Radiated power * Antenna gain, R = distance from equipment or device in meters.

Table 2.1 has been provided depicting some of the most common radios. The student is urged to add on to the table for new wireless devices that may present a problem later.

2.1.1 Conducted Emissions

These types of emissions are due to capacitive, inductive or impedance transfer (poor grounding of equipment). Tables 1.3 and 1.9 in Chapter 1 are the emission limits set by the Federal Communication Commission (FCC) in CFR 47 Regulations Part 15. This is also where the radiation limits are located the data can be downloaded free from the FCC web site in tables. The next objective is to determine how the various emissions are coupled into the devices, equipment and systems.

Electromagnetic Compatibility: Analysis and Case Studies in Transportation, First Edition.
Donald G. Baker.
© 2016 John Wiley & Sons, Inc. Published 2016 by John Wiley & Sons, Inc.
Companion website: www.wiley.com/go/electromagneticcompatibility

TABLE 2.1 Radiation Effects on Electronic Equipment

	Cell phones	Walkie talkie	Walkie talkie	Micro cellular	Comment
Power output	600 mW	3 Watt	5 Watt	1 mW	Micro cellulars are used in hospitals
Antenna gain	1.6	1.6	1.6	1.6	
ERP	0.96 W	4.8 W	8.0 W	1.6 mW	
Distance in cm			Volts/meter		
300	1.8	4	5.2	0.07	(a) Very few electronics will fail at 1 V/m
200	2.7	6	7.7	0.11	(b) About 50% of the victim circuits will fail at 20–50 V/m
150	3.5	8	15.5	0.14	(c) Most victim circuits will malfunction at 100 V/m
100	5.4	12	15	0.22	
75	7.2	16	21	0.29	
50	11	24	31	0.49	
30	19	40	52	0.73	
20	27	60	77	1.1	
15	35	80	103	1.4	
10	54	120	155	2.2	
5	107	240	310	4.4	

2.2   VARIOUS TYPES OF EMISSION COUPLING

The various emission couplings are as follows:

1. Inductive coupling
2. H (field) coupling
3. Capacitive coupling
4. E (field) coupling
5. Transfer impedance due to poor grounding
6. Combinations of all the others.

The various types of emission couplings are described in Figure 2.1. These are only a few of the methods that allow noise to get on signals and power lines. Figure 2.1a depicts the coupling between adjacent wire capacitors, C_b. They are connected to different circuits. The two capacitors C_a and C_c are the capacitance to ground. Capacitors C_a and C_b form a capacitance transformer for all of the capacitance in Farads/meter $\left[C_b / (Cb + C_a) \right] \times E_{\text{Culprit Noise}} = E_{\text{Victim Noise}}$.

Figure 2.1b is a representation of a wire with shielding. This shows the capacitance between the shield and the single wire (coaxial or power lead that may require some shielding). C_1 will couple in noise from the shield if the length of the wire and shield is longer than $1/20\lambda$. If longer than this value the shield should be grounded at both ends. Inductive coupling is represented in Figure 2.1c. This is the mutual (M) inductance between two wires. The inductance shown is considered the inductance/ meter of each wire. The coupling is the inductance due to magnetic flux in the Culprit inducing noise on the Victim wire. It is similar to a one-turn transformer. The diagram in Figure 2.1d is a common mistake made by many engineers and technicians, that is daisy chaining the grounds of cabinets, equipment or devices as shown in the diagram. A possibility is that the cabinet ground is isolated from the signal or power ground and capacitance must be added to the diagram showing this capacitance that may produce a ground loop. Note, the inductance or resistance of the ground wire may lift cabinet C off the ground at certain frequencies. The best solution is to ground each cabinet to a central ground bus bar.

Figure 2.1e is an example of the use of a wireless devices, cell phone or two-way radio operated in front of an open rack or an enclosure with the door open. The devices or equipment has printed circuit boards such as SCADA, signals, audio public address, video, fire alarm, fire suppression and security or communication equipment. These types of equipments and devices are commonly found in most transportation communication bungalows. The E field leakage from the devices can enter through vents, doors without seals, holes cut for wiring and any other openings in the cabinet. Operating with the doors open will occasionally occur to the untrained person. The E field entering through small openings may form an aperture antenna that can even produce gain. If the longest dimension of the aperture is the about $1/2\lambda$ or $1/4\lambda$ (0.5λ or 0.25λ) this could produce some serious problems. Open racks are often used when shielding of the equipment is considered more than adequate;

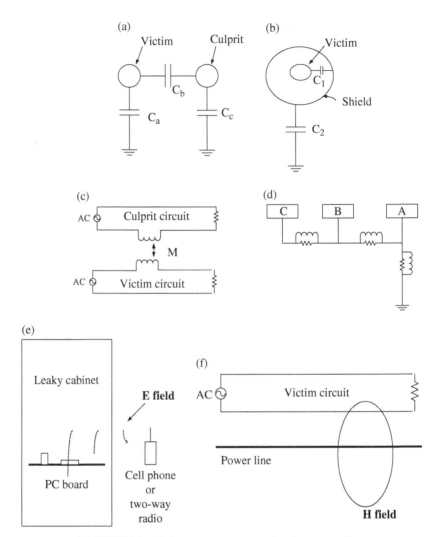

FIGURE 2.1   Various common types of emission coupling

however, the wiring to the equipment power and ground may be a problem over-looked during the system design.

Figure 2.1f is a description of how a power line 60 Hz hum can enter a signal circuit such as a signal line of an audio circuit. The H field = $I/(2\pi\, r)$ A/m where r = radius of the H field coupled into the loop formed by the victim circuit. $B = \mu H$ webers/m^2 is the magnetic field density. The $\psi = B \times$ (area of the loop) = Weber's magnetic field and $V = \omega\,\psi$ Volts induced on the loop of the culprit circuit. Figure 2.2 is a practical example of how a culprit extraneous field can enter a circuit. The EMI can be a combination of E and H fields. Since both are vectors the fields must be broken down into the fields along each axis. Maxwell's equations are often employed to isolate the

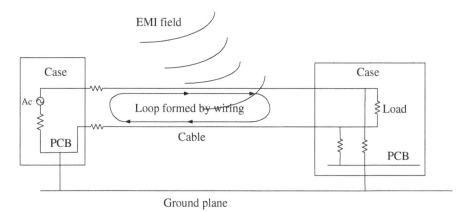

FIGURE 2.2 Ground loop in a circuit from an extraneous source

E and H fields that are difficult to separate to solve the necessary equation for each axis; however, if the divergence of Maxwell's vector equations is taken and the resulting wave equations derived, solutions can be extracted as shown in Chapter 3. This will decouple the E and H fields.

Example 2.1

If Figure 2.2 has an H field perpendicular to the loop that is generated by a power source such an inverter will generally have harmonics of 60 Hz. Harmonics from subway DC power supplies are generally odd harmonics: 5th, 7th, 9th and 11th. If the field is from a power feed near the loop that supplies 100 Amps to a device or system at 60 Hz with a 5% third harmonic calculate the effect on the loop at 2 meters. The space between the wires is 1 cm; the length of the wire is 10 meters. This is a general example and does not necessarily represent a subway application. Later in Chapter 4 a subway application is presented. The fundamental approach will be expanded upon to provide the audience with the complex view of the problems. Use Equations 2.2–2.5 for one-dimensional problems.

$$H = I / 2\pi R \tag{2.2}$$

$$B = \mu_{air} H \tag{2.3}$$

$$\Psi = B * (\text{area of loop A}) \tag{2.4}$$

$$V = \omega (\text{rad/sec}) * \Psi \tag{2.5}$$

1. $H \text{ field} = 7.95\,A/m\,\text{at}\,60\,Hz$
2. $H \text{ field} = 0.397\,A/m\,\text{at}\,180\,Hz$

3. Area $= 0.01 \times 10 = 0.1 \ \text{m}^2$ area of the loop

4. $B_{60\text{Hz}} = 7.95 * 4\pi \times 10^{-7} \ \text{webers/m}^2 = 1.9 \times 10^{-5} \ \text{webers/m}^2 \text{at } 60\text{Hz}$

5. $B_{180\text{Hz}} = 0.397 * 4\pi \times 10^{-7} \ \text{webers/m}^2 = 5.0 \times 10^{-7} \ \text{webers/m}^2 \text{at } 180\text{Hz}$

6. $\Psi_{60\text{Hz}} = 1.9 \times 10^{-5} \ \text{webers/m}^2 \times 0.1 \text{m}^2 = 1.9 \times 10^{-6} \ \text{webers}$

7. $\Psi_{180\text{Hz}} = 5.0 \times 10^{-7} \ \text{webers/m}^2 \times 0.1 \text{m}^2 = 5.0 \times 10^{-8} \ \text{webers}$

8. $V_{60\text{Hz}} = 1.9 \times 10^{-6} \ \text{webers} \times 2\pi \times 60 = 0.715 \text{mV}$

9. $V_{180\text{Hz}} = 5.0 \times 10^{-8} \ \text{webers} \times 2\pi \times 180 = 0.0.056 \text{mV}.$

Example 2.2

If the line above is an audio line for a telephone, the thermal noise on the line is = –174 dBm/Hz. For a 3000 Hz bandwidth the noise level is –174 + 10Log 3000 = –139.2 dBm. The noise due to 60 and 180 Hz for a 600-ohm line is calculated below.

1. $W_{60\text{Hz}} = \left(10\text{Log}\, 0.715 \text{mV}^2/600\right) + 30 = -60.7 \text{dBm} \left(\text{dBm ref. to } 1\text{mW}\right)$

2. $W_{180\text{Hz}} = \left(10\text{Log}\, 0.056 \text{mV}^2/600\right) + 30 = -82.8 \text{dBm}$

3. Audio will be –37 to –42 dBm

4. SNR for 60 Hz –42 + 60.7 = 18 dB, adequate for speech >15 dB.

If the length of the wiring is 100 meters with all other parameters the same, the voltage impressed on the loop area is 10 times larger. The voltage is also 10 times larger or 20 dB for 60 Hz; for 7.15 mV the noise power at 60 Hz is –40 dBm. Thus the noise power is larger than the signal power. The simplified approach to loop analysis can be used because the separation between wiring is very small. For larger separation dimensions calculus must be used to get closer approximations to the magnetic flux Ψ. This will be covered later when dealing with ground loops in Chapter 4 in a more complex form. This is a one-dimensional approach to the problem as is often the case, due to many analyses in the far field. Chapter 4 has some case analyses that are not so straightforward. The 100 meter wiring is installed in conduit or duct banks for transportation applications. The duct banks are tagged for the type of wiring such as power, signal lines, audio, SCADA or communications, but contractors occasionally run wiring in the wrong duct. This can become a serious problem.

2.3 INTERMODULATION

Table 2.2 gives an example of intermodulation (IM) products due to unintentional mixing of radiators and radios. This is only a short list of equations. All the combinations of frequencies for the fourth order or higher are not shown. The list is complete for third and second order products. The extended list can be searched on the Internet. The objective of the table is to show the audience how to generate the equations for

TABLE 2.2 Examples of intermodulation Products

Order of IM	Frequencies (MHz)	Victim receiver	Hit
Second A−B	857.25 − 444.5	412.75	412.75 D hit 412.5 ND hit
Third A−B+C	902 − 425.25 + 428.5	905.25	905.25 D hit 905.0 ND hit
Third A+B+C	902 + 425.25 + 428.5	1755.75	1755.75 D hit Spread spectrum
Third 2A−B	2*857.25 − 902	812.5	812.5 D hit 812.25 ND hit
Third 2A+C	2*857.25 + 902	2616.5	2616.5 D hit Spread spectrum
Third 3A	3*508.25	1524.75	1524.75 D hit Spread spectrum
Fourth A+B+C−D	902 + 425.25 + 428.5−902	853.75	853.75 D hit 853.5 ND hit
2A−B+C	2*857.25 − 902 + 812.25	1624.75	1624.75 D hit Spread spectrum
3A+B	3*508.25 + 425.50	1950.25	1950.25 D hit Spread spectrum

higher order products. Direct hits on a radio receiver cannot be filtered out. Some form of separation or shielding must be used to remove the culprit or culprits from the receiver. Often a directional antenna can solve the problem. When several antennas are stacked on the same mast separating the antennas may be the only means of solving the intermodulation problem. The phrase "rusty bolt problem" came into being because un-bonded joints that exhibit corrosion will have a diode action across the joint, especially if it consists of dissimilar metals. Mixing occurs at the joint, producing the IM products.

A short list of IM products is provided below:

A−B, A+B, 2A are second order products

3A, 2A−B, A+B−C are third order products

3A−B, 2A−B−C are fourth order products

3A−B−C, A+2B−2C, A+B+C−D−E are fifth order products.

Examples of how the IM products are used to calculate hits on radios is described in Table 2.2. The hits are labeled as direct (D hits) and non-direct (ND hits). A D hit cannot be filtered because it is at the same frequency as the carrier. The ND hits can be filtered with careful filtering such as a notch filter.

In some cases shielding can reduce the hit magnitude, if it is due to ground plane anomalies or changing the antenna polarization of the culprit to something manageable.

Occasionally the antenna producing the hit can be replaced with a directional antenna that will also reduce the hit. The hits may be produced by high rise-time transients that make the culprit source difficult to identify, and it will require a storage oscilloscope to capture the waveform image.

Example 2.3

This is a typical example of how a cell phone can impact the operation of navigation equipment on an aircraft. Radios are not at these exact frequencies, so this is a rough estimate of how the calculations can be made. The third harmonic of the navigation channel was chosen it would be the strongest signal from the radio transmitters. The receivers have a minimum sensitivity of −118 dBm. Some of the newer digital radios have a minimum sensitivity of −124 dBm. Cell phones operate at nearly 1 W (30 dBm). As can be observed in Figure 2.3, the skin of the aircraft is aluminum and a good ground plane, thus the signal in the cockpit can be very large at the instrument panel for passengers operating cell phone close to a structure grounded to the fuselage skin. If multiple passengers are allowed to use cell phones the noise can be very large. It is therefore unwise to allow cell phone to operate while the aircraft is aloft or taxiing.

A cell phone operating on aircraft can produce IM products in the navigation system.

Cell phone	857 MHz
Navigation radio operates in the frequency band	325.6–335.4 MHz
ILS receiver in the pilot's compartment operates in the frequency band	108–118 MHz

Intermodulation $3A - B = 3*325 - 857 = 118$ MHz direct hit on the navigation system.

Most radios operate at a carrier frequency with a four-digit fraction of a megahertz, such as 325.4325 for example. The exact frequencies must be used to calculate IM. A direct hit is exactly on the carrier, while a small offset can be filtered out, that is 0.025 MHz from the exact carrier frequency. When doing an analysis this should

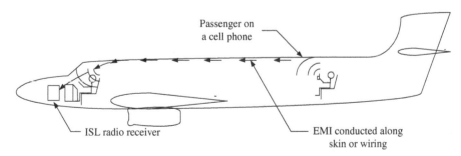

FIGURE 2.3   Example of an aircraft with a passenger operating a cell phone

be kept in mind. This is considered a narrow band radio receptor; there are a host of others. The newer aircraft have fiber optics onboard and the receivers are high gain devices and generally well shielded. The newer aircraft have composite and smart skins, so excluding the use of cell phones may not be necessary. Cell phones are not the only culprit sources onboard aircraft; wireless devices such as games and computers are also a source of culprit emissions.

A single cell phone may not be a problem but mixing can occur between harmonics and can fall within the ILS receiver or GPS bandwidth. Harmonics are generated by non-linear components such as diodes. Two dissimilar metals with slight corrosion will appear as a diode; dirty or corroded connectors are a particular problem. A problem occurred on an aircraft due to a placard $2 \times 4''$ (5×10 cm) glued to the fuselage of the plane. It formed a diode that caused erroneous ground level reading (discussed by Professor Dr Chamberlin in a problems class [3]).

A short list of victim devices, circuits and equipment that can be affected by radiation and conducted emission or both is as follows:

Victim circuits and other receptors

1. Analog sensors, audio amplifiers, instrumentation amplifiers
2. Fiber optic receivers and other high gain devices
3. Industrial controllers and microprocessors
4. Audio equipment, especially microphone leads
5. Computers and peripheral devices
6. Radio narrow band receivers
7. Human body, especially at 300 MHz (may be hazardous).

Electromagnetic interference problems have occurred that seemed rather mundane but had serious consequences. Here is a case in point: a simple problem due to a chain link fence occurred on a subway system. Fence corrosion formed a diode that resulted in the mixing of two radio waves that radiated a signal resulting in EMI to a radio in the form of crosstalk at the receiver located in the transit car. This occurred in a transit system where radio communication was seriously affected and it is considered a life and safety issue. Often radios communicate with the dispatcher via a leaky coaxial cable in the tunnels. Dirty connectors or corrosion in connectors can produce crosstalk on the cable that can impact receiver and transmitter performance.

2.4 COMMON MODE REJECTION RATIO

Figure 2.4 is an example of a common mode noise on signal and ground lines. This can be the result of circuit board traces or no central grounding point. The amplifier is single ended with R_L at the low signal side or ground. The common mode signal will appear on the low side input but not on the R_h high side.

This will produce an undesirable common mode signal at the output along with the desired signal. In audio amplifiers, this can appear as hum when close to a 60 Hz

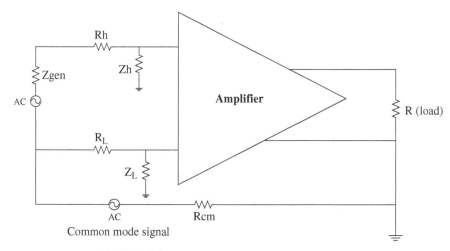

FIGURE 2.4   An amplifier with common mode noise

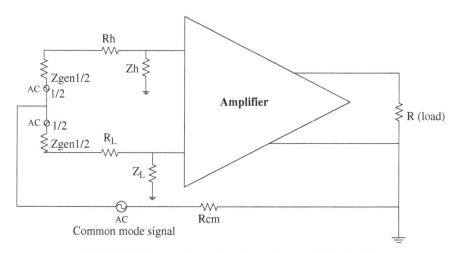

FIGURE 2.5   Amplifier with balanced input and high CMRR

power source. To reduce or eliminate the problem, a balanced amplifier input is implemented as shown in Figure 2.5. The common mode signal can be reduced or canceled to produce a clean signal at the load. When wiring an amplifier for long distance transmission, this technique is often used. A calculation is provided to demonstrate the effect of the Common mode rejection ratio on signal lines.

This technique will assist circuit designers on deciding whether to implement shielding, twisted pair or both to reduce noise on signal lines. An amplifier is shown in this application; however, balancing the input and output with and without amplifiers is also an option if no gain is necessary. This technique is shown further in this chapter.

Example 2.4

An amplifier has a single ended connection as shown in Figure 2.4. The groundside has V_{Incm} = 0.15 V of hum. The input signal is 20 mV, about the same as a telephone line. For a gain of 100 the noise and signal would be amplified. The hum on the line would be annoying to the listener and could produce crosstalk to other circuits. If the amplifier is wired as shown in Figure 2.5, Equation 2.6 applies. Equation 2.6 indicates the effect of a CMRR (80 dB amplifier specification) on the circuit.

$$CMRR = 20LogA * V_{Incm}/V_{Outcm} \qquad (2.6)$$

$$A * V_{Incm}/V_{Outcm} = anti\text{-}Log CMR/20 = 10\,000$$

$$10\,000 = 100 * 0.15/V_{Outcm}. \text{ Then } V_{Outcm} = 1.5\,mV\,(culprit\,signal\,level).$$

The signal to noise ratio (SNR) is SNR = 20Log 20 mV/1.5 mV or 25.4 dB is a reasonable value for voice communication.

The amplifiers shown are not necessarily operational amplifiers; the balance connection provides the amplifier specification with the required CMRR. Differential operational amplifiers have an input stage that is balanced with a constant current source with a small capacitance across it. As the frequency increases, the gain of each of the differential input (common emitter connection) transistors increases. This also increases the common mode gain A_{cm}. The differential gain A_{diff} remains constant. CMRR is also equal to 20 Log (A_{diff}/A_{cm}). Therefore, as the frequency of the input signal increases the CMRR will decrease.

2.5 SUSCEPTIBILITY AND IMMUNITY

The equipment or device responds to extraneous noise from various sources, for example crosstalk between radio channels or wiring in a cable. The objective of EMC is to allow the equipment or device to operate satisfactorily in a noisy environment. This is commonly called immunity. The objective of this book course is to design cabling, equipment, devices or radios to have the necessary immunity and achieve the necessary performance. The next section is devoted to achieving this goal using various methods.

The common solution techniques for EMC problems are illustrated in the flow diagram (Figure 2.6). The first item to investigate is the electromagnetic ambient for extraneous signals. The objective is to find any emissions of both E and H fields that violate FCC or MIL-STD-461 standards, depending on the application. This same process can be applied for all other standards, including CSA, EU, medical equipment, light-rail systems and so on. If a violation of a particular standard is present in the ambient electromagnetic fields, determine if the emission is from a licensed radiator or unlicensed radiators.

When emissions are of sufficient magnitude to affect the equipment installed and EMI problems are apparent, informing the FCC or equipment operator is an option

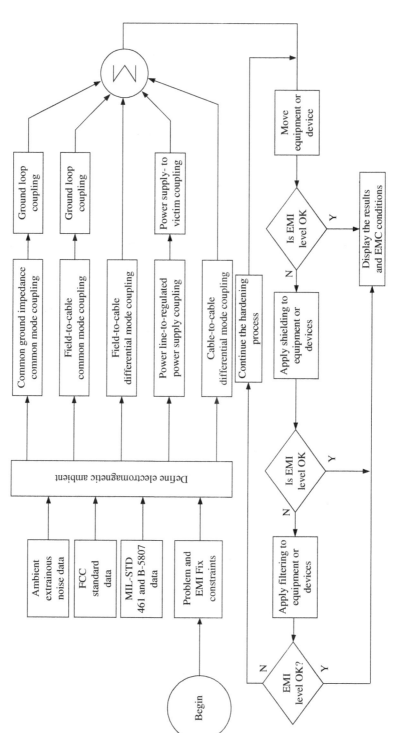

FIGURE 2.6 Flowchart of techniques for EMC

to solve the problem for both licensed and unlicensed radiators. Licensed radiators are required to fix the problem using filtering, shielding or both. The equipment operator causing the EMI to installed equipment has the responsibility to fix the problem. In some cases, reporting the EMI issue to the FCC may result in the culprit equipment being shut down until the operator fixes the problem. In some cases only a line filter to the victim's equipment may be all that is necessary. An example of EMI may be a ham radio or base station. These operators are very conscientious and will generally fix the problem without resorting to a FCC complaint. If the radiation source is undetermined, then moving equipment, shielding or filtering may be required.

EMI coupling to equipment can be one or multiple methods shown in Figure 2.6. When daisy-chained grounds and both ends are not grounded at the same point, a ground loop is formed that can capture electric and magnetic fields that will circulate noise current on the grounds. Providing a bus as a common ground point is a possible solution.

Cable to cable coupling is the result of a magnetic field around a cable carrying current that produces magnetic coupling to an adjacent cable. This is similar to a one-turn transformer with mutual inductance between the two cables. This coupling will produce crosstalk between the two cables. If the frequency is high enough so that one of the cables radiates similar to an antenna, this indicates E (field) coupling through capacitance between the cables. A wire can become an antenna if it is greater than $1/20$ of a wavelength (λ). Leads to devices or equipment should be kept as short as possible, preferably $\leq 1/20\lambda$. A future discussion will illustrate methods to combat cable to cable coupling. The separation of cables, if possible, is the simplest method to eliminate EMI. If possible, the use of steel conduit will shield cables from both electric and magnetic fields. Separating the culprit and victim, twisted pair cable, shielding or filtering are several options to consider.

Field to cable common coupling is the result of a strong extraneous H or E field that couple into a cable, single wire or multiple wires. Power lines and power supplies are typical culprit circuits but not the only source. This is addressed later in the document when means are discussed to combat the problem. The flow chart in Figure 2.6 indicates the various methodologies for controlling emissions to provide the necessary immunity to produce a viable robust system or product. The discussion that follows will provide the means to complete this task. The flowchart is meant as a guide and it does require the solutions to be in sequence. At times, the separation of culprit and victim cannot be done or shielding cannot completely solve the problem. For example, radio channels may have filtering as the only option.

The currents shown in Figure 2.7 are the Is signal current and Ir signal return. The H field is identical for each loop. Examine how the extraneous H field affects the noise coupled into the loops by the magnetic field B. Note, the first loop nearest the generator is clockwise and the next loop circulates counterclockwise; thus the two loop currents generated by the external field cancel out. All of the succeeding loop currents continue alternating between clockwise and counterclockwise circulation. However, the loops do not have exactly the same area. Assume they are within 10% of having equal area. This is a fairly good estimate.

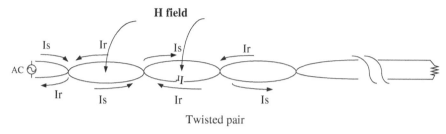

Twisted pair

FIGURE 2.7   Twisted pair as a method to reduce EMI

Example 2.5

Use the equations in Example 2.1 for the calculations.

If the loops each have an H (field) = 10 A/m at a frequency of 60 Hz,
the B field = $4\pi \ 10^{-6}$ webers/m^2.
Use Equations 2.2, 2.3.

Generally twisted pair wiring will have a pitch (twists/foot) of 6–12. The example
will use a pitch of 6. Each twist length L = 1/6 per ft (L =1/6 per 0.05 m at
0.305 m = 1 ft). The separation of the wire is approximately 4 mm. Insulation
+ Air gap area A of a loop is approximately:

$A = 0.05 * 0.004 = 2 * 10^{-4} \, m^2$, the magnetic flux $\psi = B * A = 8\pi 10^{-10}$ webers

$V = \omega * B * A$

$V = 2 * \pi * 60 * 8\pi \ 10^{-10} \, \text{Volts} = 9.495 * 10^{-7} \, \text{Volts/Loop}$

$V_{10\%} = 9.496 * 10^{-8} \, \text{Volts} = 9.496 * 10^{-2} \, \mu V$ or –20 dB μV noise injected due to
mismatch in loop areas. If the signal level is 16 mV telephone line amplitude,
the SNR $= 20 \text{Log} 16 * 10^3 / -20 \text{dB} \mu V = 84 - 20 = 64 \text{dB}$

This example depicts how twisted pair wiring can improve the signal to noise of a
telephone circuit. Increasing the frequency will reduce the isolation if the loop area
A and B field are held constant. Increasing the Pitch will reduce the noise coupled in
by the magnetic flux. The current I Amps producing the H field can be due to power
line 3.1 A/m circuit 0.5 m from the telephone line. The cable often has a copper or
aluminum shield with negligible effect on low frequency magnetic fields below
10 KHz. Steel conduit is the material of choice for keeping power line magnetic
fields under control. The calculations are only an estimate the separation in the wire;
the length of loop and the amount of untwisted cable at the ends can have a significant
effect compared to actual tests. This can be as much as 10–20 dB.

The use of twisted pair cabling is a common method to reduce feedback audio
signal coupling into the microphone line. The audio amplifier can have a 70 V output
with the microphone input in the millivolt range. The larger the number of twists per
unit length the greater the H (field) rejection. All speaker and microphone wiring is
twisted pair to prevent hum and decrease the possibility of feedback. Shielded twisted

pair wiring is often used on microphone wiring with the precaution that shielded wire has capacitance from shield to signal and shield to return lines. These lengths should be kept less than $1/20\lambda$.

2.5.1 Calculations for Shielding

Example 2.6

A typical problem is how to shield an area from RF using a screen room. The question is how thick must the copper screen be to attenuate a 125 MHz signal by at least 100 dB? The second question is what are the minimum dimensions of the screen to approximate a solid shield? The second question is easy to answer: screen mesh $< 1/20\lambda$.

$\lambda = 3.1\times10^8/1.25\times10^8 = 2.48$ m. Screen mesh is generally 0.124 m or less.

Screen comes in 0.003 m (3 mm) standard thickness.

The only loss accounted for are those within the copper itself and not due to imperfect boundaries. Copper conductivity is $\sigma = 5.8\times10^7$ S/m.

$$\alpha \approx (\pi f \mu \sigma)^{1/2} \quad \text{factor/meter} \tag{2.7}$$

Where f = frequency, μ = magnetic permeability and α = attenuation.

$$\alpha = (\pi * 125\times10^6 4\pi\times10^{-7} * 5.8\times10^7)^{1/2}$$
$$\alpha = 1.692\times10^5 \tag{2.8}$$
$$-100 = 20\text{Log}_{10}(e^{-\alpha d})$$

$$5 = \alpha d * \text{Log}_{10}(e)$$
$$d = 5/(1.692\times10^5 * 0.434)\text{m}$$
$$d = 6.8\times10^{-5}\text{m or } 0.68\text{mm}$$

The thickness of the screen used to construct the screen room must be greater than 0.068 mm to allow a little margin for error. If a screen room is to be constructed, the best procedure is to determine the highest frequency to be shielded. Then determine the shortest wavelength, this will provide the minimum mesh size. It would be prudent to calculate several frequencies between the lowest and highest frequencies using EXCEL spread sheets or perhaps graph the results of the spreadsheet for later use.

Example 2.7

A communication house is made of concrete with 0.5 inch rebar mesh 6 inches center to center on the rebar. A more detailed version of this communication house is provided in the case studies. This can be considered a Faraday shield, neglecting the effect of the concrete. Calculate the shielding effect for an f = 60 Hz on low carbon steel, use the following rebar parameters:

$$\mu_r\mu = (2000)4\pi \times 10^{-7} \text{ Where } \mu_r = 2000 \text{ and } \sigma = 8.33 \times 10^6 \text{ S/m}$$

$$d = (0.5/12) \times 0.305 = 1.271 \times 10^{-2} \text{ m}$$

$$\alpha = (\pi 60 * 2000 * 4\pi \times 10^{-7} * 8.33 \times 10^6)^{1/2} = 1.9865 \times 10^3 \text{ nep/m}$$

$$\alpha d = 1.986 \times 10^3 * 1.27 \times 10^{-2} = 25.22$$

$$\text{Attenuation} = 20\text{Log}_{10}(e^{-25.22}) = 219 \text{ dB}$$

$$\text{Attenuation} = 219 \text{ dB}$$

This may be considered a great deal of attenuation but these houses are usually near high power lines. The fields may be large enough to light florescent lamps when held in the E field. Some of these issues will be considered in the case studies.

Leaky enclosures or open rack installations of electronic equipment can present a particular class of problems, as shown in Figure 2.8. Enclosures with electronic printed circuit (PC) boards can have a leakage from an uninterrupted power supply (UPS), and there can be radiation from either circuit boards or the inverter. Long wave H fields from the inverters can be produced by magnetic fields. The apertures formed by small openings in cabinets represent small leaks which will not produce gain. However, if the UPS for some reason has a defect such as poor shielding, the

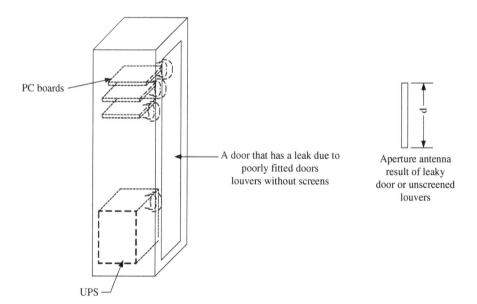

PC boards

A door that has a leak due to poorly fitted doors louvers without screens

Aperture antenna result of leaky door or unscreened louvers

UPS

FIGURE 2.8 Enclosure with leakage shown

magnetic field or high frequencies that can leak through openings in the enclosure or radiate out as emissions can violate the FCC regulations.

A detail description of leakage fields due to the aperture formed is presented in Chapter 4. The radiation from fields due to leakage openings close to one-quarter wavelength will produce gain similar to an aperture antenna. This analysis is not a fundamental concept and requires the use of vector calculus. Most of the analysis thus far has only considered the far field. The far field has the restriction shown in Equation 2.9.

$$\text{Far Field} > 2D^2 / \lambda \qquad (2.9)$$

Where D = largest dimension of radiating element, λ = wavelength.

Frequencies with wavelengths in centimeters or millimeters will radiate through the openings with gain which is dependent on the largest dimension of the opening, as illustrated in Figure 2.8. The aperture gain equations provided below are evaluated in Chapter 4 for a particular application:

$$E_\theta = C \sin \phi \left[(\sin X)/X \right] \left[(\sin Y)/Y \right] \qquad (2.10)$$

$$E_\phi = C \cos \phi \cos \theta \left[(\sin X)/X \right] \left[(\sin Y)/Y \right] \qquad (2.11)$$

Where C = $(jkabE_0 e^{-jkr})/2\pi r$, a = length, b = width, k = $2\pi/\lambda$ all in meters, r = distance from the aperture, E_0 = maximum field V/m, X = (ka/2) sinθ cosϕ, Y = (ka/2) sinθ sinϕ.

Equations 2.10 and 2.11 are provided for observation purposes as follows: (a) values for C, (b) the magnitude of the electric field is a function of 1/r distance and the equations. This implies that, if far field calculations are made for near fields, the square of the distance is used and the equations do not apply. C is inversely proportional to wavelength, that is as wavelength gets smaller C gets larger. The leakage is very susceptible to wavelengths in the GHz range in particular, which is the fundamental, second and third harmonic of many of the new computer clocks. Values of X and Y are calculated with a power series or Bessel functions and these details will be examined in more detail in Chapter 4. What is not apparent is that, as the ratio of a/b gets larger, the E field increases and the bandwidth decreases [1]. Further reading suggested for information on the development of the aperture equations will be found in Reference [2].

Circuit boards are designed with short interconnect layouts (usually 1/20 wavelength) to prevent radiation. Even a layout of 1/20 wavelength can be a Hertzian antenna, which usually do not present much of a problem; however, when searching for culprit sources check out all possibilities. As an example, a phase lock loop oscillator at 900 MHz is in a PC board layout with a wavelength of approximately one-third of a meter ($\lambda = 3.1\times10^8/9\times10^8$ m). The interconnect leads should be kept near 1/60 or 1.67 cm or less for a good design. An 8.25-cm lead is approximately 1/4λ at 900 MHz. This size interconnect can radiate fairly efficiently. Unterminated leads will also radiate, such as an unobtrusive PC board standoff. They can radiate, resulting in serious emission problems.

Braided shield calculations [2]:

$$A_{Total} = 20Log\left(1 + \mu_r t/2d_{radius\,of\,shield}\right) \tag{2.12}$$

Where μ_r = relative permeability of the metal shield, t = thickness, d = shield radius. Equation 2.12 is useful for calculating very low frequency shield effectiveness down to DC. The equation can be applied for wire harnesses, coaxial cable, twisted pair cables and conduit shield effectiveness. If braided or non-continuous materials are used for shields care must be taken to provide correction for air or other materials used in the braid or other construction.

$$\mu_r = V_t V_m \mu_m + V_{nm} + V_{air} \tag{2.13}$$

Where: V_t = total volume of shield material

V_m = decimal equivalent of magnetic material by volume V_m/V_t

V_{nm} = decimal equivalent of non-magnetic material by volume V_{nm}/V_t

V_{air} = 1 – V_t

Example 2.8

Braid surrounding a harness has a radius of 0.5 inches. It is composed of copper clad steel (3% tin, 27% copper, 50% steel) and 20% air. The shield thickness is 0.02 inches.

$$V_t = 80\%$$
$$V_m = 50/80 = 0.625$$
$$V_{nm} = 30/80 = 0.375$$

$$\mu_r = V_t V_m \mu_m, \text{the first term} >> V_{nm} \text{or} V_{air}$$
$$\mu_r = 0.8 * 0.6 * 1000 = 480$$
$$A_{Total} = 20Log\left(1 + 480 * 0.02/2 * 0.5\right) = 20.5\,dB$$

This is field attenuation. Often a dominant field is evident, either H or E attenuation.

Example 2.9

A 0.125 inch thick, 0.75 inch diameter steel 98% with 2% tin coating conduit is used to shield telephone wiring. Now find the attenuation. The tin coating can be considered negligible.

$$\mu_r = 0.98 * 1000 = 980$$
$$A_{Total} = 20Log\left(1 + 980 * 0.125/2 * 0.75\right) = 38.4\,dB$$

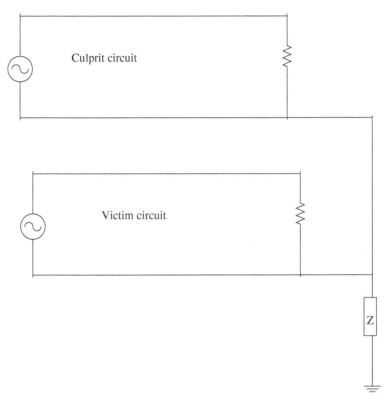

FIGURE 2.9 Common ground impedance

The assumption here is that the fields are one-dimensional vectors, that is they are radial. Most shielding problems can be reduced to scalar quantities to simplify the calculations. The calculations dealing with antenna to antenna issues are very complex and often vector calculus is required to solve field problems.

Common ground impedance is one of the most difficult EMI problems to detect; it is not always due to poor grounding of the copper wire alone. As an example, a piece of equipment may be installed on an insulator that is between the enclosure that is grounded and the case of the floating equipment. The equipment inside may be grounded to the equipment case. The case on the insulator will form one plate of a capacitor and the equipment enclosure the other capacitor plate. Some of the frequencies will be grounded to the enclosure; others will not. The case to cabinet capacitance will be a common impedance (Figure 2.9).

At high frequencies, some types of wire may be too inductive to be used as a ground. This often occurs when designing a transistor circuit with leads that are too long. This is another reason for using short traces and leads on circuit boards.

Several methods are used to isolate victim from culprit circuitry. They are as follows:

1. Use optical fiber when possible for interconnections.
2. RF float equipment, devices, cabinets and so on with RF chokes.

3. Float equipment in shielded containers.
4. Balance circuits when possible.
5. Install isolation transformers.
6. Add shielding RF or magnetic shielding.
7. Separate, that is increase spacing.
8. Add ferrite beads, chokes or filters that increase impedance.

Fiber optic circuits are one of the best means for equipment, device or enclosure isolation. Perhaps the only drawback to fiber optics is the highly sensitive trans-impedance receiver. The receivers are generally well shielded for most applications, but always be vigilant when using fiber optic components. Electric field intensity (E) can easily be shielded compared to magnetic field intensity (H). The E field radiation is the most likely culprit for high frequencies above 1 MHz.

Optical fiber is unaffected by lightning or electrical transients. It is however affected by other things such as nuclear radiation and hydrogen. Exposure to nuclear radiation can result in darkening of the fiber and increase attenuation. However, mild radiation can cause "bleaching" in the fiber and actually lower attenuation. Increased attenuation will result in poor performance do to excessive loss. Decreased attenuation has the effect of overdriving the receiver, causing saturation. Both conditions will result in degradation in performance.

Hydrogen can diffuse into fiber due to its small atomic structure. It can combine with other doping materials to produce higher attenuation or blackening of the fiber. Beware of running optical fiber through battery rooms or near areas where hydrocarbon exhaust emissions can envelop optical fiber in gases unless it has a provision for isolation for such gases. Optical fiber is also susceptible to stress corrosion when exposed to water or high amounts of water vapor. No details of fiber optic design will be given in this book but data can be found in Reference [4] that gives the details of fiber optic design and any fiber optic pitfalls.

The addition of chokes to increase ground impedance to a particular band of frequencies or frequency is common to decouple power supplies from other circuits. These techniques can be implemented to decouple a victim from a culprit circuit. A piece of equipment can be installed in a shielded container with a choke to float the container to isolate RF grounds.

Signals between equipment implemented with cables using balanced, shielded or twisted pair are methods to increase the isolation between circuits. Also, do not connect the shield to ground at both ends for short runs of cable, this will produce a ground loop with large area that can pick up a large number of currents at various frequencies on the shield. Long runs of shield cable may in some cases be grounded at both ends and this will control the capacitance of the shield to ground.

Isolation transformers are another means of isolating grounds and another method of balancing circuits. This technique is often used in audio circuits. For higher frequencies, Faraday shielded isolation transformers are often employed.

2.6 FILTERS FOR EMC

The filters are generally low pass filter elements at the filter input. Prior to selecting the other elements the first pass at filtering a power line or signal line is to try a simple low pass filter and LC, RC or RL filter. If the noise persists, use one of the more common filters such as Butterworth, Chebycheff, elliptical or notch (to remove a specific frequency). For signal lines, a host of analog active, switched capacitor and digital filters are available for wave shaping; these filters can also be implemented (Figure 2.10). The use of active, digital and switched capacitor filters save space on printed circuit boards. The Z part of the filters in Figure 2.10c–f are optional, that is, they may be omitted if low pass filtering is achieved with a single series choke, shunt capacitor or LC (Figure 2.10e) filter section.

Power line filters can be physically large and expensive; The Z section may be required to reduce the size of chokes or capacitors. A four-section filter can have a great deal more attenuation, usually four poles that can provide 24 dB of attenuation of the

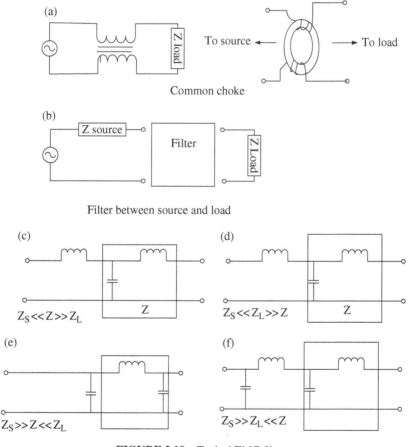

FIGURE 2.10 Typical EMC filters

culprit signal. The annular ring wound as part of Figure 2.10a is a choke, as shown in the circuit diagram. These are commonly used on power line applications. The annular ring is usually made of a ferrite which it is difficult to saturate with large currents. Filters will be covered in the case studies when necessary. Several books are available on filter theory and design. Often for a particular project the manufacturers of filter products provide application notes that have a wealth of information. Application engineers are very helpful to provide some filter products; however, always check to make sure the design is accurate because the EMC design engineer is liable not the application engineer.

Example 2.10

A DC power supply for transit vehicles produces strong 5th, 7th and 11th harmonics at 60 Hz. These harmonics must be filtered. The following parameters are required in the calculations.

$$r_{ref} = 1, R_G = R_L = 2\,\text{Ohms}, L_{ref} = R_{ref}/(2 * \pi f_{ref}) = 0.53\,\text{mH} \qquad (2.14)$$

$$C_{ref} = 1/(2 * \pi f_{ref} * R_{ref}) = 1326\,\mu F \qquad (2.15)$$

$$L_1 = 0.53 * 1\,\text{mH} = 0.53\,\text{mH}$$
$$C_2 = 1.326 * 2\,\mu F = 2562\,\mu F$$
$$L_3 = 0.53 * 1\,\text{mH} = 0.53\,\text{mH}$$

The 1, 2 and 1 multipliers are taken from tables in *"Reference Data for Engineers: Radio, Electronics and Computer Communications"*. These tables appear in most engineering handbooks (Figure 2.11).

Figure 2.12 is similar to Figure 2.10a with shunt capacitors. Compound filters are installed by subway car manufacturers for motor drive systems. These filter components are very large due to the traction motor power; these are also common power line filters that can be found in power feed circuits that provide large currents at 120 V AC. The filter in Figure 2.10 is the type that would need to be designed for a DC supply such as used for subway transit vehicles. The subway cars are supplied with 150 HP AC traction motors on each truck and pulse width modulated (PWM) drives. The power is drawn for the 600 V 800 A DC supply from two 13.5 kV substations. Two substations are required but the cars run with limited performance on one; this provides a backup to the system. These supplies have some high power harmonics to be filtered out. Each motor will take a maximum power of 112 kW. The filter requires some very large components to suppress these harmonics. The even harmonics are summed out and generally the third is filtered out using the power line choke. The 5th, 7th and 11th are still very strong.

The harmonics are wavelengths in the kilometer range. The H field is the most dominant. Long wave H fields can be filtered out to prevent them from permeating through other circuitry. Some of these EMC issues are address in case studies in high

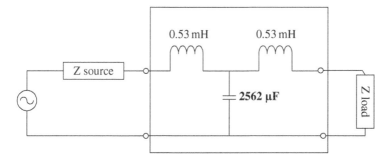

Filter between source and load completed

FIGURE 2.11   The result of calculation for Figure 2.10c

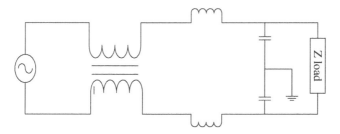

FIGURE 2.12   Compound power line filter

power examples. In particular PWM drives can be used in a host of other applications such as electric car drives. PWM drives are popular because instead of DC motors for the applications the lower cost AC motors can be implemented. The advent of ceramic permanent magnets with high permeability may add a new dimension to electric motor design reducing the size.

2.7   LIGHTNING STROKE ANALYSIS

It is not possible to address all the EMI sources listed in Table 2.1 but the most prevalent is addressed here. One of the most destructive sources is lightning, which consists of a high-energy pulse rich in harmonics. Even the if the initial stroke is not direct the harmonics can be destructive. When listening to an AM radio, the static heard is the result of harmonics generated by the visible and invisible result of field discharges.

Lightning metrics are shown in Figure 2.13 with the equation below to provide a means to calculation potential.

$$H = I/2\pi D \text{ A/m, magnetic field } f(D) \text{ from he stroke} \qquad (2.16)$$

$$V = \rho I / \left[(2\pi) * (1/D - 1/(D+x)) \right] V \qquad (2.17)$$

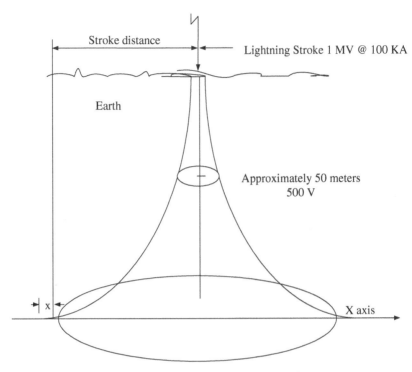

FIGURE 2.13   Lightning stroke details

Where: ρ= resistance (ohms-meter), D = distance from stroke, x = distance from edge of stroke.

Example 2.11

Find the H (field), E (field) and V potential at 10 m from the periphery of the stroke. The distance to the periphery of the stroke is 100 m. Stroke metrics are 1 MV, 100 kA and rise time of 10 μsec with ground resistivity of ρ= 1000 Ω-m.

$H(\text{field}) = 100\,\text{kA}/2\pi * 100 = 159.1\ \text{A/m using Equation 2.15}$

$$V = 1000 * 15.9\,\text{kA}/\left[(1/100) - \left\{1/(100+10)\right\}\right]$$
$$= 15.9 * 10^6 \left[0.01 - 0.0091\right] = 14.91\text{kV}$$

The importance of these calculations is that the larger x is, the greater the potential. If the ground rod spacing on a building is large, for example, a communication bungalow in a signals system with a spacing of 10 m between the power neutral and signal earth ground, a large potential can occur. The potential between the two grounds will be at 14.91 kV, if the lightning stroke is 100 m from either ground. If the protection devices are sufficient to handle the surge, equipment may not be destroyed. Similar precautions are taken for equipment during a nuclear blast. The EMP of a

TABLE 2.3 Lightning Protection Devices

Protection method	Comment
Lightning rods	This is for direct strike protection
Low impedance conductors	Large gauge wire
Low impedance earth	Condition the soil
Use conduit throughout installation	Avoid using flexible conduit
Use conduit between facilities	Between buildings and shelters
Install protection devices on external signal lines	Gas discharge arrestors
Install surge suppressors on power mains	Varistors or MOVs
Filtering, chokes, ferrite beads or sleeves	Reduce current surges
Isolation transform with Faraday shields	
Faraday shields where necessary	
Use fiber optic cables for signal leads	

nuclear blast is not included in this table and is not in the scope of this book. A list of precautions for hardening against lightning strikes is presented in Table 2.3.

Not listed here is the complete isolation from ground; this is a fairly difficult task. There always seems to be a sneak path to ground.

2.8   SKIN EFFECT IN WIRE

The skin effect for higher frequencies will result in a low current density J (A/m²) in the center of the conductor and higher current near the surface. Note that J is a vector with the z direction down the center of the wire. If Maxwell's equations are solved for other directions J can be multidirectional. Field equations are occasionally necessary to explain some of the coupling techniques; however, this is provided on a case by case basis in the case studies. The material presented will often be an approximation of the electromagnetic vector equations and reduced to one-dimensional scalar equations to reduce the complexity of a particular problem.

Equations are used when possible to provide a means for generating calculations for tables or plots. Some authors provide plots and tables that may have limitations where the mathematics break down. Generating plots and tables will provide the audience with confidence that the plots or tables are accurate. Some of the presentations are based on approximations to give a first pass estimate for a solution to a problem or case study. This provides a quick answer for an EMI anomaly with methods to correct it. Always, if possible, the answers to calculations are compare with test results to determine the accuracy of the calculations.

Example 2.12

$$\alpha \approx \left(\pi f \mu \sigma \right)^{1/2} \tag{2.18}$$

$$\delta \approx 1/\alpha \qquad (2.19)$$

A copper wire carrying current has a 0.05 mm radius and frequency of 1 GHz. Find the skin depth and percentage of the wire that carries the current.

Calculate $\alpha = \left(\pi * 10^9 * 4\pi * 10^{-7} * 5.8 * 10^7 \right)^{1/2}$

$\alpha = 4.785 \times 10^5$

$\delta = 1/4.785 \times 10^5 = 2.09$ μm skin depth.

$$A = \pi \left[\left(50 * 10^{-6} \right)^2 - \left(50 * 10^{-6} - 2.09 * 10^{-6} \right)^2 \right] m^2$$
$$A = 6.4588 * 10^{-10} m^2 \qquad (2.20)$$

$$\text{Resistance/meter } R_m = 1/A\sigma = 1/\left(6.4588 \times 10^{-10} \times 5.8 \times 10^7 \right)$$
$$R_m = 26.694 \Omega/m \qquad (2.21)$$

Percentage of cross-sectional area used:

$$P_{cs} = [(6.4588 * 10^{-10} / \pi \left(5 * 10^{-5} \right)^2] * 100$$
$$P_{cs} = 8.223\%$$

At high frequencies as shown in Example 2.12 the center of the conductor is unnecessary, due to skin effect. Hollow pipe waveguides can be implemented since only 8% of the wire carries the current. Skin effect is also the same for flat ground planes where electromagnetic waves impinge on the surface from radiation sources such as cell phones, two-way radios and spread-spectrum devices. This scenario is responsible for EMI affecting electronic circuitry such as computer PC boards. A thin ground plane will carry as much as a thick one if the depth of penetration criteria is the same for both.

Example 2.13

A power line may be tapped for measurement purposes to find the current flow or for other purposes, such as getting a little free electricity. The electric company will, of course, find the leak in the power line. The high voltage power line shown in Figure 2.14 has current flow from left to right. To determine the current magnitude in the high voltage line, a loop of N turns can be constructed on a wooden frame to capture magnetic field flux ψ from the power line. The H (field) is in amps/meter operating in air, $\mu = 4\pi \times 10^{-7}$.

In henrys/m, the flux density is B webers/meter². Note that H and B are both vectors. The total flux flowing through the loop must be found to determine the current flowing in the loop (see Figure 2.14).

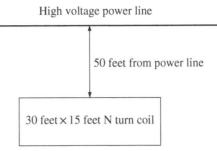

FIGURE 2.14   Using a loop to measure power line current

The equation relating ψ, B and H is as follows:

$$\Psi = \int B * dS = \int \mu H * dS \tag{2.22}$$

$$\int H = \int 1/2\pi r \ \ A/m \tag{2.23}$$

$$\Psi = \int_{0}^{9.15} \int_{15.25}^{19.8} \left(\mu I/2\pi r\right) * dy\,dx \ \text{webers} \tag{2.24}$$

Note: r = y; and change the dimensions of integration to meters.
The dot product of the two vectors in Equation 2.22 is a scalar.
In completing the integration in Equation 2.24 the result is as follows:

$$\psi = \left(4\pi \times 10^{-7} I/2\pi\right)(9.15)\left[\ln\left(19.8\right) - \ln\left(15.25\right)\right]\sin 377t \ \text{webers} \tag{2.25}$$

$$\psi = \left[0.4796 \times 10^{-6}\right] * I * \sin 377t \ \text{webers} \tag{2.26}$$

$$d\psi/dt = 1.806 * 10^{-4} I * \cos\left(2\pi 60t\right) V \tag{2.27}$$

V = ω*ψ = 0.1806*cos (377t) at power line current of 1000 A
Turns N = 115/0.1806 = 637
Then

$$V = 115\,\text{volts open circuit RMS voltage} \tag{2.28}$$

The current is dependent on the area and distance of the loop to capture the magnetic flux ψ. The voltage is dependent on the number of loop turns. If the transmission line current is increased by a factor of 10–10 000 A, the loop current will also increase by a factor of 10 and the turns ratio will go down by a factor of 10 to maintain the same 115 V.

Assumptions: A square loop was chosen, due to the ease of analysis with rectangular coordinates. If a circular loop is used cylindrical or spherical coordinates will be required, the integration process gets quite messy. The loop and transmission line are in the same x–y plane. If the loop is not in the transmission line plane, the z component will add an order of undue complexity to the problem. This is addressed in the case studies of Chapter 4.

There are several instruments that can be purchased in the US\$ 500 range to measure field strength of both E and H fields. However, if such equipment is not available while working in the field, rough measurements can be made using a loop as described previously or antennas for radiating E fields.

2.9   CONCLUSION

The power line example indicates how coupling can occur in a loop or ground loop as related to its area. If a ground loop can be measured or calculated, and the distance is known from the power source or culprit circuit with the current measured, a reasonable estimate can be made of 60 Hz hum. The example illustrates why large area ground loops produce noise. The magnetic flux from a 60 Hz line will induce current flow in the ground resulting a connection to have potential between them. This is one of the reasons grounds between enclosures should not have daisy-chained connections to ground. If possible use a single ground point.

2.10   REVIEW PROBLEMS

Problem 2.1
A transmitter from a two-way radio has an antenna gain of 1.6 and radiated power of 1 W. If it is held 1 m away from a computer: (a) Calculate the field strength in Volts/meter. (b) Is the field strength large enough to reduce computer performance or damage the computer?

Problem 2.2
The culprit noise voltage using the capacitance transformer described Figure 2.1a will produce a noise voltage on the victim circuit (for $C_c << C_a$, C_b). (a) Calculate the noise voltage on the victim circuit for $C_a = 10$ pF/ft, $C_b = 0.2$ pF/ft, cable length of 20 ft (1 ft = 30 cm) and a culprit at 2 V.

Problem 2.3
A 3 Watt radio operating 0.5 m from a enclosure (door open) with a gain of 1.6 produces an E field of 24 V/m at a printed circuit board in Figure 2.8. What is the possibility of failure or malfunction?

Problem 2.4
Use Example 2.1 to calculate the 100-m wiring B field impressed on a loop for the 5th and 11th harmonics. The current harmonics magnitude is a function of 1/5 and 1/11.

Problem 2.5
Use Example 2.2 to calculate the SNR for the 11th harmonic (B field in 2.3 is the same magnitude as the magnetic flux due to the loop area of 1 m^2). Determine if it is less than 15 dB for good audio quality.

Problem 2.6
What order is the intermodulation product 3A-B?

Problem2.7
If a cell phone is used on an aircraft, will it interfere with normal radio operation?

Problem2.8
Changing the hum to 1 V (Example 2.4), CMRR to 100 dB and amplifier gain to 100 the common mode signal for the balanced input is?

Problem2.9
Calculate SNR for Problem 2.8.

Problem 2.10
Twisted pair wiring is excellent for signal lines near power lines?

Problem 2.11
Increasing the pitch for twisted pair wiring will increase the isolation reducing noise on signal lines?

Problem 2.12
Using Example 2.6 calculate the thickness of the screen if the material is cold rolled steel instead of copper. Use data from the companion website at www.wiley.com/go/electromagneticcompatibility.

$$CR\ \mu\ steel = 150*4\pi*10^{-7}\ CR\ steel\ \sigma = 0.0384*5.37*10^{7}.\ \alpha \approx (\pi f\mu\sigma)^{1/2}.$$

Hint: $(150*0.0384)^{1/2}*$copper calculation find d

Problem 2.13
A 1 GHz oscillator is designed on a circuit board. What is the maximum lead length recommended for interconnects?

Problem 2.14
Is twisted pair wiring used to reduce magnetic and electric field coupling between wires?

Problem 2.15
Poor grounds are the result of resistance only. Is this statement true or false?

Problem 2.16
Equations 2.9 and 2.10 represent the aperture E (fields). The trigonometric functions are the phase terms. C is the magnitude of the radiating element. The length (a) in function C should be ____ for 857 MHz to prevent leakage from a cell phone from entering the enclosure.

Problem 2.17
If stainless steel is used for the braid in Example 2.8 with a permeability of 200 instead of 1000 (nickel steel) the attenuation is_____ dB.

Problem 2.18
Will increasing the coating of tin and decreasing the copper in Question 2.17 improve the attenuation?

Problem 2.19
Which of the following methods is an excellent choice to reduce EMI coupling but has a high gain element that may require shielding? (a) Fiber optics, (b) RF float equipment, devices and so on with RF chokes, (c) float equipment in shielded containers or (d) balance circuits when possible.

Problem 2.20
If the filter in Figure 2.11 has an increase in (f_{ref}), the L and C required would be: (a) larger, (b) smaller or (c) stay the same?

Problem 2.21
A communication house has a ground at two corners 5 meters apart. The edge of a lightning stroke is a one ground. Use the parameters given in Example 2.11. What is the potential between the two grounds?

Problem 2.22
If a heavy gauge wire (0.5 Ω) is connected to the two grounds, what are the potential and current?

Problem 2.23
Protection against lightning stroke is the following: (a) ground condition decreasing ground resistance, (b) increasing ground wire size, (c) complete isolation from ground or (d) all of the above?

Problem 2.24
The depth of penetration for a steel pipe instead of a copper tube is: (a) greater than copper, (b) less than copper or (c) no difference.

Problem 2.25
A thin ground plane will carry as much current as a thick one if the depth of penetration criteria is the same for both. Is the statement true or false?

Problem 2.26
Decreasing the power line current in Example 2.13 does what?

Problem 2.27
If the distance to the loop in Figure 2.14 is decreased the magnet flux through the loop will increase. Is this statement true or false?

Problem 2.28
If a connection to ground is 10 Amps and 14-gauge wire is used to a ground grid, is this adequate for lightning protection? This is within the wire size for the current according to the code.

2.11 ANSWERS TO REVIEW PROBLEMS

Problem 2.1
(a) Calculated electric field strength is 6.9 V/m. (b) This is not likely to result in failure or performance degradation.

Problem 2.2
0.784 V.

Problem 2.3
30–50%.

Problem 2.4
B $=1.9*10^{-4}/5 = 3.8*10^{-5}$ webers/m^2 and B $=1.9*10^{-4}/11 = 1.727*10^{-5}$ webers/m^2.
Since area $= 1$ m^2 the magnetic flux Ψ is equal in the magnitude of B at the 11th harmonic.

Problem 2.5

$$W_{7260\,Hz} = V^2/R = \left[\omega(rad/sec)*\Psi\right]^2/600 = \left[2*\pi*7260*1.727*10^{-5}\right]^2/600$$

$$= 1.034\,mW\left(-29.8\,dBm\right), SNR = (-42\,dBm - (-29.8\,dBm)$$
$$= 12.2\,dB < 15\,dB.$$

This will impact speech worst case.

Problem 2.6
Fourth order.

Problem 2.7
Yes.

Problem 2.8
10 mV.

Problem 2.9
SNR = 6.02 dB, very poor for voice communication.

Problem 2.10
True.

Problem 2.11
True.

Problem 2.12
d = 0.3897 mm.

Problem 2.13
Maximum length = 1.55 cm.

Problem 2.14
Yes.

Problem 2.15
False.

Problem 2.16
1.808 cm.

Problem 2.17
The attenuation is 9.3 dB.

Problem 2.18
No.

Problem 2.19
(a).

Problem 2.20
(b).

Problem 2.21
Potential = 7.57 kV.

Problem 2.22
Potential is 3.785 V, current 7.57 A.

Problem 2.23
(d).

Problem 2.24
(b).

Problem 2.25
True.

Problem 2.26
Decrease the open circuit voltage of the loop.

Problem 2.27
True.

Problem 2.28
No.

3

INTRODUCTION TO ELECTROMAGNETIC FIELDS

3.1 AN INTRODUCTION TO ELECTROMAGNETIC FIELDS

The electromagnetic field can be analyzed using Maxwell's equations but the electric or magnetic field must be known to completely analyze the unknown field. The fields cannot exist independently, for example E = σ*D; as can be observed the E and H fields are independent. Decoupling of the E and H fields will make analysis a great deal easier. The wave equation introduced does the decoupling of Maxwell's equation, but it will require the solution of six second order equations. However, not all of the equations are necessary for most analyses. The wave equation solution will begin with the Cartesian coordinate system but it will also be extended to cylindrical and spherical coordinate systems. See page 3 in Reference [1] for a complete treatment of this term. The complete derivation of all of the equations in this chapter can be found in References [1] and [3]; only the derivations are necessary. Any restrictions on the use of these equations are provided.

Maxwell's Equation Differential Form (with source charge):

$$\bar{\nabla} x \bar{E} = -\frac{\partial \bar{B}}{\partial t} + \bar{M} \tag{3.1}$$

The M term (see Reference [1]) shown in Equation 3.1 is not physically realizable but it is added to provide balance to Maxwell's equations. Most books on the subject do not have this term included.

$$\bar{\nabla} x \bar{H} = \frac{\partial \bar{D}}{\partial t} + \bar{J}_c \tag{3.2}$$

Electromagnetic Compatibility: Analysis and Case Studies in Transportation, First Edition.
Donald G. Baker.
© 2016 John Wiley & Sons, Inc. Published 2016 by John Wiley & Sons, Inc.
Companion website: www.wiley.com/go/electromagneticcompatibility

$$\bar{\nabla} * \bar{D} = \rho_q \tag{3.3}$$

$$\bar{\nabla} * \bar{B} = \rho_m \tag{3.4}$$

Maxwell's Equation Integral Form (with source charge):

$$\oint_c \bar{E} * d\bar{l} = -\int_s \left(\frac{\partial \bar{B}}{\partial t} + M \right) dS \tag{3.5}$$

Equation 3.5 is Faraday's law.

$$\oint_c \bar{H} * d\bar{l} = -\int_s \left(\frac{\partial \bar{D}}{\partial t} + \bar{J}_c \right) dS \tag{3.6}$$

Equation 3.6 is Ampere's law.

$$\oiint_s \bar{D} * d\bar{S} = \iiint_v \rho_c * dv \tag{3.7}$$

Equation 3.7 is Gauss' law.

$$\oiint_s \bar{B} * d\bar{S} = \iiint_v \rho_m * dv \tag{3.8}$$

When all charge sources are removed from Maxwell's equation the free space version is shown below. A great deal of the case study analyses are performed using the free space version of Maxwell's equations.

Maxwell's Equations Differential Form (free space):

$$\bar{\nabla} x \bar{E} = -\frac{\partial \bar{B}}{\partial t} \tag{3.9}$$

$$\bar{\nabla} x \bar{H} = \frac{\partial \bar{D}}{\partial t} \tag{3.10}$$

$$\bar{\nabla} * \bar{D} = 0 \tag{3.11}$$

$$\bar{\nabla} * \bar{B} = 0 \tag{3.12}$$

Maxwell's Equation Integral Form (free space):

$$\oint_c \bar{E} * d\bar{l} = -\int_s \frac{\partial B}{\partial t} dS \tag{3.13}$$

$$\oint_c \bar{H} * d\bar{l} = \int_s \frac{\partial D}{\partial t} dS \tag{3.14}$$

$$\oiint_s \bar{B} * d\bar{S} = 0 \tag{3.15}$$

$$\oiint_s \bar{D} * d\bar{S} = 0 \tag{3.16}$$

The derivation of Maxwell's differential or integral equations can be derived using Stokes theorem and divergence theorem, also known as Gauss divergence theorem, Equations 3.17 and 3.18 respectively. If the surface integral is taken on both sides of Equations 3.1 and 3.2 and Stokes theorem is applied to the left side of each equation the integral form is the result. The process can also be reversed to find the differential version of Maxwell's equations from the integral version. For finding the integral form of Equation 3.4 and 3.5 the volume integral is taken of both sides of each equation and the divergence theorem is applied to the lefthand side of the equation.

$$\oint_c \bar{E} * d\bar{L} = \iint_S (\bar{\nabla} x \bar{E}) * \mathrm{d}S \tag{3.17}$$

$$\oiint_s \bar{D} * \mathrm{d}\bar{S} = \oiiint_v (\bar{\nabla} * \bar{D}) * dv \tag{3.18}$$

The dell operator is a vector defined by Equation 3.19, where i, j and k are unit vectors, all of Maxwell's equations are vectors. Some examples are provided later as single dimensional analyses. These of course will be scalar equations.

$$\nabla = \frac{\partial * i}{\partial x} + \frac{\partial * j}{\partial y} + \frac{\partial * k}{\partial z} \tag{3.19}$$

The wave equation may be derived by taking the curl of Equation 3.9. The result of this operation is shown in Equations 3.21 and 3.22. Since most of the case studies are studied in the frequency domain we will replace the derivative of the time with the expression for frequency domain, Equation 3.20.

$$\frac{d}{dt} = j\omega \tag{3.20}$$

The curl of Maxwell's Equations 3.1 and 3.2, with no sources, is where $D = \epsilon E$, $J_c = \sigma E$ and $B = \mu H$. The term M is removed from Equation 3.1. With the vector identity $\nabla x \nabla x E = \nabla (\nabla * E) - \nabla^2 E$ in the lefthand side of Equations 3.21 and 3.22 the resulting equations are shown below, Equations 3.23 and 3.24.

$$\bar{\nabla} x \bar{\nabla} x \bar{E} = -j\omega\mu (\bar{\nabla} x \bar{H}) \tag{3.21}$$

$$\bar{\nabla} x \bar{\nabla} x \bar{H} = (\nabla x E)(\sigma + j\omega) \tag{3.22}$$

$$\bar{\nabla} (\bar{\nabla} * \bar{E}) - \nabla^2 \bar{E} = j\omega\mu (\sigma + j\omega\varepsilon) \bar{E} \tag{3.23}$$

$$\bar{\nabla} (\bar{\nabla} * \bar{H}) - \nabla^2 \bar{H} = (j\omega\mu)(\sigma + j\omega\varepsilon) \bar{H} \tag{3.24}$$

Insert the electromagnetic Equations 3.1 and 3.2 into Equations 3.23 and 3.24 and they are rearranged. The results are shown in Equations 3.25 and 3.26. Both of these equations have one variable in either H or E field. For free space $\nabla * E = \nabla * H = 0$ Equations 3.23 and 3.24 will reduce to the final equations, shown as Equations 3.25 and 3.26 respectively; these are second order wave equations.

$$\nabla^2 \bar{E} = j\omega\mu(\sigma + j\omega\varepsilon)\bar{E} \equiv \gamma^2 \bar{E} \tag{3.25}$$

$$\nabla^2 \bar{H} = (j\omega\mu)(\sigma + j\omega\varepsilon)\bar{H} \equiv \gamma^2 \bar{H} \tag{3.26}$$

The field equations E and H are vectors; they are shown below:

$$\bar{E} = E_x(x,y,z)i + E_y(x,y,z)j + E_z(x,y,z)k$$
$$\bar{H} = H_x(x,y,z)i + H_y(x,y,z)j + H_z(x,y,z)k$$

Inserting the vector equations $\bar{E}$ and $\bar{H}$ into Equations 3.25 and 3.26 the result is six second order scalar differential wave equations.

$$\frac{\partial^2 E_x(x,y,z)}{\partial x^2} + \gamma^2 E_x(x,y,z) = 0 \qquad \frac{\partial^2 H_x(x,y,z)}{\partial x^2} + \gamma^2 H_x(x,y,z) = 0$$

$$\frac{\partial^2 E_y(x,y,z)}{\partial y^2} + \gamma^2 E_y(x,y,z) = 0 \qquad \frac{\partial^2 H_y(x,y,z)}{\partial y^2} + \gamma^2 H_y(x,y,z) = 0$$

$$\frac{\partial^2 E_z(x,y,z)}{\partial z^2} + \gamma^2 E_z(x,y,z) = 0 \qquad \frac{\partial^2 H_z(x,y,z)}{\partial z^2} + \gamma^2 H_z(x,y,z) = 0$$

The solution of each axis of the wave equations are of the form shown below

$$E_1(x) = A_1 cos(\beta_x x + B_1 \sin(\beta_x x)$$

$$E_2(x) = C_1 e^{-j\beta_x x} + D_1 e^{j\beta_x x}$$

$$E_3(y) = A_2 cos(\beta_y y) + B_2 \sin(\beta_y y)$$

$$E_4(y) = C_2 e^{-j\beta_y y} + D_2 e^{j\beta_y y}$$

$$E_5(z) = A_3 cos(\beta_z z) + B_3 \sin(\beta_z z)$$

$$E_6(z) = C_3 e^{-j\beta_z z} + D_3 e^{j\beta_z z}$$

The six solutions to the wave equations for the electric field are shown. There are six more for the H fields. They will not be shown here; it's just a matter of changing the E fields to H fields. These are general solutions. The constants are determined by boundary conditions; therefore, the constants are different for the H fields when numerical values are calculated. Each application to EMC problems will require at most the product of three solutions for each field, due to the three axes. In most cases however, the field of interest may be along only one axis. This is considered a one-dimensional problem and the others may be ignored. The case studies have several examples in them showing the use of one and two and occasionally the third axis to be one of interest.

Calculations for propagation constant:

$$\alpha + j\beta = \pm\sqrt{j\omega\mu\sigma - \omega^2\mu\varepsilon} \tag{3.27}$$

Squaring both sides of Equation 3.27 and equating imaginary in the real parts of both sides of the equation, the result is shown as Equations 3.28 and 3.29. Solving the simultaneous Equations 3.28 and 3.29 for the attenuation and phase constants the result is Equation 3.30 and 3.31. These two constants are exact for both conductors and insulators. Table 3.1 shows the approximations for both these constants.

As one can observe from the table these are useful for calculating other expressions that are affected by boundary conditions.

$$\alpha^2 - \beta^2 = -\omega^2\mu\varepsilon \tag{3.28}$$

$$2\alpha\beta = \omega\mu\sigma \tag{3.29}$$

$$\alpha = \omega\sqrt{\mu\varepsilon}\left(\left\{\frac{1}{2}\left[1+\left(\frac{\sigma}{\omega\varepsilon}\right)^2\right]^{1/2} - 1\right\}\right)^{1/2} \quad \text{Np/m} \tag{3.30}$$

$$\beta = \omega\sqrt{\mu\varepsilon}\left(\left\{\frac{1}{2}\left[1+\left(\frac{\sigma}{\omega\varepsilon}\right)^2\right]^{1/2} + 1\right\}\right)^{1/2} \quad \text{Radians/m} \tag{3.31}$$

Next it is prudent to consider the material constants of the boundary conditions. These constants are the result of materials that have specific properties, such as air, steel, copper and so on. The first consideration is for a source free environment such as air. This will apply to many of the case studies that are communications EMC problems. In some of these cases, Cartesian coordinate systems will suffice when the electromagnetic waves are considered planer; however, in a few instances spherical or cylindrical coordinate systems must be used to completely analyze the problem. For the Cartesian coordinates, in free space analysis E and H fields are orthogonal provided that the electromagnetic wave is not in the near field of a radiating antenna.

The methods used to arrive at the approximations for good conductors and good insulator is the expansion of the attenuation and phase constant in an infinite series.

TABLE 3.1 Calculations Using Various Physical Constants for Solving EMC Field Problems

Exact expression	Good insulator (dielectric) $\left(\frac{\sigma}{\omega\varepsilon}\right)^2 \ll 1$	Good conductor $\left(\frac{\sigma}{\omega\varepsilon}\right)^2 \gg 1$	Remarks
$\alpha = \omega\sqrt{\mu\varepsilon}\left\{\left[\frac{1}{2}\left[1+\left(\frac{\sigma}{\omega\varepsilon}\right)^2\right]^{1/2}-1\right]\right\}^{1/2}$	$\cong \frac{\sigma}{2}\sqrt{\frac{\mu}{\varepsilon}}$	$\cong \sqrt{\frac{\omega\mu\varepsilon}{2}}$	Also for good conductor $\cong \sqrt{\pi f \mu\sigma}$
$\beta = \omega\sqrt{\mu\varepsilon}\left\{\left[\frac{1}{2}\left[1+\left(\frac{\sigma}{\omega\varepsilon}\right)^2\right]^{1/2}+1\right]\right\}^{1/2}$	$\cong \omega\sqrt{\mu\varepsilon}$	$\cong \sqrt{\frac{\omega\mu\varepsilon}{2}}$	Also for good conductor $\cong \sqrt{\pi f \mu\sigma}$
Intrinsic impedance $\eta = \sqrt{\frac{j\omega\mu}{\sigma+j\omega\varepsilon}}$	$\cong \sqrt{\frac{\mu}{\varepsilon}}$	$\cong \sqrt{\frac{\omega\mu}{2\sigma}}(1+j)$	Polar good conductor $\cong \sqrt{\frac{\omega\mu}{2\sigma}\angle 45^0}$
Free space vacuum $\eta_0 = \sqrt{\frac{\mu_0}{\varepsilon_0}}$	Different for other gases and gas mixtures	Different for gases	$\eta_0 = 120\pi\ \Omega \approx 377\Omega$
Velocity $v = \frac{\omega}{\beta}$	$v \cong \frac{1}{\sqrt{\mu\varepsilon}}$	$v \cong \sqrt{\frac{2\omega}{\mu\sigma}}$	c_o = speed of light $c_o = \frac{1}{\sqrt{\mu_0\varepsilon_0}}$
Wavelength $\lambda = \frac{\omega}{\beta}$	$\lambda \cong \frac{2\pi}{\sqrt{\mu\varepsilon}}$	$\lambda \cong 2\pi\sqrt{\frac{2}{\omega\mu\sigma}}$	Wavelength and air $\lambda \cong \frac{c_o}{f}$
Skin depth $\delta = \frac{1}{\alpha}$	$\cong \frac{2}{\sigma}\sqrt{\frac{\varepsilon}{\mu}}$	$\cong \sqrt{\frac{2}{\omega\mu\sigma}}$	μ_r frequency dependent from 100 kHz to greater than 1 GHz where it may have an imaginary component ε_r it may have an imaginary component above 1 GHz

When dealing with semiconductors the exact expression for the attenuation and phase constant must be used. When frequency analysis is above 1 GHz the material constants have real and imaginary parts that must be included.

The solutions to the wave equation shown previously only considered the phase constant. The complete solution for conductive materials shown in Equation 3.32. This is not the only solution since $\gamma = \pm(\alpha + j\beta)$ and $\gamma = \pm(\alpha - j\beta)$. This example represents a solution for one of four possible solutions on the X axis only. The solutions for the other axes are similar except the dependent variable is changed for the axis to be analyzed. The total solution to a three-dimensional system is shown in Equation 3.33. To perform a particular three-dimensional waveguide analysis, a model is selected to meet the criteria of the problem. The equation to do the analysis can be done for a simple case by inspection. Equations 3.34, 3.35 and 3.36 are the general equations necessary to model the fields in a rectangular waveguide (Figure 3.1). The parameters a and b are the width and height respectively for the waveguide. The width is along the x axis, height is the y axis and n and m are integer values for phase analysis using the sine and cosine terms. The length of the waveguide along the z axis is infinite. Attenuation factor α may be eliminated for short lengths of waveguide or long wavelengths. For long lengths of waveguide, there will be attenuation in some instances and standing waves and others.

$$f(x) = \left[A_1 Cos(\beta_x x) + B_1 Sin(\beta_x x) \right] e^{-\alpha x} \tag{3.32}$$

$$\text{Total solution} = f(x) * g(y) * h(z) \tag{3.33}$$

The solutions to the wave equation for the waveguide are as follows:

$$f(x) = C_1 Cos(\beta_x x) + D_1 Sin(\beta_x x)$$
$$g(y) = C_2 Cos(\beta_y y) + D_2 Sin(\beta_y y)$$
$$h(z) = A_3 e^{-(j\beta_z + \alpha)z} + B_3 e^{(j\beta_z + \alpha)z}$$

Table 3.2 shows all the possibilities of waves traveling down a transmission path.

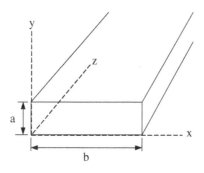

FIGURE 3.1 Rectangular waveguide.

TABLE 3.2   All Possibilities for Waves Traveling Down a Transmission Path

Wave type	Wave function solutions	Zeros of wave functions
Traveling waves	$e^{-j\beta axis}axis = (x, y, z)$ $e^{+j\beta axis}axis = (-x, -y, -z)$	axis = +infinity axis = −infinity
Standing wave	$Cos(\beta axis), axis = \pm(x, y, z)$ $sin(\beta axis), axis = \pm(x, y, z)$	β axis = $\pm(n + 1/2)\pi$ β axis = $\pm n\pi$
Evanescent waves	$e^{-\alpha^* axis}axis = (x, y, z)$ $e^{+\alpha^* axis}axis = (-x, -y, -z)$	axis = +infinity axis = −infinity
Attenuating traveling waves	$e^{-j\beta axis} * e^{-\alpha^* axis}axis = (x, y, z)$ $e^{+j\beta axis} * e^{+\alpha^* axisgame\ you}axis = (-x, -y, -z)$	axis = +infinity axis = −infinity
Attenuated standing waves	$cos(\gamma axis) = cos(\alpha^* axis)cosh(\beta axis) - j$ $sin(\alpha^* axis)sinh(\beta axis) for \pm axis$	$\pm j(n + 1/2)\pi$
Attenuated standing waves	$sin(\gamma axis) = sin(\alpha^* axis)cosh(\beta axis) +$ $jcos(\alpha^* axis)sinh(\beta axis) for \pm axis$	$\pm jn\pi$ $n = 0, 1, 2, 3 \ldots$

Example 3.1

The boundary conditions are y = 0 and y = a, $E_x = 0$, the wave is standing (x = 0, x = b) and it is unbounded in the z direction. The assumptions are that the waves are standing waves perfectly conducting and they are infinite ±z direction. These solution were taken from Reference [3]. As can be observed, examining Equation 3.34 the Sine term in the equation $E_x = 0$ is for y = 0 and y = a for m = integer. The wave is a standing wave between x = 0 and x =b with n = integer.

$$E_x(x,y,z) = E_{0x}Cos\left(\frac{n\pi}{b}x\right)Sin\left(\frac{m\pi}{a}y\right)e^{-(j\beta_z)z} \quad \text{A possible solution} \quad (3.34)$$

$$E_y(x,y,z) = E_{0y}Sin\left(\frac{n\pi}{b}x\right)Cos\left(\frac{m\pi}{a}y\right)e^{-(j\beta)z} \quad (3.35)$$

$$E_z(x,y,z) = E_{0y}Sin\left(\frac{n\pi}{b}x\right)Cos\left(\frac{m\pi}{a}y\right)e^{-(j\beta)z} \quad (3.36)$$

3.2   WAVE EQUATION SOLUTIONS FOR CYLINDRICAL COORDINATE SYSTEMS

The dell operator expression in cylindrical coordinates is Equation 3.39.The gradient of the electric field in Equation 3.37 ($\hat{a}_\rho, \hat{a}_\phi, \hat{a}_z$ with unit vectors that have a magnitude of one in the direction of each axis) is as shown below. The curl of Equation 3.37 is taken with the result provided in Equation 3.38. This term is necessary in order to derive the scaler differential equation wave equations in a cylindrical coordinate system.

$$\bar{\nabla} * \bar{E} = \frac{1}{\rho}\frac{\partial(\rho E_\rho)}{\partial \rho} + \frac{1}{\rho}\frac{\partial E_\phi}{\partial \phi} + \frac{\partial E_z}{\partial z} \tag{3.37}$$

$$\bar{\nabla}(\bar{\nabla} * \bar{E}) = \hat{e}_\rho \left\{ \frac{\partial}{\partial \rho}\left[\frac{1}{\rho}\frac{\partial(\rho E_\rho)}{\partial \rho} + \frac{1}{\rho}\frac{\partial E_\phi}{\partial \phi} + \frac{\partial E_z}{\partial z} \right] \right\}$$
$$+ \hat{e}_\phi \frac{1}{\rho}\left\{ \frac{\partial}{\partial \phi}\left[\frac{1}{\rho}\frac{\partial(\rho E_\rho)}{\partial \rho} + \frac{1}{\rho}\frac{\partial E_\phi}{\partial \phi} + \frac{\partial E_z}{\partial z} \right] \right\} +$$
$$+ \hat{e}_z \left\{ \frac{\partial}{\partial z}\left[\frac{1}{\rho}\frac{\partial(\rho E_\rho)}{\partial \rho} + \frac{1}{\rho}\frac{\partial E_\phi}{\partial \phi} + \frac{\partial E_z}{\partial z} \right] \right\} \tag{3.38}$$

$$\bar{\nabla} = \hat{e}_\rho \frac{\partial}{\partial \rho} + \hat{e}_\phi \frac{1}{\rho}\frac{\partial}{\partial \phi} + \hat{e}_z \frac{\partial}{\partial z} \tag{3.39}$$

The left side of Maxwell's differential Equation 3.1 for the electric field is Equation 3.40.

$$\bar{\nabla}x\bar{E}(\rho,\phi,z) = \hat{e}_\rho\left(\frac{1}{\rho}\frac{\partial E_z}{\partial \phi} - \frac{\partial E_\phi}{\partial z} \right) + \hat{e}_\phi\left(\frac{\partial E_\rho}{\partial z} - \frac{\partial E_z}{\partial \rho} \right) + \hat{e}_z\left(\frac{\partial E_\phi}{\partial \rho} + \frac{E_\phi}{\rho} - \frac{1}{\rho}\frac{\partial E_\rho}{\partial \phi} \right)$$
$$\tag{3.40}$$

The divergence (curl) of Equation 3.40 is shown in Equation 3.41. The evaluation of this determinant is Equation 3.42. This will be used later to derive the expression for scaler cylindrical wave equations for each axis.

$$\bar{\nabla}x\bar{\nabla}x\bar{E}(\rho,\phi,z) = \begin{bmatrix} \hat{e}_\rho & \rho\hat{e}_\phi & \hat{e}_z \\ \dfrac{\partial}{\partial \rho} & \dfrac{\partial}{\partial \phi} & \dfrac{\partial}{\partial z} \\ \left(\dfrac{1}{\rho}\dfrac{\partial E_z}{\partial \phi} - \dfrac{\partial E_\phi}{\partial z}\right) & \rho\left(\dfrac{\partial E_\rho}{\partial z} - \dfrac{\partial E_z}{\partial \rho}\right) & \left(\dfrac{\partial E_\phi}{\partial \rho} + \dfrac{E_\phi}{\rho} - \dfrac{1}{\rho}\dfrac{\partial E_\rho}{\partial \phi}\right) \end{bmatrix}$$
$$\tag{3.41*}$$

$$\bar{\nabla}x\bar{\nabla}x\bar{E}(\rho,\phi,z) = \hat{e}_\rho\left[\frac{1}{\rho}\frac{\partial^2 E_\phi}{\partial \rho \partial \phi} + \frac{1}{\rho^2}\frac{\partial E_\phi}{\partial \phi} - \frac{1}{\rho^2}\frac{\partial^2 E_\rho}{\partial \phi^2} - \frac{\partial^2 E_\rho}{\partial z^2} + \frac{\partial^2 E_z}{\partial z \partial \phi} \right] +$$
$$+ \hat{e}_\phi\left[\frac{1}{\rho}\frac{\partial^2 E_z}{\partial \phi \partial z} - \frac{\partial^2 E_\phi}{\partial z^2} - \frac{\partial^2 E_\phi}{\partial \rho^2} - \frac{1}{\rho}\frac{\partial E_\phi}{\partial \rho} + \frac{E_\phi}{\rho^2} + \frac{1}{\rho}\frac{\partial^2 E_\rho}{\partial \rho \partial \phi} - \frac{1}{\rho^2}\frac{\partial E\phi}{\partial \phi} \right] +$$
$$+ \hat{e}_z\left[\frac{\partial^2 E_z}{\partial \rho \partial z} + \frac{1}{\rho}\frac{\partial E_\rho}{\partial z} - \frac{\partial^2 E_z}{\partial \rho^2} - \frac{1}{\rho}\frac{\partial E_z}{\partial \rho} - \frac{1}{\rho^2}\frac{\partial^2 E_z}{\partial \phi^2} + \frac{1}{\rho}\frac{\partial^2 E_\phi}{\partial \phi \partial z} \right]$$
$$\tag{3.42}$$

---

* $\hat{e}_\phi, \hat{e}_\rho, \hat{e}_r$ are unit vectors with a magnitude of 1.

Inserting all the identical terms from Equations 3.41 and 3.42 into Equation 3.43 the result is an expression for the radial components cylindrical, as shown in Equation 3.44. The expression within the parentheses of Equation 3.44 is $\nabla^2 E_\rho$ with the resulting equations as shown in Equation 3.45.

$$\nabla^2 \bar{E} = \bar{\nabla}\left(\bar{E} * \bar{\nabla}\right) - \bar{\nabla}x\bar{\nabla}x\bar{E} = \beta^2 \bar{E} \tag{3.43}$$

$$\left(\frac{\partial^2 E_\rho}{\partial\rho^2} + \frac{1}{\rho}\frac{\partial E_\rho}{\partial\rho} + \frac{1}{\rho^2}\frac{\partial^2 E_\rho}{\partial\phi^2} + \frac{\partial^2 E_\rho}{\partial z^2}\right) - \frac{E_\rho}{\rho^2} - \frac{2}{\rho^2}\frac{\partial E_\phi}{\partial\phi} = -\beta^2 E_\rho \tag{3.44}$$

$$\nabla^2 E_\rho - \frac{E_\rho}{\rho^2} - \frac{2}{\rho^2}\frac{\partial E_\phi}{\partial\phi} = -\beta^2 E_\rho \tag{3.45}$$

This same technique is used to derive Equation 3.46 for the angular components in the cylindrical coordinate system, with the final result shown in Equation 3.47.

$$\left(\frac{\partial^2 E_\phi}{\partial\rho^2} + \frac{1}{\rho}\frac{\partial E_\phi}{\partial\rho} + \frac{1}{\rho^2}\frac{\partial^2 E_\phi}{\partial\phi^2} + \frac{\partial^2 E_\phi}{\partial z^2}\right) - \frac{1}{\rho^2} + \frac{2}{\rho^2}\frac{\partial E_\rho}{\partial\phi} = -\beta^2 E_\phi \tag{3.46}$$

$$\nabla^2 E_\phi - \frac{1}{\rho^2} + \frac{2}{\rho^2}\frac{\partial E_\rho}{\partial\phi} = -\beta^2 E_\phi \tag{3.47}$$

The z axis component in Equation 3.49 is derived using the same methods as used to find the radial and angular components.

$$\left(\frac{\partial^2 E_z}{\partial\rho^2} + \frac{1}{\rho}\frac{\partial E_z}{\partial\rho} + \frac{1}{\rho^2}\frac{\partial^2 E_z}{\partial\phi^2} + \frac{\partial^2 E_z}{\partial z^2}\right) = -\beta^2 E_z \tag{3.48}$$

$$\nabla^2 E_z = -\beta^2 E_z \tag{3.49}$$

The general solution for the wave equation is Equation 3.50, henceforth shown as by $\psi = \text{fgh}$. Inserting the general solution Equation 3.51 to the wave equations derivation of Equation 3.52 is a result. Dividing both sides of Equation 3.52 by $\psi = \text{fgh}$ the result in Equation 3.53.

$$\psi\left(\rho,\phi,z\right) = \text{f}\left(\rho\right)\text{g}\left(\phi\right)\text{h}\left(z\right) \tag{3.50}$$

$$\left(\frac{\partial^2 \psi}{\partial\rho^2} + \frac{1}{\rho}\frac{\partial\psi}{\partial\rho} + \frac{1}{\rho^2}\frac{\partial^2\psi}{\partial\phi^2} + \frac{\partial^2\psi}{\partial z^2}\right) = -\beta^2\psi \tag{3.51}$$

$$\left(gh\frac{\partial^2 f}{\partial\rho^2} + gh\frac{1}{\rho}\frac{\partial f}{\partial\rho} + fh\frac{1}{\rho^2}\frac{\partial^2 g}{\partial\phi^2} + fg\frac{\partial^2 h}{\partial z^2}\right) = -\beta^2 fgh \tag{3.52}$$

$$\left(\frac{1}{f}\frac{\partial^2 f}{\partial\rho^2} + \frac{1}{f\rho}\frac{\partial f}{\partial\rho} + \frac{1}{g\rho^2}\frac{\partial^2 g}{\partial\phi^2} + \frac{1}{h}\frac{\partial^2 h}{\partial z^2}\right) = -\beta^2 \tag{3.53}$$

At this point in the analysis, the z axis component is decoupled from the other terms in the equation; and all the terms are equal to a constant β^2 that is equal to the sum of two constants. At this point Equation 3.54 is a simple second order scalar differential equation.

$$\frac{d^2 h}{dz^2} + \beta_z^2 h = 0 \tag{3.54}$$

$$\beta^2 = \beta_z^2 + \beta_\rho^2, \quad \beta^2 - \beta_z^2 = \beta_\rho^2 \quad \text{or} \quad \beta^2 - \beta_\rho^2 = \beta_z^2 \tag{3.55}$$

Multiplying both sides of Equation 3.53 by ρ^2 and inserting the appropriate constant the result is Equation 3.56. The term with derivative ϕ is replaced by a constant. The result of replacing this component by a constant provides a simple second order scalar differential Equation 3.57. Inserting the constant $-m^2$ and multiplying both sides of Equation 3.56 results in Equation 3.58. The differential equations numbered in italics must be solved for the total solution to a particular problem in cylindrical coordinates.

$$\left(\frac{\rho^2}{f} \frac{\partial^2 f}{\partial \rho^2} + \frac{\rho}{f} \frac{\partial f}{\partial \rho} + \frac{1}{g} \frac{\partial^2 g}{\partial \phi^2} + \left(\beta^2 - \beta_z^2 \right) \rho^2 \right) =$$
$$\left(\frac{\rho^2}{f} \frac{\partial^2 f}{\partial \rho^2} + \frac{\rho}{f} \frac{\partial f}{\partial \rho} + \frac{1}{g} \frac{\partial^2 g}{\partial \phi^2} + \left(\beta_\rho \rho \right)^2 \right) = 0 \tag{3.56}$$

$$\frac{1}{g} \frac{\partial^2 g}{\partial \phi^2} = -m^2 \tag{3.57}$$

$$\rho^2 \frac{\partial^2 f}{\partial \rho^2} + \rho \frac{\partial f}{\partial \rho} + \left[\left(\beta_\rho \rho \right)^2 - m^2 \right] f = 0 \quad \text{Bessel differential equation} \tag{3.58}$$

$$h_2(z) = C_3 \cos\left(\beta_z z \right) + D_3 \sin\left(\beta_z z \right) \tag{3.59}$$

$$h_1(z) = A_3 e^{-j\beta_z z} + B_3 e^{j\beta_z z} \tag{3.60}$$

$$g_2(\phi) = C_2 \cos(m\phi) + D_2 \sin(m\phi) \tag{3.61}$$

$$g_1(\phi) = A_2 e^{-jm\phi} + B_2 e^{jm\phi} \tag{3.62}$$

$$f_1(\rho) = \left[A_1 J_m \left(\beta_\rho \rho \right) + B_1 Y_m \left(\beta_\rho \rho \right) \right] \quad \text{Bessel function equation} \tag{3.63}$$

$$f_2(\rho) = \left[C_1 H_m^1 \left(\beta_\rho \rho \right) + D_1 H_m^2 \left(\beta_\rho \rho \right) \right] \quad \text{Henkel function equation} \tag{3.64}$$

Bessel functions J_m and Y_m are of the first and second kind respectively. Henkel functions H_m^1 and H_m^2 are of the first and second kind respectively. Equation 3.63 is used to represent standing waves and Equation 3.64 represents traveling waves. Large tables of Bessel and Henkel functions can be found in most mathematical handbooks. Equations 3.63 and 3.64 can be reduced in complexity if the argument for these expressions is large and asymptotic estimate can be derived. This is provided in Reference [1] and more mathematical rigor is provided. The objective here is to provide solutions to Maxwell's equations in a cylindrical coordinate system.

A general solution for a typical problem $\psi(\rho,\phi,z) = f(\rho)g(\phi)h(z)$ is provided as Equation 3.65. This may appear as a very complex solution; however, when the boundary conditions are used to solve a particular problem, many of the terms may drop out and in some cases the solution will be one-dimensional. As can be observed, the solution equations are general thus far but when these equations are applied to both E and H fields this will expand the number of equations to 12 possible solutions. When boundary conditions are applied several of the terms in the equations will drop out and the complexity will be reduced somewhat. Because the boundary conditions depend on the physical properties of various materials, as the analysis of frequencies increases above 1 GHz some of the constants become complex, involving imaginary numbers. This is beyond the scope of this particular book. The analysis here is for frequencies below 1 GHz. Most of the noise components from frequencies in transportation are due to noise components below this value.

$$\psi(\rho,\phi,z) = f(\rho)g(\phi)h(z) = \left[A_1 J_m(\beta_r \rho) + B_1 Y_m(\beta_r \rho) \right] * $$
$$\left[C_2 \cos(m\phi) + D_2 \sin(m\phi) \right] * \left[A_2 e^{-jm\phi} + B_2 e^{jm\phi} \right]$$

(3.65)

The wave equation is derived in a similar way as the Cartesian coordinate system version with the exception that the axes are changed for both E and H. Taking the curl of both sides of Equations 3.1 and 3.2 and using the identity $\nabla x \nabla x E = \nabla(\nabla * E) - \nabla^2 E$ and Equations 3.21 through 3.24, the six scaler wave equations can be derived.

3.3   WAVE EQUATION SOLUTIONS FOR SPHERICAL COORDINATE SYSTEMS

The dell operator is a vector Equation 3.66 as shown below and $\hat{a}_r, \hat{a}_\theta, \hat{a}_\phi$ are unit vectors with a magnitude of one in the direction of each axis.

$$\bar{\nabla} = \hat{a}_r \frac{\partial}{\partial r} + \hat{a}_\theta \frac{1}{r} \frac{\partial}{\partial \theta} + \hat{a}_\phi \frac{1}{r \sin\theta} \frac{\partial}{\partial \phi}$$

(3.66)

$$\bar{\nabla} x \bar{E}(r,\theta,\phi) = \hat{a}_r \frac{1}{r \sin\theta} \left(\frac{\partial(E_\phi \sin\theta)}{\partial \theta} - \frac{\partial E_\theta}{\partial \phi} \right)$$
$$+ \hat{a}_\theta \frac{1}{r} \left(\frac{1}{\sin\theta} \frac{\partial E_r}{\partial \phi} - \frac{\partial r * E_\phi}{\partial r} \right) + \hat{a}_\phi \frac{1}{r} \left(\frac{\partial(r * E_\theta)}{\partial r} - \frac{\partial E_r}{\partial \theta} \right)$$

(3.67)

Solving Equation 3.43 with the same techniques as used to solve the wave equation for cylindrical coordinates systems the result is Equation 3.68. This is the general scaler solution for the wave equation.

$$\psi\left(r,\theta,\phi\right)=\frac{1}{r^2}\frac{\partial}{\partial r}\left(r^2\frac{\partial\psi}{\partial r}\right)+\frac{1}{r^2\sin\theta}\frac{\partial}{\partial\theta}\left(\sin\theta\frac{\partial(\psi)}{\partial\theta}\right)+\frac{1}{r^2}\frac{1}{\sin^2\theta}\frac{\partial^2 E\psi}{\partial\phi^2}=-\beta^2\psi$$

(3.68)

To decouple Equation 3.68 into equations for each axis Equation 3.69 is inserted into Equation 3.68 and the result is Equation 3.70. Dividing both sides of Equation 3.70 by Equation 3.69 and multiplying both sides by $r^2\sin^2\theta$ the result is Equation 3.71 (note that h is decoupled).

$$\psi\left(r,\theta,\phi\right)=f\left(r\right)g\left(\theta\right)h\left(\phi\right)=fgh$$

(3.69)

$$\psi\left(r,\theta,\phi\right)=\frac{gh}{r^2}\frac{\partial}{\partial r}\left(r^2\frac{\partial f}{\partial r}\right)+\frac{fh}{r^2\sin\theta}\frac{\partial}{\partial\theta}\left(\sin\theta\frac{\partial(g)}{\partial\theta}\right)+\frac{fg}{r^2}\frac{1}{\sin^2\theta}\frac{\partial^2 h}{\partial\phi^2}=-\beta^2 fgh$$

(3.70)

$$\frac{\sin^2\theta}{f}\frac{\partial}{\partial r}\left(r^2\frac{\partial f}{\partial r}\right)+\frac{\sin\theta}{g}\frac{\partial}{\partial\theta}\left(\sin\theta\frac{\partial(g)}{\partial\theta}\right)+\frac{1}{h}\frac{\partial^2 h}{\partial\phi^2}=-\left(\beta r\sin\theta\right)^2$$

(3.71)

The expression in Equation 3.71 for h is equal to a constant $(-m^2)$ is shown in Equation 3.72.

$$\frac{1}{h}\frac{\partial^2 h}{\partial\phi^2}=-m^2$$

(3.72)

Inserting the constant of Equation 3.72 into Equation 3.71 the result is Equation 3.73.

$$\frac{\sin^2\theta}{f}\frac{\partial}{\partial r}\left(r^2\frac{\partial f}{\partial r}\right)+\frac{\sin\theta}{g}\frac{\partial}{\partial\theta}\left(\sin\theta\frac{\partial(g)}{\partial\theta}\right)-m^2+\left(\beta r\sin\theta\right)^2=0$$

(3.73)

Dividing Equation 3.73 by $\sin^2\theta$ the result is Equation 3.74. At this point it should be noted that the decoupling is completed by removing g terms from Equation 3.74 and equating them to a constant. The result is Equation 3.75. Inserting the constant into Equation 3.74 the result is Equation 3.76.

$$\frac{1}{f}\frac{\partial}{\partial r}\left(r^2\frac{\partial f}{\partial r}\right)+\left(\beta r\right)^2+\frac{1}{g\sin\theta}\frac{\partial}{\partial\theta}\left(\sin\theta\frac{\partial(g)}{\partial\theta}\right)-\left(\frac{m}{\sin\theta}\right)^2=0$$

(3.74)

$$\frac{1}{g\sin\theta}\frac{\partial}{\partial\theta}\left(\sin\theta\frac{\partial(g)}{\partial\theta}\right)-\left(\frac{m}{\sin\theta}\right)^2=-n\left(n+1\right)$$

(3.75)

Then

$$\frac{1}{f}\frac{\partial}{\partial r}\left(r^2\frac{\partial f}{\partial r}\right)+\left(\beta r\right)^2-n\left(n+1\right)=0$$

(3.76)

Equations 3.72, 3.75 and 3.76 are the scaler equations for all three axes in the decoupled form. A summary of all the differential equations for each coordinate system is provided in Table 3.3 for Cartesian coordinate systems, Table 3.4 for cylindrical coordinate systems and Table 3.5 for spherical coordinate systems. The six general solutions for each scaler equation are also provided in these tables. Table 3.3 is a Cartesian scaler equation summary. The solutions are found in most books on differential equations. Table 3.4 is a summary of the differential scaler equations for cylindrical coordinate systems; these of course are more complex. The first entry in the table is similar to Cartesian coordinate system differential equations and is solved in a similar manner, with the solutions as shown. However, the second entry in Table 3.4 is a bit more complex differential scaler equation expressed with solutions to the Bessel function of the first and second kind, J_m and Y_m respectively. The third entry in the table is in terms of Henkel functions of the first and second kind, as H^1 and H^2 respectively. The fourth entry in the table is the scaler differential equation similar to Cartesian coordinate solutions.

TABLE 3.3   Second order Differential Solutions in Rectangular Coordinate Systems

Differential equations	Solutions	Remarks
$\dfrac{d^2f}{dx^2} + \beta_x^2 f = 0$	$f_1(x) = C_1 Cos(\beta_x x) + D_1 Sin(\beta_x x)$ $f_2(x) = A_1 e^{-(j\beta_x + \alpha)x} + B_1 e^{(j\beta_x + \alpha)x}$	Second order differential equation
$\dfrac{d^2g}{dy^2} + \beta_y^2 g = 0$	$g_1(y) = C_2 Cos(\beta_y y) + D_2 Sin(\beta_y y)$ $g_2(y) = A_2 e^{-(j\beta_y + \alpha)y} + B_2 e^{(j\beta_y + \alpha)y}$	Second order differential equation
$\dfrac{d^2h}{dz^2} + \beta_z^2 h = 0$	$h_1(z) = C_3 Cos(\beta_z z) + D_3 Sin(\beta_z z)$ $h_2(z) = A_3 e^{-(j\beta_z + \alpha)z} + B_3 e^{(j\beta_z + \alpha)z}$	Second order differential equation

TABLE 3.4   Second order Differential Solutions in Cylinderical Coordinate Systems

Differential equations	Solutions	Remarks
$\dfrac{d^2g}{d\phi^2} + m^2 g = 0$	$g_1(\phi) = C_2 Cos(m\phi) + D_2 Sin(m\phi)$ $g_2(\phi) = A_2 e^{-jm\phi} + B_2 e^{jm\phi}$	Second order differential equation
$\rho^2 \dfrac{d^2f}{d\rho^2} + \rho \dfrac{df}{d\rho} + \left[(\beta_\rho \rho)^2 - m^2\right] f = 0$	$f_1(\rho) = \left[A_1 J_m(\beta_\rho \rho) + B_1 Y_m(\beta_\rho \rho) \right]$	
$\rho^2 \dfrac{d^2f}{d\rho^2} + \rho \dfrac{df}{d\rho} + \left[(\beta_\rho \rho)^2 - m^2\right] f = 0$	$f_2(\rho) = \left[C_1 H_m^1(\beta_\rho \rho) + D_1 H_m^2(\beta_\rho \rho) \right]$	
$\dfrac{d^2h}{dz^2} + \beta_z^2 h = 0, \; \beta^2 = \beta_z^2 + \beta_\rho^2$	$h_2(z) = C_3 \cos(\beta_z z) + D_3 \sin(\beta_z z)$ $h_1(z) = A_3 e^{-j\beta_z z} + B_3 e^{j\beta_z z}$	

TABLE 3.5   Second order Differential Solutions in Spherical Coordinate Systems

Differential equations	Solutions
$\dfrac{d^2h}{d\phi^2} + m^2h = 0$	$h_1(\phi) = C_3Cos(m\phi) + D_3Sin(m\phi)$ $h_2(\phi) = A_3e^{-jm\phi} + B_3e^{jm\phi}$
$\dfrac{1}{g\sin\theta}\dfrac{d}{d\theta}\left(\sin\theta\dfrac{d(g)}{d\theta}\right) - \left(\dfrac{m}{\sin\theta}\right)^2$ $= -n(n+1)$	$P_n^m(\cos\theta)$ $Q_n^m(\cos\theta)$
$\dfrac{1}{g\sin\theta}\dfrac{d}{d\theta}\left(\sin\theta\dfrac{d(g)}{d\theta}\right) - \left(\dfrac{m}{\sin\theta}\right)^2$ $= -n(n+1)$	$P_n(\beta r) = \dfrac{1}{2n!\,d^n(\beta r)}\dfrac{d^n}{}\left[(\beta r)^2 - 1\right]^n$ $P_n(\xi) = \dfrac{1}{2n!\,d^n\xi}\dfrac{d^n}{}\left[(\xi)^2 - 1\right]^n$
$\dfrac{1}{f}\dfrac{d}{dr}\left(r^2\dfrac{df}{dr}\right) + (\beta r)^2 - n(n+1) = 0$	$f_1(r) = A_1\sqrt{\dfrac{\pi}{2\beta_r r}}J_{n+1/2}(\beta_r r)$ $+ B_1\sqrt{\dfrac{\pi}{2\beta_r r}}H_{n+1/2}(\beta_r r)$
$\dfrac{1}{f}\dfrac{d}{dr}\left(r^2\dfrac{df}{dr}\right) + (\beta r)^2 - n(n+1) = 0$	$f_2(r) = A_1\sqrt{\dfrac{2}{\pi r}}\cos\left(r - \dfrac{\pi}{4} - \dfrac{n\pi}{2}\right) +$ $+ B_1\sqrt{\dfrac{2}{\pi r}}\sin\left(r - \dfrac{\pi}{4} - \dfrac{n\pi}{2}\right)$
$J_n(\beta r) = \displaystyle\sum_{m=0}^{\infty}\dfrac{(-1)^m(\beta r/2)^{n+2m}}{m!(n+m)!(n+m+1)}$	
$H_n(\beta r) = \dfrac{(\cos n\pi)J_n(\beta r) - J_{-n}(\beta r)}{\sin n\pi}$	

A summary of the scaler differential equations for spherical coordinate systems is provided in Table 3.5. The first entry in the table is the scaler differential equation similar to Cartesian coordinate system equations and the solutions are very similar. The second entry in the table is presented as a function of Legendre functions P_n^m and Q_n^m. The equations showing the Legendre functions provided in the fourth entry as a series must be evaluated for P_n^m and Q_n^m; this is similar to ln(x). The fifth entry in Table 3.5 provides an alternative solution to the Bessel and Henkel functions in entry for large values of r. The sixth entry in Table 3.5 depicts the evaluation of the Bessel function as a series and the seventh entry is how the Bessel and Henkel functions are related. Reference [2] has excellent tables of Bessel and Henkel functions.

A summary of wave functions in Cartesian coordinate systems is given in Table 3.2.

The cylindrical and spherical coordinate systems do not have as easy table to generate as the Cartesian coordinates shown in Table 3.2. See reference [2] for more detail on standing, traveling, evanescent, attenuated standing and traveling waves. As the boundary conditions are added to the solution of the cylindrical and spherical equations, a decision will be apparent that will illustrate what type of wave is being analyzed. As the case studies are examined, it will become apparent what type of waveform is being analyzed.

This book will use some of the equations generated here; however, several of the case studies have a one- or two-dimensional analysis. The case studies generally have boundary conditions in the far field and do not have active sources within the field so that analysis is fairly straightforward. Most radiation that requires EMC analysis is not within cables such as coaxial cable. Most of the problems occur on the outside of the shield, therefore getting bogged down with cylindrical coordinate systems for coaxial analysis is unnecessary unless specifically required on a case by case basis. Spherical coordinate systems are important for EMC analysis when dealing with the radio transmission of electromagnetic waves. These again are studied and analyzed on case by case basis for case studies.

$$P_{supplied} = \frac{1}{2}\iiint_v (\bar{H}^* * \bar{M}_i + \bar{E} * J_i^*)\,dv \quad \text{Supplied complex power}(\text{W}) \quad (3.77)$$

$$P_{Poynting} = \frac{1}{2}[\text{Real}(\bar{E}x\bar{H}^*)] = \hat{E}_{time}x\hat{H}_{time} \quad \text{Exiting complex power}(\text{W}) \quad (3.78)$$

$$P_{dissipation} = \frac{1}{2}\iiint_v \sigma |E|^2\,dv \quad \text{Dissipated real power}(\text{W}) \quad (3.79)$$

$$W_{stored\_E} = \frac{1}{4}\iiint_v \varepsilon |E|^2\,dv \quad \text{Time average electric energy}(\text{J}) \quad (3.80)$$

$$W_{stored\_H} = \frac{1}{4}\iiint_v \mu |H|^2\,dv \quad \text{Time average magnetic energy}(\text{J}) \quad (3.81)$$

$$P_{stored\_E} = -j2\omega \iiint_v 1/4\varepsilon |E|^2\,dv \quad \text{Electric field power storage} \quad (3.82)$$

$$P_{stored\_H} = -j2\omega \iiint_v 1/4\mu |H|^2\,dv \quad \text{Magnetic field power storage} \quad (3.83)$$

The details for analyzing power supplied by fields are shown in References [1] and [3]. The final result is Equations 3.77 through 3.81. Checking the units for each of the equations it can be observed that volume integration is necessary for all except Equation 3.78. This should always be done when examining equations as a check for accuracy. Equation 3.78 is the radiated power from such things as antennas and obstacles that may reflect power. Equation 3.79 is the real power dissipated. Note Equation 3.80 expresses energy storage of an electric field, and Equation 3.81 energy stored in the magnetic field. Since most of the EMC studies for radiation are in the far field, Equations 3.78 and 3.79 are more frequently used. Equations 3.80 and 3.81 show energy storage that is used for near field studies or the transition between near field and far field. The near field to far field is not abrupt. The limit where the far field begins is 2D²/λ.

$$\hat{\overline{P}}_{Poynting} = \hat{\overline{E}}_{time} x \hat{\overline{H}}_{time} \tag{3.84}$$

$$\overline{P}_{Poynting} = \frac{1}{2}\left[\text{Real}\left(\overline{E}x\overline{H}^*\right)\right] \tag{3.85}$$

Reactive power is represented by Equations 3.82 and 3.83, often referred to as watt-less power. This is the power in capacitive and inductive reactance respectively. This power is due to electric fields of capacitance and stray capacitance, and inductive reactance is due to leakage inductance in transformers and in the windings of both transformers and motors. The near field of antennas stores electric and magnetic field energy; therefore, these are very important parameters when dealing with the radiated power. Storage aspects of the field storage will change the field shape for both magnetic and electric fields.

The time-varying Poynting vector expression is Equation 3.84 with the magnitude shown as peak values. Equation 3.85 is expressed in phasor form with the RMS magnitudes. The latter Poynting vector is the one most often used. When examining transients, for example catenary arcing, Equation 3.84 will be used in several the calculations. At this point, the introduction to electromagnetic fields does not provide analysis for oblique waveforms on obstacles or surfaces. On a case by case basis, further analysis will be introduced showing the impact of oblique waveforms. A reason for not discussing all the subtleties of oblique waveforms is the complexity of these analyses are better understood with the actual analysis of a particular case study problem. The objective of this book is not to produce a complete electromagnetic analysis of fields, but to use electromagnetic analysis to solve actual problems in the transportation industry. Reference [4] is easy to read as a good start on understanding field theory. Reference [1] is more difficult to read but it is a very in-depth study with several examples.

Example 3.2

Examples of field theory for solving Kirchhoff's current and voltage law are as follows. Kirchhoff's current law is that the sum of the currents flowing into a node is equal to zero. It is presented as Equation 3.86. This equation has a restriction that there is no charge buildup at the node. Equation 3.87 allows charge buildup at the node, which provides a term for stray leakage capacitance. The result is leakage capacitance at the node, as expressed by Equation 3.88.

$$\sum i = 0 \quad \text{Normal circuit analysis with no charge buildup at node} \tag{3.86}$$

$$\sum i = \oiint \overline{J}_{ic} * d\overline{S} \quad \text{Field relation of charge on note} \tag{3.87}$$

$$\sum i = -\frac{\partial Q_{node\_charge}}{\partial t} = -\frac{\partial C_{leakage} v}{\partial t} = -C_{leak}\frac{\partial v}{\partial t} \quad \text{Leakage capacitance at node} \tag{3.88}$$

Example 3.3

Kharkov's voltage law is that the sum of the voltages around the loop are equal to zero. This is presented as Equation 3.89. The left side of Equation 3.90 indicates that circulation integration of the electric field around the loop is not zero. This is due to the magnetic flux flowing through the loop resulting in a stray magnetic flux. This will result in a voltage, as shown in Equation 3.90. To account for the stray voltage. an additional inductive reactance is added for the stray voltage as the last term in Equation 3.91

$$\sum v = 0 \quad \text{Normal circuit analysis with no magnetic flux in the loop} \quad (3.89)$$

$$\oint \bar{E} * d\bar{l} = -\frac{\partial}{\partial t} \iint_s \bar{B} * d\bar{S} = -\frac{\partial \psi_m}{\partial t} \quad \text{Field relation with magnetic flux in the loop}$$

$$(3.90)$$

$$\sum v = -\frac{\partial \psi_m}{\partial t} = -\frac{\partial (Li)}{\partial t} = -\frac{L\partial(i)}{\partial t} \quad \text{Leakage inductance in the loop} \quad (3.91)$$

As can be observed, the examples indicate how the electromagnetic field analysis will impact EMC calculations.

Example 3.4

A lossy rectangular conducting strip is used to carry electric current. The current is non-uniform across the cross-section of the strip. The current rapidly decays on the upper and lower surfaces as a function of y as they approach the center of the strip. The dimensions of the strip are 0.5 mm (y) thick and 10 mm (x) wide. The current density is represented by $J(y\} = \bar{a}_z 10^5 e^{-10^6 y}$ A/m^2. The objective here is to find the current flowing through the cross-section of the strip.

Solution:

$$I = \iint \bar{J} * d\bar{S} = 2 \int_0^{2.5 \times 10^{-4}} \int_0^{10^{-2}} \left[\hat{a}_z (10^5) e^{-10^6 y} \right] * \hat{a}_z dx dy$$

$$I = 2(10^{-2})(10^5) \int_0^{2.5 \times 10^{-4}} e^{-10^6 y} dy$$

$$I = 2 * 10^3 \left. \frac{e^{-10^6 y}}{-10^6} \right|_0^{2.5*10^{-4}} = -2 * 10^{-3} \left(e^{-2.5*10^2} - 1 \right) \approx 2\,\text{mA}$$

The lossy strip in Figure 3.2 is an example that can also be configured as a ground plane for larger dimensions and smaller thickness of the material. The exponential in equation describing the current density can be due to a skin effect so this simple example can be expanded for other uses, such as circulation PC boards, stainless steel ground plane and circulation currents on enclosures.

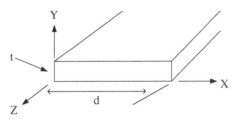

FIGURE 3.2 Lossy strip

Example 3.5

The height of a transmitting antenna for a ground to air communication system is placed at 10 m above the water (reflective material $\varepsilon_r = 81$).

$$\sin \theta_i = \sqrt{\frac{\frac{\varepsilon_2}{\varepsilon_1}\frac{\mu_2}{\mu_1}}{\frac{\varepsilon_2}{\varepsilon_1} - \frac{\varepsilon_1}{\varepsilon_2}}} \qquad \text{Brewster angle} \qquad (3.92)$$

$$\theta_c = \sin^{-1} \sqrt{\frac{\varepsilon_2}{\varepsilon_1}\frac{\mu_2}{\mu_1}}$$

$$\theta_i = \theta_B = \tan^{-1} \sqrt{\frac{\varepsilon_2}{\varepsilon_1}} = \tan^{-1} \sqrt{\frac{81\varepsilon_1}{\varepsilon_1}} = \tan^{-1}(9) = 83.72° \quad \text{Brewster angle}$$

$$(3.93)$$

If the aircraft radio transmission in Figure 3.3 is not to have a reflection from the water, the sine of the Brewster angle is 0.994 (83.72°). The angle at which the reflection coefficient is zero is referred to as the Brewster angle. The relative constants are $\varepsilon_1, = \mu_1, = \mu_2 = 1$. The relative constant is $\varepsilon_2 = 81\,\varepsilon_1$ for water. The distance d between the ground station and the aircraft is 20 km. If the aircraft is at 2212 m, the aircraft radio will not produce any reflections at the h_1 antenna. The assumptions are the water surface is flat, lossless and free of any ripples. The problem is similar to the notes in Reference [3].

$$\theta_i = \theta_B = \tan^{-1} \sqrt{\frac{81\varepsilon_1}{\varepsilon_1}} = \tan \frac{x}{10} \quad \text{Then } x = 90\,\text{m}.$$

The x direction parallel to the water is where the incident wave impinges on the water at 90 m.

$$\tan^{-1} \frac{2*10^4 - 90}{h_2} = \tan^{-1} 9, \quad \frac{2*10^4 - 90}{h_2} = 9 \quad h_2 = 2212\,\text{m}$$

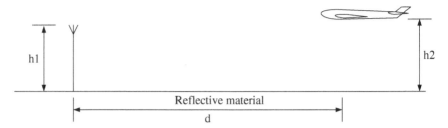

FIGURE 3.3 Aircraft landing reflection example

The height that the aircraft must be approximately 6000 ft, which is reasonable for it to be within tower control space.

Example 3.6

A perpendicular electromagnetic wave in free space is incident upon a block of material. The wave is traveling in the z direction and the reflection is in the z direction (Figure 3.4). The material is a perfect conductor.

Where $\sigma = \infty$, $\bar{E}_x^i = -\hat{a}_x E_o e^{-j\beta_0 z}$, $\bar{E}_x^r = -\hat{a}_x \Gamma E_o e^{+j\beta_0 z} = \hat{a}_x E_o e^{+j\beta_0 z}$, $\beta_0 = \omega\sqrt{\mu_0 \varepsilon}_0$

$$\bar{H}_i = -\frac{1}{j\omega\mu_o}\bar{\nabla}x\bar{E}_i = -\hat{a}_y \frac{1}{j\omega\mu_0}\frac{\partial E_x^i}{\partial z} = \hat{a}_y E_o \frac{\beta_0}{\omega\mu_0}e^{-j\beta_0 z} = \hat{a}_y E_o \sqrt{\frac{\varepsilon_0}{\mu_0}}e^{-j\beta_0 z} \quad (3.94)$$

$$\bar{H}_r = -\frac{1}{j\omega}\bar{\nabla}x\bar{E}_r = -\hat{a}_y \frac{1}{j\omega\mu_0}\frac{\partial E_x^r}{\partial z} = \hat{a}_y E_o \frac{\beta_0}{\omega\mu_0}e^{+j\beta_0 z} = \hat{a}_y E_o \sqrt{\frac{\varepsilon_0}{\mu_0}}e^{+j\beta_0 z} \quad (3.95)$$

$$\bar{J}_s = \bar{n}x\bar{H}_m = -a_z x\hat{a}_y\left(H^i + H^r\right) = \hat{a}_x\left(H^i + H^r\right) = \hat{a}_x E_o\sqrt{\frac{\varepsilon_0}{\mu_0}}\left(e^{-j\beta_0 z} + e^{+j\beta_0 z}\right)_{z=0}$$

$$(3.96)$$

$$\bar{J}_s = 2\hat{a}_x E_o\sqrt{\frac{\varepsilon_0}{\mu_0}} = \hat{a}_x \frac{2E_o}{377} = \hat{a}_x E_o 5.305 mA/m,\left(Where\sqrt{\frac{\mu_0}{\varepsilon_0}} = 377\Omega\right) \quad (3.97)$$

The electromagnetic wave is incident on the x, y plane and it is traveling in the −z direction. The reflected wave is traveling in the z direction. Using Maxwell's Equation 3.1 with no sources, M = 0, the phasor form of Maxwell's equation and $\bar{B} = \mu\bar{H}$, Equation 3.94 can be derived for the incident magnetic field. Since the electric field is in the x direction. The H field must be in the y direction because these two fields must be orthogonal in the far field and they form the x, y plane and z is normal to this plane. Observing the phase term in all equations, that is the exponent of e requires −z in the positive direction (incident) and +z in the negative direction (reflective).

The current density J at the surface of the material, that is the perfect conductor, is normal to the H field as presented in Equation 3.96. Note that the cross product of

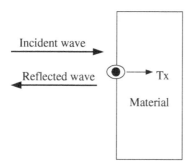

FIGURE 3.4   Perpendicular electromagnetic wave in free space incident upon a block of material (z direction perpendicular to the surface, x direction vertical and y direction out of the page)

the unit for z and y is in reverse order. This makes the cross product negative. Therefore the negative of a negative provides a positive value for the unit vector in x. The incident and reflected wave sum is equal to two, that is this produces a standing wave because the wave is 100% reflected. The waves are traveling through air that has a resistance of 377 ohms. The current density in the x direction is E_0* 5.305 mA/m. If E_0 = 3 V/m the current density would be 15.91 mA/m.

Example 3.7

The free space incident and reflected fields of an oblique wave is incident on a perfect conductor material, as shown in Figure 3.5.

Solution: $\bar{H}_i = -\dfrac{1}{j\omega\mu}\nabla x\bar{E}_i$. Solving this version of Maxwell's equation for the incident version of the H_i field that has no sources is Equation 3.98. Note in this solution, Equation 3.98 is a two-dimensional solution because of the incident wave angle.

$$\bar{H}^i = \left(\hat{a}_y x\hat{a}_z\right)E_0\frac{\partial}{\partial x}\left(e^{-j\beta_0(x\sin\theta_i+z\cos\theta_i)}\right)+\hat{a}_z)E_0\frac{\partial}{\partial z}\left(e^{-j\beta_0(x\sin\theta_i+z\cos\theta_i)}\right) \qquad (3.98)$$

$$\bar{E}^r = \hat{a}_y E_0\Gamma_{rc}e^{-j\beta_0(x\sin\theta_i+z\cos\theta_i)} \qquad (3.99)$$

The two equations above are phasor representations of incident H field and reflected E field waves. The reflected wave has the reflection coefficient (Γ_{rc}) included. Along the interface between air and material tangent to y axis z is equal to zero. Find the H field, and electric current intensity J for the value of the incident electric field at 3 V/m. Where $\beta_0 = \omega\sqrt{\mu_0\varepsilon}_0$, and $\sqrt{\dfrac{\mu_0}{\varepsilon_0}} = 377\,\Omega$.

$$E^i + E^r = 0 = \bar{E}^r = \hat{a}_y E_0\left(1+\Gamma_{rc}\right)e^{-j\beta_0(x\sin\theta_i)},\left(1+\Gamma_{rc}\right)=0,\Gamma_{rc}=-1$$

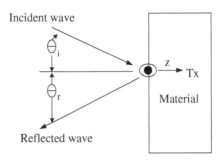

FIGURE 3.5 Free space incident and reflected fields of an oblique wave incident on a perfect conductor material (z direction horizontal, x direction vertical and y direction out of the page)

Using Maxwell's equations below to find the H fields and performing the function in Equation 3.98, the resulting calculations are Equations 3.100 and 3.101, the magnetic fields.

$$\bar{H}_i = -\frac{1}{j\omega\mu_o}\bar{\nabla}x\bar{E}_i, \bar{H}_r = -\frac{1}{j\omega\mu_o}\bar{\nabla}x\bar{E}_r$$

$$\bar{H}^i = H_0\left(-\hat{a}_x\cos\theta_i + \hat{a}_z\sin\theta_i\right)e^{-j\beta_0(x\sin\theta_i + z\cos\theta_i)} \tag{3.100}$$

$$\bar{H}^r = -H_0\left(\hat{a}_x\cos\theta_i + \hat{a}_z\sin\theta_i\right)e^{-j\beta_0(x\sin\theta_i - z\cos\theta_i)} \tag{3.101}$$

Where $H_0 = \dfrac{3\,\text{V/m}}{377} = 7.96\,\text{mA/m}$.

The surface current density normal to the H field is represented by Equation 3.102. The final result is Equation 3.103.

$$\bar{J}_s = \bar{n}x\left(\bar{H}^i + \bar{H}^r\right) = -\hat{a}_z x\left[\hat{a}_x x\left(\bar{H}^i + \bar{H}^r\right) + \hat{a}_z x\left(\bar{H}^i + \bar{H}^r\right)\right]_{z=0} \tag{3.102}$$

$$\bar{J}_s = -\hat{a}_z x\left[\hat{a}_x x\left(\bar{H}^i + \bar{H}^r\right)\right]_{z=0} = \hat{a}_y 2H_0\cos\theta_i\left(e^{-j\beta_0 x\sin\theta_i}\right)$$

$$\bar{J}_s = -\hat{a}_z x\left[\hat{a}_x x\left(\bar{H}^i + \bar{H}^r\right)\right]_{z=0} = \hat{a}_y 2H_0\cos\theta_i\left(e^{-j\beta_0 x\sin\theta_i}\right)$$

$$\bar{J}_s = \hat{a}_y 15.91\cos\theta_i\left(e^{-j\beta_0 x\sin\theta_i}\right)\text{mA}/m \tag{3.103}$$

Example 3.8

Find a critical phase angle in Example 3.7

$\theta_c = \sin^{-1}\sqrt{\dfrac{\varepsilon_2}{\varepsilon_1}\dfrac{\mu_2}{\mu_1}}$ Where $\theta_i \le \theta_C$ most conductors $\varepsilon_1 = \varepsilon_2$, μ_1 (air) $< \mu_2$ (good conductor)

Sin $(\theta_c) \gg 1$. With this condition electric field will be a surface wave.

3.4   REVIEW PROBLEMS

Problem 3.1
A lossy rectangular conducting strip is used to carry electric current. The current is non-uniform across the cross-section of the strip. The current rapidly decays on the upper and lower surfaces as a function of y as they approach the center of the strip. The dimensions of the script are 0.5 mm (y) thick and 5 mm (x) wide (Figure 3.6). The current density is represented by $J(y) = \bar{a}_z 10^5 e^{-10^6 y}$ Amps/m^2. The objective here is to find the current flowing through the cross-section of the strip.

Problem 3.2
The height of a transmitting antenna for a ground to air communication system is placed at a height of 10 m above the water (reflective material ε_r = 81). If the aircraft radio transmission in Figure 3.3 is not to have a reflection from the water, the angle at which the reflection coefficient is zero is the Brewster angle. The relative constants are ε_1, = μ_1, = μ_2 =1. The relative constant is ε_2 = 81 ε_1 for water. The distance d between the ground station and the aircraft is 10 km (Figure 3.7). The assumptions are the water surface is flat and lossless. The problem is similar to the notes in Reference [3]. Find the sine of the Brewster angle and distance h_2.

Problem 3.3
Refer to Example 3.6 Replace the value ε_0 with $\varepsilon_r \varepsilon_0$ = 3.3 ε_0 (snow) in the example and find the value J_s. The same calculation can be done for our humidity with a great deal of water vapor or other gases in place of air. This same technique can be used on certain solids such as Plexiglas.

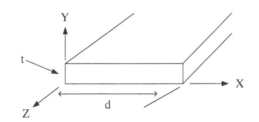

FIGURE 3.6   Lossy rectangular conducting strip

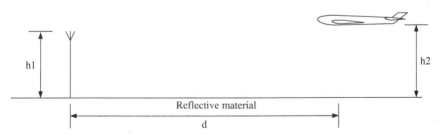

FIGURE 3.7   Aircraft landing reflection

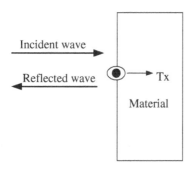

FIGURE 3.8   Perpendicular electromagnetic wave in free space incident upon a block of material

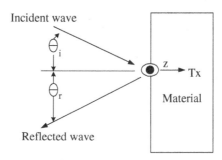

FIGURE 3.9   Free space incident and reflected fields of an oblique wave incident on a perfect conductor material (z direction horizontal, x direction vertical and y direction out of the page)

Problem 3.4
Solve Example 3.6 with the assumption that copper is a perfect conductor (Figure 3.8).

Problem 3.5
Find a critical phase angle in Example 3.7 (Figure 3.9).

3.5   ANSWERS TO REVIEW PROBLEMS

Problem 3.1
$I = 10^{-4}$ A.

Problem 3.2
Brewster angle= 82.73°, h_2 = 1106 m.

Problem 3.3
$J = 9.637 \times 10^{-3}$ A/m

Problem 3.4
$\bar{J}_s = \hat{a}_y 15.91 \cos\theta_i \left(e^{-j\beta_0 x \sin\theta_i} \right)$ mA/m.

Problem 3.5
$\sin(\theta_c) \gg 1$. With this condition the electric field will be a surface wave.

4

CASE STUDIES AND ANALYSIS IN TRANSPORTATION SYSTEMS

4.1 BACKGROUND INFORMATION FOR SUBWAY SYSTEMS

4.1.1 Types of equipment and bungalows/houses

The communications houses have the largest percentage of equipment sensitive to electric and magnetic fields. The audience is introduced to rail transportation in this chapter. Chapter 5 also has rail case studies but those are generally radio. Table 4.1 lists the equipment that is installed in most communications houses. The equipment is usually installed in cabinets or open racks. SCADA equipment and occasionally radios are installed in open racks. Fire and intrusion alarm systems are wall mounted in special cabinets with telephone line connected to the 26 pair cable or two pair cable that runs between houses in a daisy-chain fashion to a central control room. The hard wired phone system is a backup in the event the network fails. The alarm is sent directly to the fire department. A limited version of Table 4.1 equipment is installed at or near station platforms in hardened stainless steel cabinets for SCADA, video and PA system components. Tunnels have an expanded set of equipment due to tunnel equipment, such as large vent fans, sump pumps and environmental controls that are more elaborate than communication house equipment.

Radio communications has several narrow band frequencies. The fire department has multiple narrow band frequencies in the lower UHF band. Emergency medical services (EMS) can have VHF and low UHF band frequencies, local police will have low and high UHF band frequencies, state police have VHF and low band UHF and the transit authority has upper UHF radio to communicate with transit cars. Not all the narrow band radios are installed in the communication houses. Generally only the radios for internal communication are installed. The 902–928 MHz radios and other wireless

Electromagnetic Compatibility: Analysis and Case Studies in Transportation, First Edition.
Donald G. Baker.
© 2016 John Wiley & Sons, Inc. Published 2016 by John Wiley & Sons, Inc.
Companion website: www.wiley.com/go/electromagneticcompatibility

TABLE 4.1   Communication House Equipment

Item	Description	Frequency range	Fields of sensitivity	Remarks
1	Radio communications	150–928 MHz	E field	Narrow band
2	PA system equipment	300–3200 Hz	E and H fields	Near stations
3	Video equipment DS-3	Digital 4.5 MHz	E and H fields	Parking lots and stations
4	SCADA equipment	1.0–3.2 MHz Clk	E and H fields	Monitors all subsystems
5	Network equipment	OC-192 Clk 10 GHz	E field	Fiber optics
6	Telephone equipment	T-1 1.5 MHz	E and H fields	Telephone T-1
7	Fire alarm	1–2 MHz Clk	E and H fields	Monitored by control room
8	Intrusion detection	1–2 MHz Clk	E and H fields	Monitored by control room
9	Fire suppression	Control fire alarm	E and H fields	Monitored by control room
10	Environmental	SCADA control	E and H fields	Vent fan and temperature control

devices are unlicensed bands are used at the wayside but the data is sent to the communication house and relayed to the control room via the communication network.

The PA and video equipment at stations are controlled and monitored by the central control room or from a local cabinet located at the station. The local control is used in the event of an emergency. SCADA equipment is often installed close to the PA. The SCADA system is composed of dry contact monitor points, control contacts and dual processors. The secondary processor shadows the primary as a backup for processor failure. These computers are industrial grade and hardened for temperature and other environmental factors. The clock frequency is not very high by today's computer standards; however, it is high enough to monitor all the local points in a communications house. These devices are programmable logic computers. They use ladder logic similar to older relay networks as this was the method of monitor and control prior to computer logic. The SCADA computer processor can be programmed locally using a laptop computer. This can also monitored by maintenance personnel.

The signal house/bungalow has a suite of equipment as shown in Table 4.2. These are generally located at track interlockings. This is the equipment that provides position data to the main control room and provides signaling along the rail right of way. Communications between the signal houses and the control room is via the communications houses. Short runs usually have 12 pair copper cable as the communication medium and longer runs such as 200 meters have fiber optic cable modems at each end.

TABLE 4.2 Signal Bungalow/house Equipment

Item	Description	Frequency range	Fields of sensitivity	Remarks
1	Track circuits	100 Hz	H fields	Supplied for relay circuits
2	Audio freq. track ckts	2–10 KHz	H fields	Local oscillators
3	Non-vital logic	1–3 MHz clock	E and H fields	Ladder logic (PLC)
4	Vital logic and backup	1–3 MHz clock	E and H fields	Industrial grade computers
5	Telephone	T-1 (1.5 MHz clock)	E and H fields	
6	Network F.O. or copper	T-1 (1.5 MHz clock)	E and H fields	Same as telephone
7	Switch controllers	Relay operated		
8	Fire alarm	1–2 MHz Clk	E and H fields	Monitored by control room
9	Intrusion detection	1–2 MHz Clk	E and H fields	Monitored by control room

Often the conversion of data such as RS 232 or RS 485 are converted to a fiber optic format in the signals house and back to copper at the communications house.

Interlockings are where crossovers between the inbound and outbound tracks can be done in emergencies, maintenance or to reroute traffic. These have switches that can be set by either the central control room or locally in the signal houses/bungalows with a wall screen to show the operator the setting. Item 1 in Table 4.2 is track circuits using locally generated 100 Hz to provide a balanced relay system with power. As a train enters the track circuit, the relay points open due to the axle and wheels shorting out the 100 Hz power to the relay. This occurrence will allow the central control room to observe the train entering the interlocking area. As can be observed a power outage opens the relays, warning the control room that a problem exists.

Audio frequency track circuits have a transmitter and receiver tuned for a particular frequency for a section of track that operates in a similar fashion to the 100 Hz insulated joint track circuits. When there is no train in the track section, the receiver gets a continuous audio frequency. As a train enters the section of track an open set of contacts will indicate a train is present due to the wheels and axle short circuit of the receiver. The control room will then have data for the train's location. In today's environment, subway cars are equipped with GPS systems to monitor train activity; however, tunnels and overpasses will not have this position information. The track circuits are a safety measure that will give the control room a positive fix on a car's position at all times. Most track sections are 41 meters or more long with isolation bonds between sections or tuned circuits for audio frequency track circuits. Each section has a different audio frequency so that several frequencies can be used multiple times along the tracks.

Non-vital logic is programmed using programmable logic control (PLC) units and need not be programmed in failsafe mode. Vital logic must be programmed for failsafe operation; this is due to safety issues. The PLC requires dual hardened processors with a secondary that shadows the primary. The latter logic is also in-programmed for failsafe operation in the event of a power failure. The PLC has a fairly low frequency clock as compared to the processors in the communications house. The data sent by the processors may have a heartbeat signal sent out to insure that the equipment is operational. These types of signals will transmit a digital word to the communication house computer that will process it for the central control room. If the heartbeat does not occur, a warning is sent to the central control room operator station or the central control room may process the loss of heartbeat. The processing decision is dependent on whether the authority wants distributed or central processing for the heartbeat.

Items 5 and 6 in Table 4.2 may be in some cases the same T-1 which will handle the telephone and other functions in the signal house/bungalow such as environmental, fire and intrusion alarms. This is all dependent on the overall design of the system. Often, the 26 pair copper cable is used for communications. This again is dependent on the authority specifications and design of the system. Not discussed in this book is the transit system power supply which gives the 680 V DC at 800 Amps used for traction motors. The experience of most authors is with the communication aspect of transit systems; however, any issues raised that are due to electromagnetic radiation from the DC power supply or traction motors is analyzed. The internal workings of the power supply house itself are not examined here, except how they affect telephone service or SCADA points.

The subsequent introduction in Chapter 4 is to provide the reader with information relating to subway systems. It gives an overview of the aspects of subway communications design. This information will be alluded to as the case studies are examined. The EMC aspects of the DC power supply and the traction motors are discussed as necessary to discuss a particular point.

4.2   CASE STUDIES

Case Study 4.1

A communication house commonly used on railway and subway systems is depicted in Figure 4.1.These houses are generally located 13–15 m from the center of the track. Poles (13–15 ft, 4.0–4.5 m, above the grade) or duct banks along the track carry power lines for the DC subway car, communication and signal house/bungalows (Figure 4.2). This power is three phase 13.5 kV from substations. The duct banks are designed for the H fields. The worst case is that poles that may be required where burial of duct banks is not possible. The substation power for the DC transit supply is brought in from two substations. They have a transfer trip mechanism that provides power when the main substation fails. If the

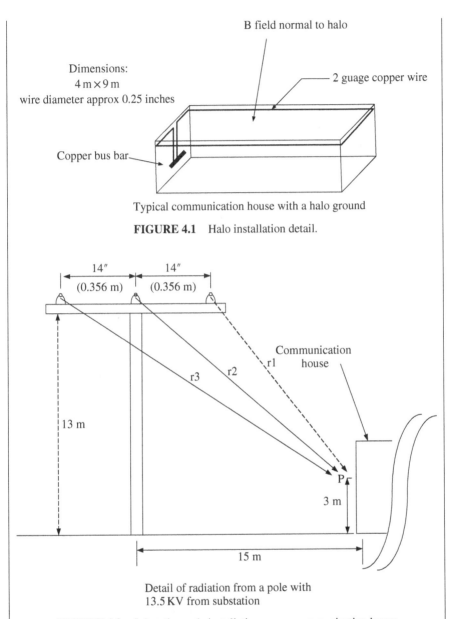

B field normal to halo

Dimensions:
  4 m × 9 m
wire diameter approx 0.25 inches

— 2 guage copper wire

Copper bus bar

Typical communication house with a halo ground

FIGURE 4.1 Halo installation detail.

14″
(0.356 m)

14″
(0.356 m)

Communication
house

r1

r3   r2

13 m

P

3 m

15 m

Detail of radiation from a pole with
13.5 KV from substation

FIGURE 4.2 Substation pole installation near a communication house

line is carrying 500 Amps from the local substation, the H field center of the communication house is calculated below. The calculations are for a communication house with and without rebar. Some have been constructed with aluminum alloys that are not good shields for magnetic fields at low frequencies, that is at 60 Hz and harmonics of 60 Hz. The author had one occasion to observe how the latter resulted in a very poor design that resulted in all of the equipment performing poorly. The cost to fix the problem was very high. Some of the equipment required mu-metal shielding which is very expensive. To improve the performance, a great deal of rework was required changing the position of equipment to reduce the effects of the low frequency magnetic fields and its harmonics. The effects were particularly noticeable when monitors changed colors and a high amount of distortion occurred as subway cars approached the station where the communication house was located. In some cases, filtering was required in order to solve the problems. This of course was an extreme case that is very rare but one should be vigilant when doing an analysis of a particular problem. It was obvious at first glance that this was to be a difficult problem to solve as an EMC engineer. EMC engineers should always look for the obvious when analyzing a problem because some designs meet the authority's expectations, that is requirements, but may have built-in EMC issues that affect the performance of systems or equipment.

A calculation is made to find the linear charge on the conductor using Equation 4.1. The substation voltage on the line is 13.5 kV. The distance r_1 is calculated at 17.7 m. The integral expression in Equation 4.1 is from 3 m to 17.7 m, as shown in the equation. Using Equation 4.2 to calculate the charge on the conductor gives a result in microcoulombs (μC)

$$V = -\int E * dr = -\frac{\rho_L}{2\pi\varepsilon_0} \int_{17.7}^{3} \frac{dr}{r} = -\frac{\rho_L}{2\pi\varepsilon_0} Ln3 - Ln17.7 = \frac{\rho_L}{2\pi\varepsilon_0}(1.77) \quad (4.1)$$

$$\rho_L = \frac{13.5kV * 2 * \pi * 8.854 * 10^{-12}}{1.77} = 0.13\,\mu C/m \quad (4.2)$$

The electric field is calculated using Equation 4.3 and the value r_1 (17.7 m). Distances r_2 and r_3 are estimated is approximately the same as r_1, due to the small differences in angles between r_1, r_2 and r_3. If the distances between the conductors is large, the x and y vector components would need to be considered. The H field is calculated by dividing the E field by 377 Ω, the impedance of air.

$$E = \frac{0.13\,\mu C/m}{2 * \pi * 8.854 * 10^{-12}\,F/m * 17.7} = 132\,V/m \quad (4.3)$$

$$H = 0.35\,A/m$$

The magnetic flux density is calculated using Equation 4.4 with the result in microwebers per meter.

$$B = \mu_0 H = 4\pi * 10^{-7} * 3.44\,Webers/m^2 = 0.439\,\mu W/m^2 \quad (4.4)$$

The area of the ground halo in Figure 4.1 is 36 m² and a Cos 377t term is necessary to calculate the magnetic flux using Equation 4.6. The derivative of Equation 4.5 results in the voltage at the bus bar.

$$\Psi = 0.\ 4.39 \times 10^{-6} \times 36 = 1.55 \times 10^{-5} \operatorname{Cos} 2\pi f \ \text{Webers} \qquad (4.5)$$

$$d\Psi / dt = -5.86\,\text{mV Sin } 377t \text{ at the bus bar } f = 60\,\text{Hz} \qquad (4.6)$$

The mass transit light rail system has a 680 V, 800 A DC power supply that provides traction motor power. Each car truck (two sets of wheels and axle) has a 150 hp (112 kW) AC traction motor. The frequency range of these motors is 3–1100 Hz using pulse width modulated (PWM) drives.

Case Study 4.2 analyzes the effects of transit cars and rails on the communication house.

Calculations for the shielding effect of communication house in Figure 4.1 is analyzed. This analysis will be used later in Case Study 4.2. The construction is as follows: the walls are concrete 4 inches thick with rebar 3/8 to 1/2 inch in diameter that form a 4 × 6 inch matrix. This forms a Faraday shield enclosure around the communication house. λ >> the largest opening in the mesh; therefore, it is equivalent to a solid sheet of steel at frequencies of approximately 100 KHz.

Intrinsic impedance of steel, concrete and air:

$$Z_{steel} = \sqrt{\frac{f*2000*4\pi*10^{-7}}{0.34*5.7*10^{7}}}\,\lfloor 45° = 1.139*10^{-5}\sqrt{f}\,\Omega\lfloor 45° \qquad (4.7)$$

$$Z_{concrete} = \sqrt{\frac{f*3*4\pi*10^{-7}}{5*10^{-4}}}\,\lfloor 45° = 6.667*10^{-2}\sqrt{f}\,\Omega\lfloor 45° \qquad (4.8)$$

$$Z_{air} = 377\,\text{Ohms}$$

Transmission and reflections

These equations are for low frequencies below 1 MHz that have long wavelengths and electromagnetic waves perpendicular to the surface of the shielding. The transmission coefficient in Equation 4.8 is from Equation 4.10 and is the reflection from the concrete in the wall to air concrete interface. The transmission coefficient in Equation 4.11 is the transmission to the steel matrix within the wall. The reflection coefficient Equation 4.12 is the reflection at the interface between the concrete in the wall and the steel matrix. The transmission coefficient Equation 4.13 is through steel to the concrete on the other side of the steel matrix in the wall. The reflection coefficient Equation 4.14 is the reflection at the steel concrete interface on the other side of the steel matrix. Equation 4.15 is a transmission coefficient through the concrete on the other side of the steel matrix to the air interface inside the communications house and the reflection coefficient Equation 4.16 is the reflection due to the concrete air interface at the inside wall of the communications house.

Calculations for Equations 4.7 and 4.8 are provided in Table 4.3. The resistance values are very small compared to air (377 Ω). The impedances are calculated as harmonics of 60 Hz. This table will use later when dealing with rail harmonic

TABLE 4.3 Impedance of Concrete and Steel

Frequency (Hz)	Z, concrete (Ω)	Z, steel (Ω)
60	0.516 42	0.000 09
120	0.730 33	0.000 12
180	0.894 47	0.000 15
240	1.032 85	0.000 18
300	1.154 76	0.0002
360	1.264 97	0.000 22
420	1.366 33	0.000 23
480	1.460 67	0.000 25
540	1.549 27	0.000 26
600	1.633 07	0.000 28
660	1.712 78	0.000 29
720	1.788 94	0.000 31
780	1.861 99	0.000 32
840	1.932 28	0.000 33
900	2.0001	0.000 34
960	2.065 69	0.000 35
1020	2.129 27	0.000 36
1080	2.191	0.000 37
1120	2.2312	0.000 38

issues. The harmonics above 420 Hz are generally found as emanating from traction motors. These calculations only pertain to incident waves that are normal to the surface of the communication house wall. Later in the text incident waves are examined at angles less than 90°. The E and H fields are generally in the near field at these frequencies instead of the 1/r function of fields in the near field. These relationships can be at higher powers of r.

$$T_{air/concrete} = \frac{2*Z_{air}}{Z_{air} + Z_{concrete}} \quad \text{Transmission coefficient} \quad (4.9)$$

$$\Gamma_{air/concrete} = \frac{Z_{concrete} - Z_{air}}{Z_{air} + Z_{concrete}} \quad \text{Reflection coefficient} \quad (4.10)$$

$$T_{concrete/steel} = \frac{2*Z_{concrete}}{Z_{steel} + Z_{concrete}} \quad \text{Transmission coefficient} \quad (4.11)$$

$$\Gamma_{concrete/steel} = \frac{Z_{concrete} - Z_{steel}}{Z_{steel} + Z_{concrete}} \quad \text{Reflection coefficient} \quad (4.12)$$

TABLE 4.4 Transmission Coefficients (T) at each interface

Frequency (Hz)	T (air/concrete)	T (concrete/steel)	T (steel/concrete)	T (concrete/air)
60	1.997	1.9997	0.000 3416	0.002 7359
120	1.996	1.9997	0.000 3416	0.003 867
180	1.995	1.9997	0.000 3416	0.004 734
240	1.995	1.9997	0.000 3416	0.005 4643
300	1.994	1.9997	0.000 3416	0.006 1073
360	1.993	1.9997	0.000 3416	0.006 6883
420	1.993	1.9997	0.000 3416	0.007 2223
480	1.992	1.9997	0.000 3416	0.007 7190
540	1.992	1.9997	0.000 3416	0.008 1853
600	1.991	1.9997	0.000 3416	0.008 6262
660	1.991	1.9997	0.000 3416	0.009 0453
720	1.991	1.9997	0.000 3416	0.009 4456
780	1.990	1.9997	0.000 3416	0.009 8294
840	1.990	1.9997	0.000 3416	0.010 1986
900	1.989	1.9997	0.000 3416	0.010 5546
960	1.989	1.9997	0.000 3416	0.010 8989
1020	1.989	1.9997	0.000 3416	0.011 2324
1080	1.988	1.9997	0.000 3416	0.011 5562
1120	1.988	1.9997	0.000 3416	0.011 767

$$T_{steel/concrete} = \frac{2*Z_{steel}}{Z_{steel} + Z_{concrete}} \quad \text{Transmission coefficient} \quad (4.13)$$

$$\Gamma_{steel/concrete} = \frac{Z_{steel} - Z_{concrete}}{Z_{steel} + Z_{concrete}} \quad \text{Reflection coefficient} \quad (4.14)$$

$$T_{concrete/air} = \frac{2*Z_{concrete}}{Z_{air} + Z_{concrete}} \quad \text{Transmission coefficient} \quad (4.15)$$

$$\Gamma_{concrete/air} = \frac{Z_{concrete} - Z_{air}}{Z_{air} + Z_{concrete}} \quad \text{Reflection coefficient} \quad (4.16)$$

The transmission coefficients are calculated in Table 4.4 as a function of frequency. It should be noted that the incident waves have a coefficient of almost 2.0 (a standing wave). These are not quite considered standing waves. As can be observed from the table, both concrete and steel are fairly good conductors.

TABLE 4.5 Reflection Coefficient at each interface

Frequency (Hz)	Γ (air/concrete)	Γ (concrete/steel)	Γ (steel/concrete)	Γ (concrete/air)
60	−0.997	0.9997	−0.9996584	−0.997 2641
120	−0.996	0.999658	−0.9996584	−0.996 1330
180	−0.995	0.999658	−0.9996584	−0.995 2660
240	−0.995	0.999658	−0.9996584	−0.994 5357
300	−0.994	0.999658	−0.9996584	−0.993 8927
360	−0.993	0.999658	−0.9996584	−0.993 3117
420	−0.993	0.999658	−0.9996584	−0.992 7777
480	−0.992	0.999658	−0.9996584	−0.992 2810
540	−0.992	0.999658	−0.9996584	−0.991 8147
600	−0.991	0.999658	−0.9996584	−0.991 3738
660	−0.991	0.999658	−0.9996584	−0.990 9547
720	−0.991	0.999658	−0.9996584	−0.990 5544
780	−0.990	0.999658	−0.9996584	−0.990 1706
840	−0.990	0.999658	−0.9996584	−0.989 8014
900	−0.989	0.999658	−0.9996584	−0.989 4454
960	−0.989	0.999658	−0.9996584	−0.989 1011
1020	−0.989	0.999658	−0.9996584	−0.988 7676
1080	−0.988	0.999658	−0.9996584	−0.988 4438
1120	−0.988	0.999658	−0.9996584	−0.988 233

Table 4.5 is the calculations for the reflection coefficients at all the interfaces. A reflection coefficient of one means all of the electromagnetic energy is reflected. This would be the case for a perfect conducting shield, that is $\sigma = \infty$.

The result as incident electromagnetic waves travel through the various interfaces and at various interfaces reflections occur is shown in Table 4.6. Reflections in particular occur at the air conductor boundaries. The equation for transmission coefficients at the various frequencies is shown in this table. The calculation for the transmission coefficient is shown as Equation 4.17. The reflection coefficients shown in the equation represent the internal reflections at the various interfaces.

The total reflection coefficient is almost 1.0. This means that the total electromagnetic wave is reflected as expected. The attenuation of the transmission is provided in dB for the residual incident wave after the reflection at the air/concrete interface. Some examples of how these tables were used in the case study of the halo in the communication house are provided here.

$$T_{\text{total}} = T_{\text{air/concrete}} * \Gamma_{\text{concrete/steel}} * \Gamma_{\text{steel/concrete}} * T_{\text{concrete/air}} \tag{4.17}$$

TABLE 4.6 Reflection and Transmission Coefficients at
Various Frequencies from Traction Motors

Frequency	T (total)	Γ (total)	T (attenuation)
60	–0.00546	–0.993856286	–45.26
120	–0.00771	–0.991603182	–42.26
180	–0.00944	–0.989877776	–40.50
240	–0.01089	–0.988425521	–39.26
300	–0.01217	–0.987147822	–38.29
360	–0.01332	–0.985994114	–37.51
420	–0.01438	–0.984934357	–36.84
480	–0.01537	–0.983948981	–36.27
540	–0.01629	–0.983024392	–35.76
600	–0.01716	–0.982150692	–35.31
660	–0.01800	–0.981320409	–34.90
720	–0.01879	–0.980527737	–34.52
780	–0.01955	–0.979768061	–34.18
840	–0.02028	–0.979037638	–33.86
900	–0.02098	–0.978333383	–33.56
960	–0.02166	–0.977652717	–33.29
1020	–0.02233	–0.976993461	–33.02
1080	–0.02296	–0.97635375	–32.78
1120	–0.02338	–0.975937331	–32.62

Example 4.1

The communications, signal/bungalow and track power houses in subway systems are situated so that doorways do not face the track. This allows the full effect of the shielding on all internal circuits. Doorways will generally have some leaks that would allow electromagnetic radiation to enter the houses. The subway wayside along the right of way has several sources of electromagnetic radiation that can affect the internal circuitry such as catenary, track circuits and subway car onboard circuitry. Catenary and hot rail arcing are rich in harmonics due to the 10 μsec rise time and high current. These are transients that will not always occur at the same place or time, so they may be very difficult to analyze. The transit cars have strict limits on allowed emissions but transient radiation cannot always be predicted. A car starting up near a station will have inrush currents that will make any arcing the worst case as compared to arcing as a car passes a point away from a station.

Maintenance personnel have 3 W two-way radios and cell phones. These are used while performing routine and emergency maintenance. The effect of these radios on the halo are examined in this example. The halo is located about 1.5 m from the maintenance person's head height the electric field data for both cell phone and two way radio are taken from Table 2.1 in Chapter 2. The electric fields 3.5 V/m for the cell phone at 1.5 m and 8 V/m for the 3 W two-way radio. These are the values of the main beam of each of the antennas since the halo is not in a plane with the main beam but almost directly overhead 40 dB less than main beam. The H fields are 19.7 and 22.51 µA/m respectively for the cell phone and radio using Equation 4.18 for the calculations. The halo for both the cell phone and the two-way radio is in the far field, so near field and restrictions do not apply. The E field for the cell phone and radio used to make the calculations for the H fields using Equation 4.18 are 35 and 80 mV respectively. Since the medium in the communication house is air permeability of air is used for calculating both flux densities (B) in pW/m².

$$H = \frac{E}{377*2*\pi*1.5} \quad \text{H fields for the cell phone and 3 W radio} \quad (4.18)$$

$B = \mu_0 H$

$B_c = 4\pi * 10^{-7} * 19.7 * 10^{-6} = 22.475$ pW/m² for the cell phone

$B_r = 4\pi * 10^{-7} * 22.51 * 10^{-6} = 28.29$ pW/m² for the two-way radio

$\Psi = B*$(Loop area of the halo) magnetic field flux

$\Psi_c = 22.475 * 10^{-12} * 36 = 809.1$ pW flux magnitude for the cell phone

$\Psi_r = 28.29 * 10^{-12} * 36 = 1018.4$ pW flux magnitude for two-way radio

$\Psi_c = 809.1$ pW $*$ Sin $(\omega_c t)$ nW 857 MHz carrier is sinusoidal

$\Psi_r = 1018.4$ pW $*$ Sin $(\omega_r t)$ nW 467 MHz carrier is sinusoidal

$$V = \frac{d\Psi}{dt} = \text{Magnitude} * \omega * \cos(\omega t) \quad (4.19)$$

$V_c = 4.35$ V around the halo for the cell phone

$V_r = 2.99$ V around the halo for the radio

$E_c = 4.35/26$ m (length of halo wire) $= 0.167$ V/m

$E_c = 2.99/26$ m (length of halo wire) $= 0.1149$ V/m

The magnitude of magnetic flux enclosed by the halo loop is calculated as shown in the equations above. Since the carrier for both cell phone and radio are sinusoidal the last two equations for magnetic flux include the sinusoidal term. The frequency of cell phones are in the 857 MHz range and one of the normal radio frequencies for two-way radios is 467 MHz. Equation 4.19 calculates the voltage generated by the cell phone or radio wire length of the halo is 26 m; that is the perimeter of the loop. Dividing the voltage around the loop by the length of the wire around the loop the two voltages are established by the cell phone and radio as 0.167 and 0.1149 V/m respectively.

Conclusion

As can be observed, the safety ground of the enclosures should be as close together as possible to eliminate cell phone and radio noise on grounds. It is assumed, that cell phone and radio were distributed evenly across the loop due to the halo however, and this is not the usual case. All other wireless devices such as pagers, blood-pressure monitoring devices and so on should be analyzed on a case by case basis. Any device that has switching power supplies may also result in noise on the ground due to the switcher frequency appearing at the neutral of a three phase ground connection.

Case Study 4.2

Access broadband over power line (BPL) may be implemented over the power lines on subway systems. These must be considered as possible noise sources that may affect equipment performance. The following are some of the uses for a BPL "smart grid": (i) to manage electricity from suppliers, (ii) to utilize two-way communications, (iii) to control consumer appliances and (iv) to even out peak demands on electrical energy. If demand can be evened out, the energy cost can be lowered due to lower capital equipment costs for the supplier. The full text of BPL can be found in FCC 11-160 (see Example 4.2).

The potential for harmful interference in transit systems due to BPL is investigated in this case study. Transit systems have a host of subsystems that can be affected by noise produced by BPL on power lines and radio system grounds. Transit systems have wireless equipment along the way side that can be affected by extraneous noise on power lines or radiated from power lines. There are several areas where signals that appear as noise can mix with radio to produce intermodulation products. Any interference problems discovered due to BPL in transit equipment must be removed by the BPL supplier this is standard practice by the FCC. Some excerpts have been taken from FCC 11-160 and presented below.

Because access BPL devices are mounted on overhead power lines and the measurement antenna is at a lower distance closer to the ground, the actual distance from the power line to the measurement antenna is greater than the horizontal distance from the pole on which the BPL device is mounted to the measurement antenna. The correct distance for measurement is therefore the "slant range" diagonal distance measured from the center of the measurement antenna to the nearest point of the overhead power line carrying the Access BPL signal being measured.

The Commission agreed with ARRL that access BPL on overhead lines is not a traditional point-source emitter, but not with its argument that access BPL devices would cause power lines to act as miles of transmission lines, all radiating RF energy along their full length. In this regard, the Commission observed that the Part 15 emission limits for carrier current systems have proven very effective

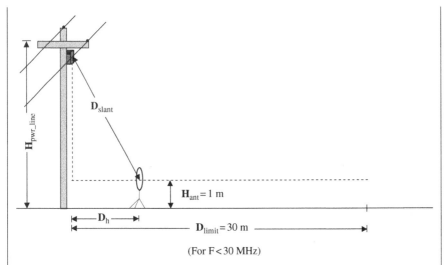

(For F < 30 MHz)

FIGURE 4.3   Measurements needed to calculate slant range. D_{slant} is the slant range distance, in meters. D_h is the horizontal (lateral) distance between the center of the measurement antenna and the vertical projection of the overhead power line carrying the BPL signals down to the height of the measurement antenna, in meters. D_{limit} is the distance at which the emission limit is specified in Part 15 (e.g. 30 m for frequencies below 30 MHz). H_{pwr_line} is the height of the power line, in meters. H_{ant} is the measurement antenna height, in meters.

at controlling interference from such systems. Also, it indicated that the design and configuration of access BPL systems would be inconsistent with the development of cumulative emission effects for nearby receivers. The Commission further concluded that, because the BPL emissions level decreases significantly with distance perpendicular from the line, the potential for interference also decays rapidly with distance from the line.

Slant Range Distance: the diagonal distance measured from the center of the measurement antenna to the nearest point of the overhead power line carrying the access BPL signal being measured (Figure 4.3).

This distance is equal to the hypotenuse of the right triangle as calculated in the formula below. The slant range distance shall be calculated as follows:

$$d_{slant} = \sqrt{\left(h_{pwr\_line} - h_{ant}\right)^2 + \left(d_h\right)^2} \qquad (4.20)$$

Where:

d_{slant} is the slant range distance, in meters;

d_h is the horizontal (lateral) distance between the center of the measurement antenna and the vertical projection of the overhead power line carrying the BPL signals down to the height of the measurement antenna, in meters;

h_{pwr_line} is the height of the power line, in meters;

h_{ant} is the measurement antenna height, in meters.

Extrapolated emission level: because the radiated emission limit for Access BPL devices operating below 30 MHz is specified at the reference distance of 30 m, when making emission measurements at a distance closer than 30 m, the measured result must be extrapolated to the specified measurement distance (i.e. 30 m) to determine compliance with the Part 15 emission limit. The extrapolated emission level (in dBµV/m) is calculated using the following Equation 4.21:

$$E_{extrap} = E_{meas} N \cdot Log_{10}\left(\frac{d_{limit}}{d_{slant}}\right) \qquad (4.21)$$

Where:

N is the distance extrapolation factor, for example 40 for frequencies below 30 MHz;

d_{limit} is the horizontal measurement distance corresponding to the Part 15 emissions limits, for example 30 m for frequencies below 30 MHz;

d_{slant} is the slant range distance, in meters;

E_{meas} is the measured electric field strength at a horizontal distance, d_h, in dBµV/m;

E_{extrap} is the electric field strength value after applying the distance extrapolation factor, in dBµV/m.

If the extrapolated electric field strength value exceeds the Part 15 emission limit (i.e. 29.54 dBµV/m for access BPL operating below 30 MHz as specified for a horizontal measurement distance of 30 m), the access BPL device must lower its emission level to comply with the limit.

Example 4.2   Extrapolated emission levels for Access BPL devices installed on typical overhead power line heights

Broadband emissions from power lines are not likely to be a problem that communications or signal houses/bungalows. The most likely areas where broadband noise may present a problem is at grade crossings, station parking lots and along the right of way. These areas have an assortment of video cameras, audio equipment, sensors and public address systems. Some of this equipment is installed on utility poles or in close proximity to utility poles. Some of the signals equipment installed along the right of way vital logic that have clocks operating below 30 MHz frequencies. Noise effect on vital logic is a safety issue to be considered.

TABLE 4.7   Power Line Height of 10 m

Frequency (MHz)	Horizontal distance from **nearest point of overhead** power **line carrying the BPL signals** (m)	Power line height (m)	Calculated slant-range distance (m)	Extrapolated radiated emission maximum allowable level (dBμV/m)
1.705–30.0	3	10	9.49	49.54
1.705–30.0	10	10	13.45	43.47

TABLE 4.8   Power Line Height of 11 m

Frequency (MHz)	Horizontal distance from **nearest point of overhead** power **line carrying the BPL signals** (m)	Power line height (m)	Calculated slant-range distance (m)	Extrapolated radiated emission maximum allowable level (dBμV/m)
1.705–30.0	3	11	10.44	47.88
1.705–30.0	10	11	14.14	42.60

TABLE 4.9   Power Line Height of 12 m

Frequency (MHz)	Horizontal distance from **nearest point of overhead** power **line carrying the BPL signals** (m)	Power line height (m)	Calculated slant-range distance (m)	Extrapolated radiated emission maximum allowable level (dBμV/m)
1.705–30.0	3	12	11.40	46.35
1.705–30.0	10	12	14.87	41.74

The tables below show examples of extrapolated maximum allowable emission levels for access BPL devices installed on typical overhead power line heights (10–12 m in the USA). If the power line height involved is different than the illustrated heights below, the slant range distance and applicable extrapolated emission level must be calculated using Equations 4.20 and 4.21 above (Tables 4.7, 4.8, 4.9).

For access BPL devices operating below 30 MHz, measurements shall be performed at the 30-meter reference distance specified in the regulations whenever possible. Measurements may be performed at a distance closer than that specified in the regulations if circumstances such as high ambient noise levels or geographic limitations are present. When performing measurements at a distance which is

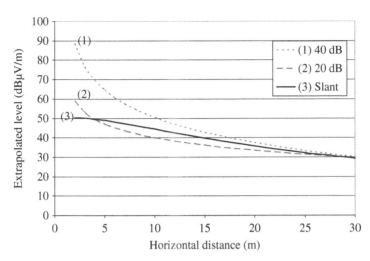

FIGURE 4.4 Extrapolated levels comparison with 40 dB/decade extrapolation factor applied to slant-range distance. Power line height at 10 m

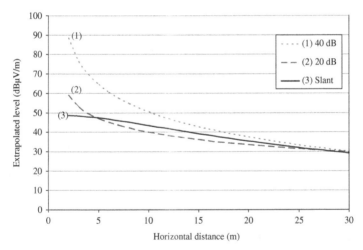

FIGURE 4.5 Extrapolated levels comparison with 40 dB/decade extrapolation factor applied to slant-range distance. Power line height at 11 m

closer than specified, the field strength results shall be extrapolated to the specified distance by using the square of an inverse linear distance extrapolation factor (i.e 40 dB/decade) in conjunction with the slant range distance defined in Section 15.3(h) of this part (Figures 4.4, 4.5, 4.6). As an alternative, a site-specific extrapolation factor derived from a straight line best fit of measurements of field strength in dBμV/m versus logarithmic distance in meters for each carrier frequency, as determined by a linear least squares regression calculation from

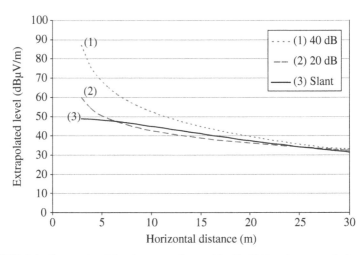

FIGURE 4.6   Extrapolated levels comparison with 40 dB/decade extrapolation factor applied to slant-range distance. Power line height at 12 m

measurements for at least four distances from the power line, may be used. Compliance measurements for access BPL and the use of site-specific extrapolation factors shall be made in accordance with the Measurement Guidelines for Access BPL systems specified by the Commission. Site-specific determination of the distance extrapolation factor shall not be used at locations where a ground conductor is present within 30 meters if the access BPL signals are on the neutral/grounded line of a power system.

For frequencies below 30 MHz, when a notch filter is used to avoid interference to a specific frequency band, the access BPL system shall be capable of attenuating emissions within that band to a level at least 25 dB below the applicable Part 15 limits.

FCC 11-160   (APPENDIX D}

Measurement Guidelines for Broadband over Power Line (BPL) Devices or Carrier Current Systems (CCS) and Certification Requirements for Access BPL Devices

This appendix is intended to provide general guidance for compliance measurements of Broadband over power line (BPL) devices and other carrier current systems (CCS). For BPL systems, the measurement principles are based on the Commission's current understanding of BPL technology. Modifications may be necessary as measurement experience is gained.

General Measurement Principles for Access BPL, In-House BPL and CCS

1. Testing shall be performed with the power settings of the Equipment Under Test (EUT) set at the maximum level.
2. Testing shall be performed using the maximum RF injection duty factor (burst rate). Test modes or test software may be used for uplink and downlink transmissions.
3. Measurements should be made at a test site where the ambient signal level is 6 dB below the applicable limit. (See ANSI C63.4-2003, section 5.1.2 for alternatives, if this test condition cannot be achieved.)
4. If the data communications burst rate is at least 20 burst per second, quasi-peak measurements shall be employed, as specified in Section 15.35(a) of the rules. If the data communications burst rate is 20 bursts per second or less, measurements shall be made using a peak detector.
5. For frequencies above 30 MHz, an electric field sensing antenna, such as a biconical antenna is used. The signal shall be maximized for antenna heights from 1 to 4 meters, for both horizontal and vertical polarizations, in accordance to ANSI C63.4-2003 procedures. For Access BPL measurements only, as an alternative to varying antenna height from 1 to 4 meters, these measurements may be made at a height of 1 meter provided that the measured field strength values are increased by a factor of 5 dB to account for height effects.
6. For frequencies below 30 MHz, an active or passive magnetic loop is used. The magnetic loop antenna should be at 1 meter height with its plane oriented vertically and the emission maximized by rotating the antenna 180 degrees about its vertical axis. When using active magnetic loops, care should be taken to prevent ambient signals from overloading the spectrum analyzer or antenna pre-amplifier.
7. The six highest radiated emissions relative to the limit and independent of antenna polarization shall be reported as stated in ANSI C63.4-2003, section 10.1.8.2.
8. All operational modes should be tested including all frequency bands of operation, as required by Section 15.31(i) of the rules.

1. Access BPL Measurement Principles
   a) Test Environment
   1. The Equipment Under Test (EUT) includes all BPL electronic devices *e.g.,* couplers, injectors, extractors, repeaters, boosters, concentrators, and electric utility overhead or underground medium voltage lines.
   2. *In-situ* testing shall be performed on three typical installations for overhead line(s) and three typical installations for underground line(s).

Radiated Emissions Measurement Principles for Access BPL on Overhead Line Installations

1. Measurements should normally be performed at the horizontal reference distance as specified in Sections 15.209 and 15.109 of the rules (*i.e.*, 30 meters for frequencies below 30 MHz and 10 meters for frequencies 30–88 MHz.) If necessary, due to ambient emissions, for frequencies below 30 MHz, measurements may be performed at a closer distance such as 10 meters (or 3 meters if necessary for safety or because measurements cannot practically be performed at 30 meters or 10 meters) from the overhead line. Distance corrections are to be made in accordance with paragraph (4), below.

2. Testing shall be performed at distances of 0, ¼, ½, ¾, and 1 wavelength down the line from the BPL injection point on the power line. Wavelength spacing is based on the mid-band frequency used by the EUT. In addition, if the mid-band frequency exceeds the lowest frequency injected onto the power line by more than a factor of two, testing shall be extended in steps of ½ wavelength of the mid-band frequency until the distance equals or exceeds ½ wavelength of the lowest frequency injected. (For example, if the device injects frequencies from 3 to 27 MHz, the wavelength corresponding to the mid-band frequency of 15 MHz is 20 meters, and wavelength corresponding to the lowest injected frequency is 100 meters. Measurements are to be performed at 0, 5, 10, 15, and 20 meters down line—corresponding to zero to one wavelength at the mid-band frequency. Because the mid-band frequency exceeds the minimum frequency by more than a factor of two, additional measurements are required at 10-meter intervals until the distance down-line from the injection point equals or exceeds ½ of 100 meters. Thus, additional measurement points are required at 30, 40, and 50 meters down line from the injection point.)

3. Testing shall be repeated for each Access BPL component (injector, extractor, repeater, booster, concentrator, etc.)

4. The distance correction used to calculate the applicable extrapolated emission levels for the measurements that are closer than the specified reference distance in Section 15.209 of the rules shall be based on the slant-range distance, which is the diagonal distance from the center of the measurement antenna to the nearest point of the overhead power line carrying the BPL signals being measured, as defined in Section 15.3(hh) of the rules. Calculations of the slant-range distance and the applicable extrapolated emission levels are made according to Equations (1) and (2) in Section 6, below.

5. For Access BPL devices operating below 30 MHz, if the site-specific alternative extrapolation method is selected, the extrapolation factor is determined by fitting a straight line to measurements of field strength in

dBμV/m vs. logarithmic distance in meters from the nearest conductor carrying BPL emissions. Site-specific determination of the extrapolation factor is not permitted for BPL devices that inject signals on the neutral/ grounded line of a power system if a grounding conductor (typically located at each pole) is located within 30 meters of any of the measurement locations.

a. Measurements shall be made for at least four horizontal distances from the overhead line, at no less than 3 meters from the lateral plane and differing from each other by at least 3 meters. If these measurements allow a straight line with a negative slope to be calculated or drawn with reasonable fit (the minimum regression coefficient of multiple correlation would be 0.9), the best straight line fit would be used to calculate field strength at the 30-meter standard measurement distance in the rules.

b. If the four measurements do not satisfy the regression coefficient requirement specified above, measurements at one or more additional distances shall be added until the regression coefficient is satisfied. If the regression coefficient is not satisfied, a site-specific extrapolation rate may not be used.

Note: In cases where Access BPL devices are coupled to low-voltage power lines (*i.e.*, In-House BPL or modem boosters), apply the overhead-line procedures as stated above along the low-voltage power lines.

b) **Radiated Emissions Measurement Principles for Access BPL in Underground Line Installations**

1. Underground line installations are those in which the BPL device is mounted in, or attached to a pad-mounted transformer housing or a ground-mounted junction box and couples directly only to underground cables.

2. Measurements should normally be performed at the horizontal reference distance as specified in Section 15.209 of the rules (*i.e.*, 30 meters for frequencies below 30 MHz and 10 meters for frequencies 30-88 MHz.) If necessary, due to ambient emissions, for frequencies below 30 MHz, measurements may be performed at a closer distance such as 3 meters or 10 meters from the in-ground transformer. Distance corrections are to be made in accordance with Section 15.31(f) in the rules.

3. Measurements shall be made at positions around the perimeter of the in-ground power transformer where the maximum emissions occur. ANSI C63.4-2003, section 8.1, specifies a minimum of 16 radial angles surrounding the EUT (in-ground transformer that contains the BPL device(s)). If directional radiation patterns are suspected, additional azimuth angles shall be examined.

c) **Conducted Emissions Measurement Principles**

Conducted emissions testing is not required for Access BPL.

2. **In-House BPL and Carrier Current Systems Measurement Principles**

1. In-House BPL devices are typically composite devices consisting of two equipment classes (Carrier current system and personal computer peripheral (Class B)). While carrier current systems require Verification, personal computer peripherals require Declaration of Conformity (DoC) or Certification, as specified in Section 15.101 of the Rules. Appropriate tests to determine compliance with these requirements shall be performed.

2. *In-situ* testing is required for testing of the carrier current system functions of the In-House BPL device.

3. If applicable, the device shall also be tested in a laboratory environment, as a computer peripheral, for both radiated and conducted emissions tests per the measurement procedures in C63.4-2003.

a) **Test Environment and Radiated Emissions Measurement Principles for In-House BPL and CCS *In-Situ* Testing**

1. The Equipment under Test (EUT) includes In-House BPL modems used to transmit and receive carrier BPL signals on low-voltage lines, associated computer interface devices, building wiring, and overhead or underground lines that connect to the electric utilities.

2. *In-situ* testing shall be performed with the EUT installed in a building on an outside wall on the ground floor or first floor. Testing shall be performed on three typical installations. The three installations shall include a combination of buildings with overhead-line(s) and underground line(s). The buildings shall not have aluminum or other metal siding, or shielded wiring (e.g.: wiring installed through conduit, or BX electric cable).

3. Measurements shall be made at positions around the building perimeter where the maximum emissions occur. ANSI C63.4-2003, section 8.1, specifies a minimum of 16 radial angles surrounding the EUT (building perimeter). If directional radiation patterns are suspected, additional azimuth angles shall be examined.

4. Measurements should normally be performed at the horizontal reference distance as specified in Sections 15.209 and 15.109 of the rules (*i.e.*, 30 meters for frequencies below 30 MHz and 3 meters for frequencies 30-88 MHz.) If necessary, due to ambient emissions, for frequencies below 30 MHz, measurements may be performed at a closer distance such as 3 meters or 10 meters around the building perimeter as outlined in step 3) above. Distance corrections are to be made in accordance with Section 15.31(f) of the Rules.

b) **Additional Measurement Principles for In-House BPL and CCS** *In-Situ* **Testing With Overhead Lines**

1. In addition to testing radials around the building, testing shall be performed at three positions along the overhead line connecting to the building (*i.e.* the service wire). It is recommended that these measurements be performed starting at a distance 10 meters down the line from the connection to the building. If this test cannot be performed due to insufficient length of the service wire, a statement explaining the situation and test configuration shall be included in the technical report.

2. Measurements should normally be performed at the horizontal reference distance as specified in Sections 15.209 and 15.109 of the rules (*i.e.,* 30 meters for frequencies below 30 MHz and 10 meters for frequencies 30-88 MHz.) Measurements may then be performed at a closer distance such as 3 meters or 10 meters from the overhead line. Distance corrections are to be made in accordance with paragraph 2.b.4 above.

c) **Measurement Principles for Testing In-House BPL and CCS as a Computer Peripheral**

1. The data rate shall be set at the maximum rate used by the EUT. Test modes or test software may be used to simulate data traffic.

2. For In-House BPL devices operating as unintentional radiators below 30 MHz, the conducted emissions shall be measured in the 535 – 1705 kHz band as specified in Section 15.107(c). For In-House BPL devices operating as unintentional radiators above 30 MHz, the conducted emissions shall be measured as specified in Section 15.107(a). Conducted emissions measurements shall be performed in accordance with ANSI C63.4-2003 (Section 7 and Annex E).

3. For In-House BPL devices operating as unintentional radiators either below 30 MHz or above 30 MHz, the radiated emissions limits of Section 15.109(a) apply. The radiated emissions from the computer peripheral shall be measured at an Open Area Test Site (OATS) in accordance with the measurement procedures in C63.4-2003 (Section 8 and Annex D)

3. **Certification Technical Report Requirements for Access BPL Devices**

1. Certification applications shall be accompanied by a technical report in accordance with Section 2.1033 of the Rules. Each device used in an Access BPL system requires its own Certification.

2. For Access BPL devices, the statement describing how each device operates shall include the following information: modulation type, number of carriers, carrier spacing, channel bandwidth, notch capability/control, power settings/control, and range of signal injection duty factors.

3. For Access BPL devices, the measurement report shall include representative emissions spectrum plot(s) of the reported data.

4. For Access BPL devices operating below 30 MHz, if the site-specific method for determining the extrapolation factor was used, the measurement report shall include detailed information on the calculations and the data points taken.

4. Responsibility of BPL operator

It is recommended that a BPL operator perform initial installation and periodic testing of Access BPL systems on his power lines. These tests shall be performed to ensure that the system in conjunction with the installation site complies with the appropriate emission limits using the measurement procedures outlined in Section 3 of this document. The BPL operator should use typical installation sites within his service area as outlined in section 2(a) of this document. Selection of typical sites shall be made according to the characteristics of the installation as a whole. The BPL operator is not required to submit the test results. In the instance that the Access BPL system was tested on the operator's network for certification purposes, the initial installation tests do not need to be repeated. However, periodic testing of installed Access BPL systems is recommended to ensure that the system maintains compliance with Part 15 emission limits.

5. Calculation of Slant-Range Distance and Extrapolated Emission Level for Access BPL operating on overhead power lines

Figure 1 shows an example of the testing configuration of Access BPL devices operating below 30 MHz on overhead power line installations. It is provided to illustrate the calculation of extrapolated emission levels using the slant range method when the device cannot be tested at the required reference distance of 30 meters specified in the rules. As explained below, when making emission measurements at a distance closer than 30 meters, the measured result must be extrapolated to the Part 15 specified measurement distance (*i.e.*, 30 meters) to determine compliance with the Part 15 emission limit, because the radiated emission limit for Access BPL devices operating below 30 MHz is only specified at the reference distance of 30 meters.

FCC 11–160   (APPENDIX E}

Extrapolated Emission Maximum Allowable Levels Using Slant-Range Method With Various Power Line Heights

I SLANT-RANGE METHOD FOR MEASURING BPL EMISSIONS ON OVERHEAD POWER LINES

The Commission adopted a slant-range method for measuring BPL emissions on overhead power lines in Appendix C of the *BPL Order*. With the slant-range method, the distance correction for the overhead-line measurements of Access BPL emissions is based on the slant-range distance, which is the diagonal distance from the center of the measurement loop antenna to the overhead power line, illustrated in Figure 2.1 of the amended Measurement Guidelines, *supra*. Slant-range distances are calculated based on the height of the power line and the horizontal (lateral) distance (D_h) between the center of the measurement antenna and the vertical projection of the overhead power line carrying the BPL signals down to the height of the measurement antenna (see Equation 2.1, below). Slant-range distance corrections are made in accordance with Section 15.31(f) (3) (*e.g.*, using 40 dB/decade extrapolation factor for frequencies below 30 MHz).

II EXTRAPOLATED LEVELS COMPARISONS WITH TYPICAL POWER LINE HEIGHTS

Typical medium-voltage power line heights in the United States are 10-12 meters. Figures 2.2-2.4 illustrate a comparison between the extrapolated emission levels with respect to horizontal (lateral) distance when using extrapolation factors of 40 dB (trace 1), 20 dB (trace 2) and when using the Commission's slant-range method calculated using 40 dB/decade extrapolation factor (trace 3).

As illustrated, the resulting extrapolated maximum allowable emission level applied to Access BPL devices is more stringent than for other Part 15 devices operating below 30 MHz by virtue of using the slant-range distance in the calculation of extrapolated emission levels. This extrapolated maximum allowable level is very close (within a 5 dB range for typical medium-voltage power line heights) to what ARRL is requesting, *i.e.*, using 20 dB/decade extrapolation factor with straight horizontal (lateral) distance.

A. Extrapolated Levels using 40 dB/decade and Slant-range distance From a Power Line Height of 10 meters

Figure 2.2 illustrates a comparison between the extrapolated maximum allowable emission levels with respect to horizontal (lateral) distance when using extrapolation factors of 40 dB/decade (trace 1), 20 dB/decade (trace 2) and when using the Commission's slant-range distance method calculated with 40 dB/decade extrapolation factor (trace 3). Calculations are made according to Equations 2.1 and 2.2 below, using a power line height of 10 meters. This results in a measurement height of 9 meters, because the measurement loop antenna is at 1 meter from the ground.

As illustrated (see Figure 2.2) for power line heights of 10 meters, the Commission's slant-range method using a 40 dB/decade extrapolation factor applied to slant-range distance provides an extrapolated emission level graph that reveals the following.

• At measurement distances greater than 5 meters, the extrapolated maximum allowable emission levels for slant-range distance based on the existing 40 dB/decade factor (trace 3) are LESS stringent than extrapolated emission levels based on 20 dB/decade extrapolation factor for horizontal (lateral) distance (trace 2), by a **maximum of 4.4 dB**. This maximum difference between the two traces (trace 2 and trace 3) is found at a horizontal distance of 9 meters from the nearest point of the overhead power pole carrying the BPL signals.

- At these same measurement distances (greater than 5 meters), the extrapolated maximum allowable emission levels for slant-range distance based on the existing 40 dB/decade factor (trace 3) are MORE stringent than extrapolated emission levels based on 40 dB/decade extrapolation factor for horizontal (lateral) distance (trace 1), by as much as 12.6 dB.

- However, at measurement distances less than 5 meters, the extrapolated maximum allowable emission levels for slant-range distance based on the existing 40 dB/decade factor (trace 3) are MORE stringent than extrapolated emission levels for horizontal (lateral) distance based on either 20 dB/decade (trace 2) or 40 dB/decade extrapolation factor (trace 1).

As illustrated (see figure 2.3) a comparison between the extrapolated maximum allowable emission levels with respect to horizontal (lateral) distance when using extrapolation factors of 40 dB/decade (trace 1), 20 dB/decade (trace 2) and when using the Commission's slant-range distance method calculated with 40 dB/decade extrapolation factor (trace 3). Calculations are made according to equations 2.1 and 2.2, using a power line height of 11 meters. This results in a measurement height of 10 meters, because the measurement loop antenna is at 1 meter from the ground.

As illustrated in Figure 2 for power line heights of 11 meters, the Commission's slant-range method using a 40 dB/decade extrapolation factor applied to slant-range distance provides an extrapolated emission level graph that reveals the following.

- At measurement distances greater than 5 meters, the extrapolated maximum allowable emission levels for slant-range distance based on the existing 40 dB/decade factor (trace 3) are LESS stringent than extrapolated emission levels based on 20 dB/decade extrapolation factor for horizontal (lateral) distance (trace 2), by a **maximum of 3.5 dB**. This maximum difference between the two traces (trace 2 and trace 3) is found at a horizontal distance of 10 meters from the nearest point of the overhead power line carrying the BPL signals.

- At these same measurement distances (greater than 5 meters), the extrapolated maximum allowable emission levels for slant-range distance based on the existing 40 dB/decade factor (trace 3) are MORE stringent than extrapolated emission levels based on 40 dB/decade extrapolation factor for horizontal (lateral) distance (trace 1), by as much as 14 dB.

- However, at measurement distances less than 5 meters, the extrapolated maximum allowable emission levels for slant-range distance based on the existing 40 dB/decade factor (trace 3) are MORE stringent than extrapolated emission levels for horizontal (lateral) distance based on either 20 dB/decade (trace 2) or 40 dB/decade extrapolation factor (trace 1).

Figure 2.4 illustrates a comparison between the extrapolated maximum allowable emission levels with respect to horizontal (lateral) distance when using extrapolation factors of 40 dB (trace 1), 20 dB (trace 2) and when using the Commission's slant-range distance method calculated with 40 dB per decade extrapolation factor (trace 3). Calculations are made according to Equations 1 and 2 below, using a power line height of 12 meters. This results in a measurement height of 11 meters, because the measurement loop antenna is at 1 meter from the ground.

As illustrated in Figure 2.4 for power line heights of 12 meters, the Commission's measurement method using a 40 dB/decade extrapolation factor applied to slant-range distance provides an extrapolated emission level graph that reveals the following.

- At measurement distances greater than 5 meters, the extrapolated maximum allowable emission levels for slant-range distance based on the existing 40 dB/decade factor (trace 3) are LESS stringent than extrapolated emission levels based on 20 dB/decade extrapolation factor for horizontal (lateral) distance (trace 2), by a *maximum of 2.7 dB*. This maximum difference between the two traces (trace 2 and trace 3) is found at a horizontal distance of 10 meters from the nearest point of the overhead power line carrying the BPL signals.

- At these same measurement distances (greater than 5 meters), the extrapolated maximum allowable emission levels for slant-range distance based on the existing 40 dB/decade factor (trace 3) are MORE stringent than extrapolated emission levels based on 40 dB/decade extrapolation factor for horizontal (lateral) distance (trace 1), by as much as 15.3 dB.

- However, at measurement distances less than 5 meters, the extrapolated maximum allowable emission levels for slant-range distance based on the existing 40 dB/decade factor (trace 3) are MORE stringent than extrapolated emission levels for horizontal (lateral) distance based on either 20 dB/decade (trace 2) or 40 dB/decade extrapolation factor (trace 1).

III EXTRAPOLATED EMISSION LEVEL CALCULATION FORMULA

If the extrapolated electric field strength value exceeds the Part 15 emission limit (i.e. 29.54 dBμV/m for Access BPL operating below 30 MHz as specified for a horizontal measurement distance of 30 meters), the Access BPL device must lower its emission level to comply with the limit.

4.3 TUNNEL RADIATION FROM A TEMPORARY ANTENNA INSTALLED ON THE CATWALK IN A TUNNEL

The radio bay station that provides power to antenna shown in Figure 4.7 is located in a cross passage near the antenna installation. This same installation will be duplicated elsewhere in the tunnel to allow temporary communications. The tunnel consists of two tubes one outbound and the other for inbound traffic and each has two lanes with standard width only a 1 ft shoulder. The antenna has an ERP of 50 W.

The objective of this study is to determine if a trucker or passenger car has an exposure limit that can become a safety issue over a period of time. If a vehicle travels this tunnel and it is exposed to traffic at a particular time of the day, traffic may be stopped at close to the same point where the temporary antenna is mounted. The temporary antenna over the period time it is to be installed may present a health issue due to the radiation.

The tunnel construction consists of an inner layer of concrete over 1.25 inch thick stainless steel and the wall construction behind the catwalk is 4 inches of concrete with the steel rebar mesh. This wall will distort the antenna E and H fields due to wall reflections. The wall is not in the near field of the radios. The near field for the VHF radio at 150 MHz is $2*(1/4\lambda)^2/\lambda = 6.25$ cm. The radios users are the police, fire and emergency medical service (EMS). The radios are in the low band UHF region and VHF. The wavelengths are 0.62–0.669 m and 2 m respectively. The antenna is actually in the center of the 1 m wide catwalk.

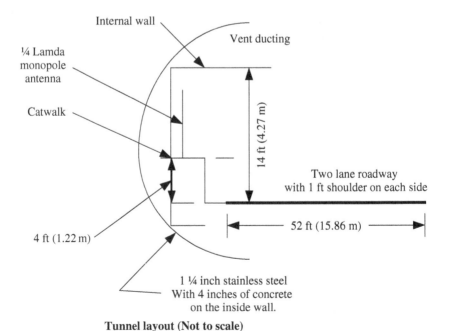

Tunnel layout (Not to scale)

FIGURE 4.7   The layout of the antenna on the catwalk in a tunnel

Antennas are placed strategically along the tunnel catwalk to provide radio minimum coverage in the tunnel until the leaky radiating cable can be installed. The installation of the leaky radiating cable will support up to 30 communication channels. These cables will be installed on both sides of the roadway. The radiation from these cables is at a very low level and will not present a health hazard to people and vehicles whether they are in trucks or cars. The case study involving examples of leaky radiating cable are presented in Section 4.5 in this chapter with more metrics on details of leaky radiating cable. The one-quarter wavelength (0.25 λ) monopole antenna at 150 MHz VHF is 0.517 m with a 3 dB beam width of 87°; and at 450 MHz with one wavelength (1.0 λ) the beam width is 47.8°. The radiation level between these frequencies must not exceed 50 W. The electric field at 50 W as calculated using Equation 4.22.

$$E = \frac{\sqrt{30 * ERP}}{R} \tag{4.22}$$

Where ERP is equivalent radiated power and R is distance from the antenna (in meters). Any radio connected to the antenna on the catwalk must not have an ERP greater than 50 W. The distance from the antenna to the roadway is 0.5 m from the antenna to the edge of the catwalk and 2 ft (0.61 m) from the catwalk to the roadway where a vehicle may be standing. The total distance R is 1.11 m. The maximum electric field E is 35 V/m and the power level is 3.2 W/m^2 (0.32 mW/cm^2). The electric field of 35 V/m represents a harsh environment for electronic circuits as shown in Table 2.1 in Chapter 2. Power levels at 0.32 mW/m^2 present a health hazard that is discussed in Chapter 7 in more detail. There is a time element that must be considered when using these radios for long periods of time with exposures as stated above. Another factor that must be considered is that of maintenance personnel working near the catwalk where the radio antenna is located. Some statistical information must be gleaned from traffic information and a hazard will be presented in Chapter 7.

The electric field and radiated power from the antenna to the roadway is the worst-case scenario because the powers taken from the center of the radiating beam. Passenger cars may have an extra margin of safety they are lower than the approximate 1.5 m from the center of the beam to roadway, where the window of a vehicle may be in line with the center of the beam. A truck for example may have a window that is in line with the antenna beam and get the full impact of the radiation in the truck cab. At one time 50 W portable radios were installed on police motorcycles. These of course were extremely hazardous to the person riding a motorcycle; the radio antenna was barely 1 m from the person riding on the motorcycle. The radios used at that time were VHF 150–300 MHz. The human body absorbs wavelengths in the 300 MHz range that are much more destructive than any of the other frequencies. Often people with portable radios such as walkie-talkies clipped the transmitter to their belt, which was a very dangerous practice that resulted in cancer in those areas where the antenna was close to the body. These practices of using radio antennas close to the body are no longer in use, the only exception being cell phones and portable phones and at home. People should always be vigilant about using wireless devices with power levels that are harmful to the human body. These are discussed in more detail in Chapter 7.

The solution to preventing overexposure of maintenance personnel from the catwalk antenna radiation requires a barrier to be installed on the catwalk with a warning that radios must cease to operate while maintenance personnel are working in the area. Fortunately the radios are not always operating; there may be a burst of activity with long periods of silence. However, during heavy traffic periods such as rush hour or emergencies in the tunnel a vehicle may be standing as long as one hour at the same position. The exposure time for safety purposes to radiation varies from 6 to 30 min depending on the frequency and the power level and whether the exposure is to the public or to maintenance personnel. This is discussed in more detail in Chapter 7.

Using the Friis transmission formula, Equation 4.23, a calculation is made for the minimum receive power at –100 dBm and transmit power at 46 dBm with a total free space loss no greater than 146 dB. The distance or length the tunnel is 2.2 km; if the radio antenna is installed in the middle of the tunnel the coverage will be adequate for either end. However, a large fade margin must be included due to the reflections from the walls of tunnel; the free space loss without fade margin included is 91 dB. The fade margin is 55 dB. This is more than adequate for the tunnel installation if it has only a slight curvature that does not exacerbate fade margin as a problem.

$$\text{Free\_Space\_Loss} = -32.44 - 20\log_{MHz} - 20\log R_{km} - \text{Fade\_Margin} \qquad (4.23)$$

A problem that can occur is the radio emanations from the tunnel entrances can interfere with simulcast radio outside the tunnel. A remedy for this of course is to reduce the ERP, that would also decrease the possibility of health hazards from radiation. At this point, the Authority decided not to pursue the analysis any further since it was decided that, due to the options provided in the analysis, all contingencies were covered. When doing an EMC analysis these types of situations occur and the Authority asking for the analysis is satisfied that EMC contingencies can be circumvented. There may be a contingency that cannot be circumvented, that is where the tunnel acts as a waveguide itself at a particular frequency. The Japanese working on a particular tunnel used a dish at one end of the tunnel to transmit radio through the tunnel which itself acted as a waveguide.

When traffic is flowing through the tunnel, trucks, buses, cars, oversize SUVs and oversize trucks, all of these vehicles will affect the transmission properties in the tunnel. The EMC analysis considering all of the aspects of traffic in the tunnel would be an insurmountable task. Since the fade margin is so high that these difficulties may be circumvented, the only sure way to know is to do field measurements tests in traffic under actual conditions. A spectrum analyzer can be connected to the antenna where the Bay Station is located and a person with a 5 W two-way radio can transmit signals at various distances from the catwalk antenna. The distances from the catwalk antenna and the receive power are recorded and the spectrum analyzer records the results. This must be done on both sides of the tunnel catwalks to get the full impact of fading. Two-way radio operation by a person on the catwalks is a complete test and will provide necessary information to make corrections at a later date. The best condition for the actual test is to use a police car or emergency vehicle and take the results as it passes through the tunnel during a traffic episode. If a single radio is used and results are taken accurately, extrapolations may be made for other frequencies.

The tunnel the analysis was conducted on was to have two leaky coaxial cables running the length of the tunnel, one on each side of the tunnel with 34 radios on each coaxial. It was found through testing and analysis that the maximum number of radios that can be installed on leaky coaxial is about 17. This is bidirectional communications on each leaky coaxial. The problem that occurred with 34 radio channels installed on one leaky coaxial cable is tuning all of the combiners in the combiner network. Once connected to the leaky coaxial cable the tuning was disturbed in the combiner network. The combiner had to be removed and taken back to the laboratory for further tuning refinement. The cost to tune and retune the combiner network became astronomical, close to one half million dollars. Through negotiations the prospect of putting 34 radio channels on one leaky coaxial cable was discarded. A reduction in the number of channels and dividing the radios into two groups each connected to a different leaky coaxial cable was one solution. This did solve the problem. The reason for 34 channels on each leaky coaxial cable was to have one as the primary cable with the other as a backup.

The 17 channel leaky coaxial cable installation is in Honolulu, Hawaii. The person who did this installation helped negotiate not to install 34 channels in the tunnel. The person that did the work in Hawaii had mentioned that crosstalk could be heard regardless of which channel was used. However, newer techniques for radio systems have made filtering much sharper and improved over the time since this was done, however, and it may be possible now with digital P 25 radios techniques to increase the number of channels on leaky coaxial cable as compared to 10 years ago.

4.4 SIMULCAST INTERFERENCE AT THE END OF THE CUT AND COVER SUBWAY TUNNEL

The tunnel is designed as a large trench covered over by highway roadway and where the subway passes underneath the highway it is considered an underpass. It is referred to as the tunnel in this document because many tunnels are constructed as cut and cover construction. The ends of this tunnel have leaky radiating cable along the walls until the subway tracks are near grade level. Communications between the subway radiating leaky cable and local radios are simulcast. This presents a problem at the ends of the tunnel where the local radios and leaky radiating cable can have an interference problem. The objective of this case study is to determine if an interference problem occurs and to find a solution if the problem exists. The subway cars have precedence over the local radio transmission.

The diagram in Figure 4.8 is a plan view of this project. The two entrances are at Flowers Street and Exposition Boulevard, respectively. The two large buildings and expressway reduce local radio reception significantly. The college buildings and college area have a great deal of open space. However, the college area along

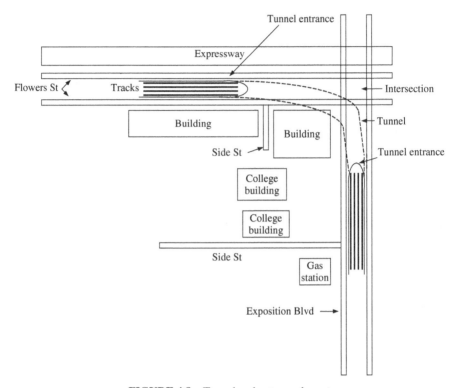

FIGURE 4.8   Tunnel and entrance layout

Exposition Boulevard has a four foot high wrought iron fence, which will provide some shielding effect at the tunnel entrance.

Methodology   A survey is conducted along the tunnel entrance areas with a portable spectrum analyzer mounted in a vehicle with an antenna mounted on the roof top. It is tuned to the various frequencies between 160 and 800 MHz. Each local radio is keyed, allowing measurements at the roadway paralleling the tunnel entrances that are Flowers Street and Exposition Boulevard.

The radiating leaky cable apertures are designed to appear as stacked horizontal dipole antennas along the length of the leaky radiating cable. The drive end of the cable has a lower leakage than the far end. This is done in a progressive manner to provide a constant signal along the length of the cable. This cable can be installed on either side of tunnel entrance. The analysis will allow the Authority communication design group to make a decision on which side of the tunnel entrance is best to install the cable.

After the data is taken, the knife edge effect is analyzed to determine the amount of the attenuation that will occur at the worst-case areas at the cable installation. This analysis examines the effects of reflections from buildings that may be along the right of way and any other factors that may result in attenuation of signals from the local radios. Structures along the right of way that result in attenuation of the local radio signals are an expressway that is very close to the right of way, a college campus and a wrought iron fence surrounding the campus. The wrought iron fence, for example, will resemble a sheet iron, due to the spacing of the pickets.

The calculations are done using a combination of Matlab and hand calculations by the author. Matlab is used to generate electric or magnetic field plots. A Google survey was taken online to determine what the structures are around the area of the tunnel entrances and these are used to predict the signal strength as compared to the leaky radiating cable at the tunnel entrances. If the two signals are close in magnitude, a prediction can be made indicating a possible interference problem. One problem that is evident is at the end of the leaky radiating cable just before grade level. Extending the concrete wall at the entrance to the tunnel may alleviate the problem, but may result in a dead spot in radio communications. These are some of the aspects that the analysis investigated.

In some cases, perhaps simple foliage can be added that will attenuate signals from the local radios, such as shrubs or trees. Other possible fixes are putting up a fence on top of the entrance wall that will attenuate the local radios to solve the problem. Other possibilities are: (i) boosting the signal level with amplifiers neared entrance of the tunnel, so as to make signal levels stronger than any incoming local radios, or (ii) reducing the power of the local radio Bay station. The last option however is very difficult because each base station is owned by a different Authority. The various Authorities are the Sheriff's Department, the local police, the state police, the Department of motor vehicle registration and the Metropolitan Transit Authority (MTA). All of these local and municipal Authorities may not require omnidirectional radio coverage. The MTA at their own expense may replace the culprit radio antenna to solve any particular problem.

Calculations when compared to a survey of the area at entrance of the tunnel
Table 4.10 is a result of this calculation for the VHF 160 MHz at one of the tunnel entrances. Item 2 is the strongest downlink signal which would be the most concerning level. But it occurs over the tunnel covered section on the site street away from the tunnel entrance. Also, the uplink is examined in this study to determine if it impacts local radio transmission. Longitude and latitude coordinates are provided in Table 4.10 so that, opening Google, the distances may be calculated. However, when the signal strength is required to augment the study, but not done here, an EDX plot can be done for the particular coverage of the areas involved. The author has performed these studies in the past and they are corroborated by actual survey results within ± 5–10 dB.

The tunnel entrance at Flowers Street is shown in Figure 4.9. The downlink are measurements taken along Flower Street with the range of –110 to –75 dBm. As measurements were taken for each particular transmitter, the worst cases are inserted in Tables 4.10 and 4.11. Corrections in the table are a result of the knife edge effect at the edge of the tunnel entrance. An additional feature shown in Figure 4.8 is the reflections off a large building that runs the length of the entrance of the tunnel. These reflections will be analyzed because the cable may be installed on the other

TABLE 4.10   Transmitter Coverage and Radio Signal Strength 160 MHz (dBm) at Flowers Street Entrance*

Item	Address	Site location (long./lat.)	Power level downlink (dBm)	Label uplink (dBm)
1	320 Santa Fe Ave	118°13′56″ W 34°3′42″ N	−110 dBm − 8dB = −118 dBm cor Tx = 47.78 dBm	−121 dBm
2	Mt Washington	118°12′54.3″ W 34°6′19″ N	−75 dBm − 8dB = −83 dBm cor Tx = 47 dBm	−121 dBm
11C	3560 Figueroe	GPS elevation 52.8 m	Uplink from cable →	−56 dBm

Note: The worst-case signal level at the shallow end of the tunnel is −83 dBm from the donor and −56 dBm from the cable at 10 m.
*These analyses were conducted to extrapolate the result of the initial survey to search for EMC anomalies.

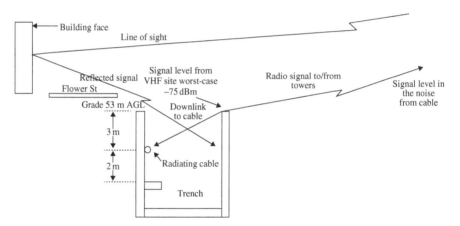

FIGURE 4.9   Signal reflection from buildings and knife edge effect at tunnel entrance

side of the tunnel, or possibly two cables will be installed one on each wall in the tunnel. These are the contingencies that must be investigated because the tunnel has not been constructed. As construction of the tunnel progresses further, measurements can be taken before the tracks are installed to allow for design modifications of the leaky radiating cable installation. Any installation of new radio sites must be analyzed in a similar manner. Thus it is evident that the design of the communications within the tunnel is an ongoing effort. However, many times when working with various Authorities, the communications aspects of tunnel design are delayed to near the end of the hardware construction phase. The leaky radiating cable is usually installed near the completion of the project. This is why the EMC analyst must be vigilant in determining contingencies for best case installation.

The various worst-case measurements for UHF radios along the roadway paralleling the mouth of the tunnel are provided in Table 4.11. Also provided in Table 4.11

TABLE 4.11   Transmitter Coverage and UHF Radio Signal Strength 460/800 MHz (dBm) at Flowers Street Entrance*

Item	Address	Site location (long. and lat.)	Power level dBm downlink	Label uplink
1	MTA 707 Wilshire Blvd f = 450 MHz	118°15′25.3" W 34°2′58" N	−50.5 dBm −39.3–12 w Corr = −102 Tx = 50.78 dBm	−101 dBm
2	One Gateway Plaza, LA Sheriff f = 450 MHz	118°13′57.3" W 34°3′23" N	−50.18dBm–40.l–12 w Corr = − 102.1 Tx = 53. 32 dBm	−139 dBm
3	Mt Lukens LAPD f = 450 MHz	118°14′19.3" W 34°16′9" N	−62.6dBm–40–12 with Corr = − 114 Tx = 51.9dBm	−139 dBm
4	LAFD City Hall f = 800 MHz	118°14′33.3" W 34°3′13" N	−49.2dBm–58.8 w Corr = −108 Tx = 50 dBm	−139 dBm
5	406 Exposition Blvd at 3590 Flowers St. used 3560 Figueroe	118°16′49.1" W 34°1′7.5" GPS Elev. 47.2 m	Uplink from cable →	−56 dBm

* These analyses were conducted to extrapolate the result of the initial survey to search for EMC anomalies.

TABLE 4.12   VHF 160 MHz Exposition Blvd Transmitter Location, Power and Power Level at Ground Level (tunnel Enterance) and Uplink Power Level*

Item	Address	Site location (long. /lat.)	Power level dBm downlink	Label uplink
1	320 Santa Fe Ave	118°13′56" W 34°3′42" N	−110 dBm–8 dB −31 dB −149 dBm cor Tx = 47.78 dBm	−121 dBm
2	Mt Washington	118°12′54.3" W 34°6′19" N	−75 dBm − 8 dB −31 dB −114 dBm cor Tx = 47 dBm	−121 dBm
11C	3560 Figueroe	GPS elev. 52.8 m	Uplink from Cable →	−56 dBm

Note: The worst-case signal level at the shallow end of the trench is −1 14 dBm from the donor and −56 dBm from the cable at 10 m.
*These analyses were conducted to extrapolate the result of the initial survey to search for EMC anomalies.

is the longitude and latitude coordinates of the intersection of the two streets that run parallel to the tunnel. Thus when a bend in the cut and cover subway tracks occurs, it must be a gentle bend in the track area. Note all uplink values of the leaky radiating cable are adjusted to −56 dBm at the amplifier driving the cable. This does not include coupling loss to a receiver of a two-way radio on subway cars or handheld mobile on tunnel catwalks.

TABLE 4.13   UHF 460/800 MHz Transmitter Location on Exposition Blvd, Power and Power Level at Ground Level (tunnel Enterance) and Uplink Power Level*

Item	Address	Site location (long. and lat.)	Power level dBm downlink	Label uplink
1	MTA 707 Wilshire Blvd f = 450 MHz	118°15′25.3″ W 34°2′58″ N	−50.5 dBm −35 −16 with Corr. = −101.5 Tx = 50.78 dBm	−101 dBm
2	One Gateway Plaza, LA Sheriff f = 450 MHz	118°13′57.3″ W 34°3′23″ N	−50.18 dBm −35 −16 with Corr. = −101.18 Tx = 53.32 dBm	−139 dBm
3	Mt Lukens LAPD f = 450 MHz	118°14′19.3″ W 34°16′9″ N	−62.6 dBm −35 −16 with Corr. = −111.6 Tx = 51.9 dBm	−139 dBm
4	LAFD City Hall f = 800 MHz	118°14′33.3″ W 34°3′13″ N	−49.2 dBm −39 −16 with Corr. = −104.2 Tx = 50 dBm	−139 dBm
5	406 Exposition Blvd at 3590 Flowers St. used 3560 Figueroe	118°16′49.1″ W 34°1′7.5″ N GPS elev. 47.2 m	Uplink from cable →	−56 dBm

Note: The worst-case signal level at the shallow end of the trench is −101.18 dBm from the donor and −56 dBm from the cable at 10 m.
*These analyses were conducted to extrapolate the result of the initial survey to search for EMC anomalies.

Along Exposition Boulevard (see Tables 4.4.3 and 4.4.4) that runs parallel with the other entrance to the tunnel college buildings are set back about 200 m from the roadway and a wrought iron fence runs along the property. One reason for the large disparity between Flowers Street and Exposition Boulevard power levels is an expressway runs parallel to Flowers Street that provides a measure of shielding. Exposition Boulevard has a great deal of open space due to the college campus and the buildings are low level in height. This allows stronger signals along the tunnel entrance on Exposition Boulevard. Some of the area has houses and a gas station that are objects that could shield the local radio signals. The radio signals along the right of way seem to be the worst-case scenario due to the lack of obstructions that may shield the entrance of the tunnel from local radio. Covered sections of the tunnel do not need to be considered because the leaky radiating coaxial cable is completely shielded by the covered area.

Table 4.14 gives the four worst-case values that are used for calculations of the shielding effects of the various obstructions to local radio signal transmission. The Authority is planning to install a new antenna near the intersection of Flowers Street and Exposition Boulevard; it will have an antenna for an 800 MHz radio. The antenna tower may include wireless devices to serve as a repeater for other local radios. The antenna to be installed will be highly directional and may or may not be a hindrance to the leaky radiating cable communications. This is a future project to analyze after

TABLE 4.14   VHF and UHF Worst-case Power Levels at Tunnel Enterance and Uplink at Transmitter*

Authority	Physical location	Power level end of the tunnel	Frequency	Street parallel to tunnel entrance
Mt Washington	118°12′54.3″ W 34°6′19″ N	−75 dBm − 8 dB −83 dBm cor Tx = 47 dBm	160 MHz	Flowers Street
MTA 707 Wilshire Blvd	118°15′25.3″ W 34°2′58″ N	−50.5 dBm −39.3 −12 with Corr. = −102 Tx = 50.78 dBm	470 MHz	Flowers Street
Mt Washington	118V12′54.3″ W 34°6′19″ N	−75 dBm −8 dB −31 dB −114 dBm corr. Tx = 47 dBm	160 MHz	Exposition Boulevard
One Gateway Plaza, LA Sheriff	118°13′57.3″ W 34°3′23″ N	−50.18 dBm −35 −16 with Corr. = −101.18 Tx = 53.32 dBm	470 MHz	Exposition Boulevard

*These analyses were conducted to extrapolate the result of the initial survey to search for EMC anomalies.

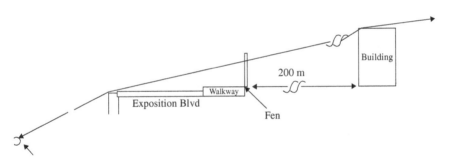

FIGURE 4.10   Transmission obstructions from roof top installation

the tunnel project is completed. As explained to the Authority, antenna fields are fairly predictable, but there is always a possibility of interaction between the antennas on the tower that can result in harmonics resulting in interference problems.

It should be noted that each of the worst-case values are at the bottom of each table. These are the cases considered in the analysis. These readings are taken from the worst-case areas near the leaky radiating cable end.

End of cable radiation   This analysis is where the cable end radiates and the surface coverage overlaps the same area. Table 4.15 is taken directly from the manufacturer's specifications. Note that the coupling loss is very close to a constant along the leaky radiating cable. An analysis of leaky radiating cables is presented in the Case study 4.6.

TABLE 4.15   Leaky Radiating Cable Attenuation per 100 ft, Coupling Loss and Total Loss at 1600 ft

Item	Loss per 100 ft (dB)	Coupling loss	Frequency (MHz)	Loss at 1600 ft
1	0.32	68/79	160	5.12
2	0.62	68/75	460	9.92
3	1.13	62/69	860	18.08

TABLE 4.16   Leaky Radiating Cable Attenuation and Signal Power at 1600 ft (at 2 m from Cable)

Item	Longitudinal loss	Coupling loss	Frequency (MHz)	Loss at 1600 ft (system loss)	Power (dBm)
1	5.12	79.88	160	85	−55.00
2	9.92	75.08	460	85	−55.00
3	18.08	66.92	860	85	−55.00

Note: System loss is constant. The cable has diminishing coupling loss as the signal approaches the far end of the cable. Aperture patterns increase with distance. The test distance is 2 m from the cable.

TABLE 4.17   Leaky Radiating Cable Attenuation and Signal Power at 1600 ft (at 10 m) Aperture Adjusted by mfg

Item	Longitudinal loss	Coupling loss	Frequency (MHz)	Loss at 1600 ft (system loss)	Power (dBm)
1	5.12	79.88	160	85	−67.00
2	9.92	75.08	460	85	−67.00
3	18.08	66.92	860	85	−67.00

Note: System Loss is constant. The cable has diminishing coupling loss as the signal approaches the far end of the cable. Aperture patterns increase with distance. Adjusted for 10 m.

Calculations made using the manufacturer's leaky radiating cable test results to make Tables 4.16 and 4.17. Due to inconsistencies in the manufacturer's data, the worst-case coupling loss is used to assure compliance with the requirements. The coupling loss includes horizontal to vertical polarization of the cable. This loss is approximately 20 dB, dependent on cable orientation.

Diffraction loss (knife edge) calculations   Mount Washington at Flowers Street is considered the worst-case VHF 160 MHz transmission. It has a small correction, due to an obstacle. The worst-case UHF radio transmission occurs from the Sheriff's Department at One Gateway Plaza. The signals were all taken at ground level with a side street entered Flowers Street. Table 4.18 is used to make the calculations for diffraction (knife edge effect) distance d in Figure 4.11.

TABLE 4.18   Radio Distances from Tunnel Entrance

Address	Location		Distance	
Mt Lukens LAPD	34°16'9" N	118°14'19.3" W	11.30 miles	18.07 km
LAFD City Hall	34°3'13" N	118°14'33.3" W	2.61 miles	4.17 km
One Gateway Plaza, LA Sheriff	34°3'23" N	118°13'57.3" W	2.90 miles	4.64 km
Mt Washington	34°6'19" N	118°12'54.3" W	6.15 miles	9.84 km
406 Exposition Blvd at 3590 Flowers St	34°1'7.5" N	118°16'49.1" W		

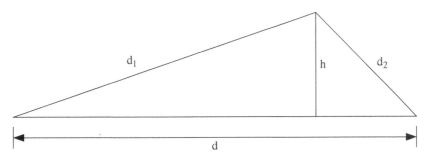

FIGURE 4.11   Knife edge loss calculations

The d distances are calculated using planar trigonometry. Longitude values are one nautical mile per degree North and latitude values are 0.444 nautical miles for each degree West. Each nautical mile is 1.125 statute miles. Table 4.18 has the distances in both miles and kilometers, because often kilometers (km) is the preferred unit of measure. Most of the calculations are completed using km because the mks system is the convention used for electrical and electronics calculations.

Fade can be estimated using Friis transmission, Equation 4.24 (Table 4.19).

$$\text{Free space in attenuation} = -32.44 - 20\log\left(f_{MHz}\right) - 20\log\left(d_{km}\right) - \text{fade}_{dB} \quad (4.24)$$

Distance calculation for d_1 and d_2:

The assumption is that $d_2 < 5\%$ d_1, $d_1 \approx d$. The distance from a building to the edge of the tunnel entrance is approximately 30 m. The loss for diffraction is calculated below. Where h = 30 m (three-story building) and d_2 = 42 m then, using Equations 4.25 and 4.26, $\Delta d \approx 42$ m and $v = 4.65$ at 160 MHz, 7.99 at 470 MHz and 10.4 at 800 MHz. The losses at the various frequencies are:

$$L\left(v\right) = 26.37\,\text{dB at }160\,\text{MHz}, 31.02\,\text{dB at }470\,\text{MHz and }33.28\,\text{dB at }800\,\text{MHz}.$$

$$\Delta d = d - \left(d_1 + d_2\right) \quad (4.25)$$

TABLE 4.19   Friis Transmission Evaluation, including Knife Edge Effects

Authority	Tx – survey Rx free space loss (dB)	$20\log(f_{MHz})$ loss (dB)	Distance loss (dB)	Fade loss (dB)
Mt Washington[a]	–122	–44.02	–19.85	–58.13
MTA 707 Wilshire Blvd	–101.28	–53.44	–10.88	–36.96
Mt Washington[a]	–122	–44.02	–19.85	–58.13
One Gateway Plaza, LA Sheriff	–103.5	–53.44	–13.33	–36.73

[a] Distance is approximately the same for Flowers Street and Exposition Boulevard.

$$v = 2\sqrt{\frac{\Delta d}{\lambda}} \tag{4.26}$$

$$L(v) = 6.9 + 20\log\left[\sqrt{v^2+1} + v\right] \tag{4.27}$$

A suggestion was made to install a radio tower behind one of the college buildings to use as a repeater station that objective here is to determine whether the repeaters will impact the leaky radiating cable at the tunnel entrance. A sketch of the proposed layout is shown in Figure 4.10. The distance from the two-story building to the edge of the tunnel is 200 + 15 m, a total of 215 m. At the fence, line of sight local radio signal from the tower is approximately 1.3 m above ground level. The fence height is approximately 1.4 m, the leaky radiating cable is 3 m below ground level. The total obstacle height to line of sight transmission is 4.4 m. The distance d = 215 m, obstacle height h = 4.4 m, $d_1 \approx 200$ m and $d_2 = 15.63$ m, $\Delta d/\lambda \approx 0.63/0.65$ and $v = 1.95$ at 470 MHz. The loss due to the obstacle is L (v) = 23.51 dB.

To calculate the free space loss, use the Friis transmission Equation 4.24.

Free space in attenuation $= -32.44 - 20\log(f_{MHz}) - 20\log(d_{km}) - fade_{dB} - L(v)$
Free space in attenuation $= -32.44 - 20\log(470) - 20\log(0.215) - 36.96 - 23.51$
Free space in attenuation $= -32.44 - 53.44 - 13.35 - 36.96 - 23.51 = -159.7\,dB$

If a 10 W repeater with a dipole antenna gain approximately 2.2 dB is installed on or near the roof top of the building (see Figure 4.10), the power level at the leaky radiating cable is –117.5 dBm (42.2 dBm – 159.7dB), but this power level would not present a problem at the leaky radiating cable. Another possibility to improve the shielding of the entrance of the tunnel is to insert it against the tunnel wall closest to the college campus.

Details of the cut and covered section of the project are shown in Figure 4.12. Note in the details leaky radiating cable is installed on both sides of the tunnel and

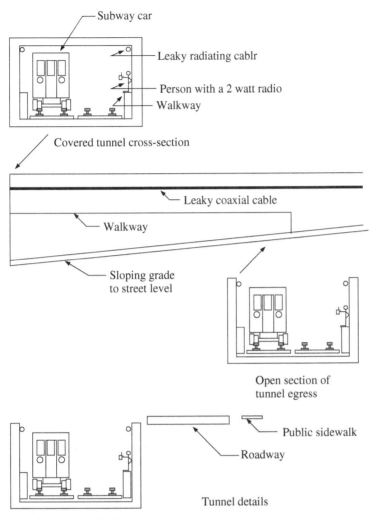

FIGURE 4.12 Tunnel and entrance details

the entrances. This may actually be the case if the power levels are insufficient to reach both transit cars if the noise levels near the leaky radiating cable are high. Another possibility where two leaky radiating cables may be installed is to use one for transmit and the other as receive. This is sometimes done in tunnels, as a matter of safety. Transmit and receive leaky radiating cables can be made bidirectional in the event that one of cables is damaged; the other may serve as a backup to the system. Roadway tunnels generally have washers that periodically clean the tunnel. These machines can damage or destroy the cable. Also shown in this diagram are maintenance persons on the catwalk with two-way 2 or 3 W portable radios. These people conduct track inspections and communicate with work crews that are doing track work. The minimum power level, as specified by the Authority, is –90 dBm at the leaky radiating cable receive end. This includes the 85 dB system loss (see Table 4.16). The distance from the leaky radiating cable to the center of the subway car is approximately 2 m.

The system power level at 2 m is –55 dBm, as shown in Table 4.16, item 2. The system loss at the end of the 1600 ft leaky radiating cable is 85 dB. The amplifier driving the cable is 30 dBm (1 W). A 2 W portable two-way radio will produce a –48 dBm signal at the receiving end.

As may be observed by the covered or tunnel layout, two subway cars side by side can produce some unique problems. The sensitivity of these radios, both on board the subway car and a portable two-way radio have a receive sensitivity of –118 dBm.

Wall reflections   The concrete surface of the tunnel and the entrance to the tunnel of covered highway absorbs most of the signal and reflects a very small portion of it. Tables 4.20, 4.21 and 4.22 show the effects of wall reflects at the entrance of the tunnel. The worst-case reflections are shown in Table 4.22. The calculations for these reflections are made using Equations 4.28 through 4.30. The result is shown under Equation 4.30. The assumption is that the center of the beam is reflecting off the concrete, and no oblique angle is used. This would be considered the worst-case reflections. It should also be noted the majority of the signal is absorbed by the concrete. Mount Washington has the largest signal, as shown in Table 4.22, at –108 dBm. This signal can appear as a noise component at the mobile radio or the subway radio.

The calculations for power in dBm (Table 4.20) have the electric field data taken from Figures 4.13–4.15. When the field contour is correct the accuracy will be improved if the geometric values are more accurate. The plot assumes that the apertures are aligned at 90° with the wall, that is the zero axes on the plots. All of the fields are in the far field. Only when the radio antenna is directly under the cable in the hand

TABLE 4.20   Wall Reflections from Tunnel for VHF 160 MHz Radios at Blvd Entrance

Item	Address	Site location longitude/latitude	Power level dBm downlink	Wall reflection
1	320 Santa Fe Ave	118°13'56" W 34°3'42" N	–110 dBm –8 dB –31 dB –149 dBm cor Tx = 47.78 dBm	–27.96 dB for concrete negligible
2	Mt Washington	118°12'54.3" W 34°6'19" N	–75 dBm – 8 dB –31 dB –114 dBm cor Tx = 47 dBm	–27.96 dB for concrete negligible
11C	3560 Figueroe	GPS elevation 52.8 m		

Note: The worst-case reflection is (–114 –27.96) = –141.96 dBm.

TABLE 4.21   Wall Reflections from Tunnel for UHF 460/800 MHz Radios at Blvd Entrance

Item	Address	Site location, longitude and latitude	Power level dBm downlink	Wall reflection
1	MTA 707 Wilshire Blvd, f = 450 MHz	118°15'25.3" W 34°2'58" N	−50.5 dBm −35 −16 with Corr. = −101.5 Tx = 50.78 dBm	−27.96 dB for concrete −129 dBm
2	One Gateway Plaza, LA Sheriff, f = 450 MHz	118°13'57.3" W 34°3'23" N	−50.18 dBm −35 −16 with Corr. = −101.18 Tx = 53.32 dBm	−27.96 dB for concrete −129 dBm
3	Mt Lukens LAPD, f = 450 MHz	118°14'19.3" W 34°16'9" N	−62.6 dBm −35 −16 with Corr. = −111.6 Tx = 51.9 dBm	−27.96 dB for concrete negligible
4	LAFD City Hall, f = 800 MHz	118°14'33.3" W 34°3'13" N	−49.2 dBm −39 −16 with Corr. = −104.2 Tx = 50 dBm	−27.96 dB for concrete negligible
5	406 Exposition Blvd at 3590 Flowers St used 3560 Figueroe	118°16'49.1" W 34°1'7.5" N GPS elevation 47.2 m		

Note: The worst-case reflection is −129 dBm.

of a 6-ft person is the E field in the near field. The wall will divert some of the energy away from the radio; next to the wall it appears as a director. Due to the steel rebar in the concrete, this can be observed as the field drops off. The field plots agree with what is expected, that is longer wavelengths propagate better than shorter ones.

The plots also match plots in the book by Constantine Balanis on advanced antenna design [2]. If the cable is lifted off the wall, it will become directive; this is an expected result. The equations used for the calculation of reflection and transmission coefficients are Equations 4.28 and 4.29 respectively. To determine whether the material is a good insulator or good conductor use Equation 4.30 due to the parameters considered and evaluation of the equation, much less one. The parameters for concrete are as follows $\mu_r = 3$, $\varepsilon_r = 3.5$ and $\sigma = 5*10^{-4}$. The wave impedances for air and concrete are 377 and 349 Ω, respectively. The reflected power is the absolute value of the reflection coefficient squared times the incident power; the remainder is transmitted into the concrete as absorption. The attenuation is $20*\log(0.04) = -27.96$ dB.

Wall reflection. Calculations

$$\Gamma_{Concrete} = \frac{Z_C - Z_{Air}}{Z_C + Z_{Air}} \tag{4.28}$$

TABLE 4.22   Worst-case Reflections at Both Tunnel Entrances

Authority	Location	Power level at end of tunnel	Frequency	Street parallel to tunnel entrance
Mt Washington	118°12'54.3" W 34°6'19" N	−75 dBm −8 dB −83 dBm corr. Tx = 47 dBm	160 MHz	Flowers Street correction for concrete reflections −27 dBm −108 dBm
MTA 707 Wilshire Blvd	118°15'25.3" W 34°2'58" N	−50.5 dBm −39.3 −12 with Corr. = −102 Tx = 50.78 dBm	470 MHz	Flowers Street correction for concrete reflections −27 dBm −129 dBm
Mt Washington	118°12'54.3" W 34°6'19" N	−75 dBm −8dB −31dB −114 dBm corr. Tx = 47 dBm	160 MHz	Exposition Boulevard −27.96 dB for concrete negligible
One Gateway Plaza, LA Sheriff	118°13'57.3" W 34°3'23" N	−50.18 dBm −35 −16 with Corr. = −101.18 Tx = 53.32 dBm	470 MHz	Exposition Boulevard −27.96 dB for concrete negligible

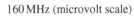

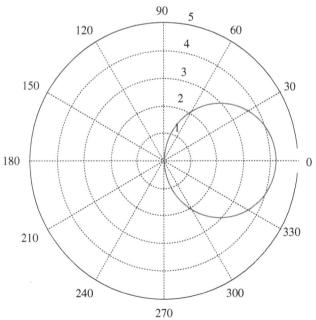

FIGURE 4.13   Tunnel wall reflections for 160 MHz

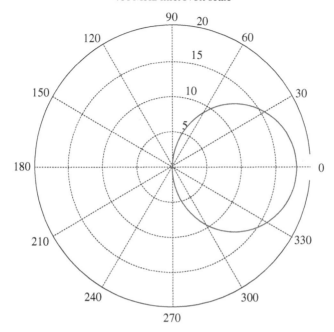

FIGURE 4.14 Tunnel wall reflections for 460 MHz

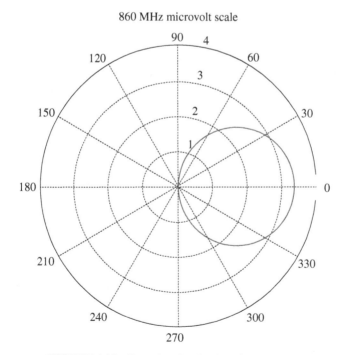

FIGURE 4.15 Tunnel wall reflections for 800 MHz

$$T = 1 + \Gamma_C \qquad (4.29)$$

$$\left(\frac{\sigma}{\omega\varepsilon_c}\right)^2 = \left(\frac{5*10^{-4}}{2\pi*8.6*10^8*3.5*8.854*10^{-12}}\right)^2 = 8.8*10^{-5} \ll 1 \qquad (4.30)$$

$$\text{Where } Z_{Air} = 377\ \Omega, Z_C \approx \sqrt{\frac{\mu}{\varepsilon}} = \sqrt{\frac{3*4\pi*10^{-7}}{3.5*8.854*10^{-12}}} = \sqrt{\frac{3}{3.5}}*377 = 349\ \Omega$$

$$\Gamma_C = -0.04, T = 1 - 0.04 = 0.96$$

The reflected power is the absolute value of the reflection coefficient squared times the incident power the remainder is transmitted into the concrete as absorption. The attenuation is $20*\log(0.04) = -27.96$ dB.

As fields propagate further from the cable they go through several transitions. The first is the near field it acts as a reactance. It is less than $2d^2/\lambda$, approximately 20 cm for 160 MHz, 3 cm at 460 MHz and 0.75 cm at 860 MHz. Therefore, only the far field is considered in the analysis. The far field electric and magnetic fields are in phase, wave impedance is real and power is the radiated. All other fields are damped out.

Figures 4.13–4.15 are the results of the program written in MetLab. The wall alters the pattern significantly into the patterns shown. The patterns shown and signal strength are calculated by the program for 10 m from the wall as the beam angle varies between $+60°$ and $-60°$. It is very easy to change parameters to larger or smaller distances from the wall. Ten meters is selected because it is close to the normal span of the lanes, usually 12 ft with a 3-ft shoulder on each side of the right of way. Figure 4.16 is a description of how the calculations were made showing the beam width but with the beam angle increased past the beam width to $+60°$ and $-60°$.

Table 4.23 is the power calculations in dBm for the patterns shown in Figures 4.13–4.15. It should be noted that the calculations only apply for the field

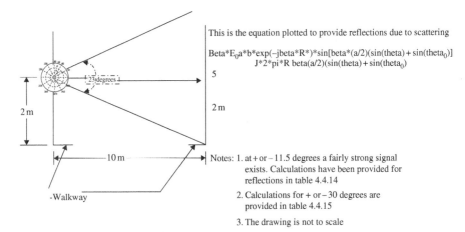

This is the equation plotted to provide reflections due to scattering

$$\text{Beta}*E_0*a*b*\exp(-j\text{beta}*R*)*\sin[\text{beta}*(a/2)(\sin(\text{theta}) + \sin(\text{theta}_0)]$$
$$J*2*pi*R\ \text{beta}(a/2)(\sin(\text{theta}) + \sin(\text{theta}_0))$$

5

2 m

2 m

Notes: 1. at + or – 11.5 degrees a fairly strong signal exists. Calculations have been provided for reflections in table 4.4.14

2. Calculations for + or – 30 degrees are provided in table 4.4.15

3. The drawing is not to scale

23 degrees

10 m

-Walkway

FIGURE 4.16 Wall reflection

TABLE 4.23 Power Reflections from the Wall, Taken from Figures 4.13–4.15

Item	E field (V/m)	Frequency (MHz)	Distance (m)	Phase (degrees)	Power (dBm)	Comments
Part 1: Reflected power is the absolute value of the reflection coefficient squared times the incident power						
1	2.00E-05	160	1.15	-60	-88.5288568	Data corrected for losses is approximated from from the graph in Figure 4.13
2	3.00E-05	160	2.38	-40	-78.6894493	
3	3.50E-05	160	3.46	-30	-74.1005306	
4	4.00E-05	160	7.46	-15	-66.2674371	
5	4.30E-05	160	10.0	0	-63.0940444	
6	4.00E-05	160	7.46	15	-66.2674371	
7	3.50E-05	160	3.46	30	-74.1005306	
8	3.00E-05	160	2.38	40	-78.6894493	
9	2.00E-05	160	1.15	60	-88.5288568	
Part 2: Power Levels at various points on the roadway and wall at 460 MHz						
1	8.00E-06	460	1.15	-60	-96.487657	Data corrected for losses is approximated from the graph in Figure 4.14
2	1.30E-05	460	2.38	-40	-85.9530073	
3	1.51E-05	460	3.46	-30	-81.4023526	
4	1.75E-05	460	7.46	-15	-73.447876	
5	1.80E-05	460	10	0	-70.6579634	
6	1.75E-05	460	7.46	15	-73.447876	
7	1.51E-05	460	3.46	30	-81.4023526	
8	1.30E-05	460	2.38	40	-85.9530073	
9	8.00E-06	460	1.15	60	-96.4876570	

(Continued)

TABLE 4.23 (Continued)

Item	E field (V/m)	Frequency (MHz)	Distance (m)	Phase (degrees)	Power (dBm)	Comments
Part 3: Power Levels at various points on the roadway and wall at 860 MHz.
1	1.50E-06	460	1.15	-60	-111.0276315	Data corrected for losses is approximated from the graph in Figure 4.15
2	2.50E-06	460	2.38	-40	-100.2730742	
3	2.90E-06	460	3.46	-30	-95.7339316	
4	3.20E-06	460	7.46	-15	-88.2056374	
5	3.50E-06	460	10	0	-84.8820526	
6	3.20E-06	460	7.46	15	-88.2056374	
7	2.90E-06	460	3.46	30	-95.7339316	
8	2.50E-06	460	2.38	40	-100.2730742	
9	1.50E-06	460	1.15	60	-111.0276315	

patterns in the figures. They are in microvolts and not an electric field units. The calculations were provided to estimate the leaky radiating cable field at 1.15 m from the wall. The drop off in signal level is fairly high as the wall is approached where the radiating cable is mounted. Figure 4.16 is a description of the beam width leaky radiating cable as a function of distance to the wall. A summary of the wall reflections is provided in Table 4.24 as a function of the 0° with the electric field electric field and the reflections from the wall.

Table 4.24 represents the worst-case scenario for reflections at the wall for maintenance personnel using a mobile two-way radio while on the catwalk. Using Friis transmission Equation 4.24 and the leaky cable radiating power of –55 dBm.

Free space in attenuation = –32.44 – 53.25 +40 = –45.69 dB, where frequency = 460 MHz and distance = 10 m. The leaky radiating cable power at the catwalk is –100.69 dBm. The maintenance channel that is commonly used is within the low band UHF range about 460 MHz. For good voice communication, the carrier to noise ratio should be about 15 dB the power level at –115 dBm reflection is adequate for good maintenance radio reception.

The shadow effect for two subway cars side by side Maintenance personnel operating two-way, 460 MHz mobile radio on the catwalk will communicate with the central office through the leaky radiating cable or the subway train crew. The worst-case scenario is when two subway cars are parked side by side obscuring the leaky radiating cable line of sight maintenance person during communication with central office. The power of the signal coming from the central office via the leaky radiating cable is –55 dBm with free space loss at –45.69 dB (not including fade) for a 10 m distance. The total line of sight signal strength is –100.69 dBm. Assume the average height of a maintenance person is 5'8" (1.72 m) and the subway car is 0.5 m above this value from the catwalk.

Use Equations 4.25–4.27 to calculate the knife edge effect where $d_1 + d_2 = 13.34$ m, $d = 12.25$ m, $\lambda = 0.6739$ m and $\Delta d = 1.09$ m.

$$\text{Attenuation } L(\nu) = 6.9 + 20\log\left[\left(2.54^2 + 1\right)^{1/2} + 1\right] = 18.33\,\text{dB, where } \nu = 2.54.$$

The signal received at the two-way radio by the maintenance person is –119.02 dBm. It has no phase margin and reception will be poor and mostly digital radios are used. P25 digital radios for example have a minimum sensitivity of –124 dBm.

The transmitter of the 2 W mobile radio has a radiated power of 33 dBm, the loss due to free space. The knife edge effect and coupling loss is 131 dB (including coupling loss of 67 dB). The total power at the input of leaky radiating cable is –98 dBm with a margin of 20 dB for fade and other anomalies.

Conclusion Data has been gathered from several sources to generate the plots and tables presented in this document. Maps generated by Google are used to observe all the streets along the right of way. The maps provided a great deal of information about the obstructions such as the fence along Exposition Boulevard and the Harbor Freeway along Flowers Street. Figueroa Street represents a gap in the obstructions that would shield the exits but it is in a covered section of the tunnel. The signal levels

TABLE 4.24   Power Reflections from the Wall for Maintenance Persons

Wall reflected power, including wall losses at 160 MHz		Wall reflected power, including wall losses at 460 MHz		Wall reflected power, including wall losses at 860 MHz	
μV	Power, dBm	μV	Power, dBm	μV	Power, dBm
1.0	−132.5178	1.0	−128.0149	1.0	−130.8607
2.0	−129.5178	2.0	−125.0149	2.0	−127.8607
3.0	−127.1178	3.0	−122.6149	3.0	−125.4607
4.0	−126.5178	4.0	−122.0149	4.0	−124.8607
5.0	−112.3822	5.0	−116.8851	5.0	−114.0393
6.0	−111.3822	6.0	−115.8851	6.0	−113.0393
7.0	−110.6822	7.0	−115.1851	7.0	−112.3393
8.0	−109.1822	8.0	−113.6851	8.0	−110.8393

TABLE 4.25   Attenuation due to Shrubs Around Tunnel Entrances Reducing External Radio Signals

Frequency (MHz)	Shrub height (ft)	Attenuation (dB)
160	2	8
450	2	9
850	2	10
160	3	9
450	3	10
850	3	12

should not present a problem if the receive cable is on the south-east side of the exit. As a precaution, the Authority is urged to plant shrubs around the entrance of the exit. A simple two- or three-foot shrub would attenuate a downlink signal, as shown in Table 4.25 below.

A wall or other obstruction is more effective but this is likely the most cost-effective. Rounded obstructions are more effective than a knife edge. They can add 3–5 dB of attenuation to the data in Table 4.25. The buildings along Exposition Boulevard are estimated at 500 m from the entrance and three stories high in Flowers Street at the Harbor. The Freeway is estimated at 7 m high and 10 m from the entrance. The underpass is very close to the covered section of the tunnel.

Measurements are necessary to refine the model for the entrance coverage. The best method is to measure coverage close as possible to ground level. This should be done carefully at different time of the day to include fading. All of the estimated data excludes fading. The EDX program is implemented without clutter included.

Measurements will be taken at 20–30 m increments along both street locations. This data can be used to further refine the estimates of performance. The measurements should be taken by keying the transmitter as various sights and observing the resulting frequency at the tunnel entrances. If possible, a two-way two-Watt radio can be keyed behind the obstructions the fence on Exposition Boulevard and Harbor Freeway along Flowers Street to determine how effective these two obstructions are as a shield.

One recommendation is to install leaky radiating cable on both sides of the tunnel entrance, due to the two-car scenario that will impact communications. It can also serve as a backup in the event of that one of the leaky radiating cables is damaged. Another possibility is to increase the power of the leaky radiating cable with amplification. This is not always a good trade-off when multiple channels of communication are used that may require larger combiner filtering.

During track maintenance that often occurs in subway systems, a two-cable solution is more desirable because maintenance personnel working on tracks at ground level will not have communications blocked by transit cars. The tunnel will no doubt be a good area to place interlocking with a signal house for crossovers in traffic for maintenance, further down the line. The tunnel is in an area, where a station, SCADA cabinets, signals equipment or other subsystems may be installed.

When excavation and concrete is poured for the foundations and walls, a radiation survey will be conducted for the area. The FCC can be contacted to determine radio stations and other licensed radiators that may affect the tunnel exit and entrance communications. It must be kept in mind that this survey is a rather double-edged sword, that is any new licensed or unlicensed radiating equipment must not affect existing installed equipment.

After the surveys are conducted, an EMC engineer will determine how the interactions between existing radiators and new equipment will affect communications in the area. Subways near airports, for example, may be exposed to weather radar, ground control radios, instrument landing equipment and so on. Some of the radiation from these installations is quite harsh. Vital logic in subway systems must be protected at all costs because it is a life and safety issue. Other equipment such as video, safety, alarm and SCADA systems may require filtering or some form of shielding to prevent corruption of communications.

An additional survey is generally conducted after all the equipment is installed in the running subways. This final survey will fine-tune the installation and the report is written and signed off by an EMC engineer. It is particularly important to note that the EMC engineer has a heavy responsibility and he/she or the corporation is generally liable for life and safety issues.

4.5 TRACKS SURVEY

The EMC track survey is an extremely important aspect of subway construction. It can be done after the route is identified or after the right of way is established, that is the land is cleared of brush and the roadbed is established. It is generally necessary to have several people to do this survey, although one may do the survey. But it will be very time-consuming, for example, the person setting up the equipment may need an assistant to record the following:

1. Equipment type oscilloscope, spectrum analyzer, laptop computer, attenuator, antenna and any other specialty equipment required
2. The serial numbers of equipment used in the survey
3. Make and model of each piece of equipment
4. Calibration date and no longer than one year since last calibration date
5. GPS location of each site where the survey is conducted approximately every 15–30 m
6. Elevation of each survey site
7. Record the distance the various obstructions within 30 m of the track
8. Antenna direction since the antennas are directional eight 360° survey must be taken at each site
9. Special anomalies such as intermittent radiation at a particular time of day
10. Time of day when the measurements are taken
11. A record of each sweep the spectrum analyzer is taken with the settings of its controls; sufficient information should be provided to duplicate the test.

As can be observed the effort required to take the survey is quite lengthy and even a two-man crew may be hard-pressed to do the survey in a timely manner. Table 4.26 gives a guide for the engineer or technician to generate a timely record of the tests. The engineer or technician conducting the tests may or may not be the engineer of record; this person is generally a professional engineer with a PE seal. The temperature and time of day for conducting the tests is very important, as well as the cloud cover. These will have an impact on the radiation from both licensed and unlicensed radiators. Tests should not be conducted after sundown because certain licenses for radio stations must reduce power, thus reducing background noise. Equipment data is self-explanatory.

Site data will include GPS location and bearing angle in degrees from North toward the easterly direction in the northern hemisphere. The antenna elevation is required and ground conditions are necessary such as sandy soil, clay, rocky or grassy area for taking a roadbed survey. If taking a survey after the track is laid, the land is usually cleared back from the roadbed and has a gravel road alongside the track to a minimum of 15 m from the center of the track. The three antennas used for the tests are monopole from 9 30 MHz, biconical from 30 230 MHz and log periodic from 230 MHz to 1 GHz. See Figure 1.1 in Chapter 1 for the various antennas used for the survey. The remarks column should be used for identifying structures and obstructions to the testing or other anomalies.

Figure 4.17 is a sketch of a survey taken after track has been laid; however, most surveys are taken initially before track is laid. In some cases, the survey may be taken in grassy areas without trees. However if trees do exist in the area 30 m from the

TABLE 4.26 Track Survey Record Sheet

Personnel conducting test and environmental data				
Name of engineer or technician			Date_____	
Signature of the engineer of record			Date_____	
Temperature		Cloud cover in percent		
Time of day and date	Date		Time	
Equipment Data				
Equipment	Model	Mfg	Serial No.	Calibration
				Date_____
Spectrum analyzer				
Attenuator				
Cable				
Biconical antenna				
Log periodic antenna				
Monopole antenna				
Laptop computer				
GPS unit				
Miscellaneous equip.				
Site Data				
GPS Location and bearing angle	Antenna Bearing Angles			
N				
W				
Biconical antenna				
Elevation_____ ft/m				
Log periodic antenna				
Elevation_____ ft/m				
Monopole antenna				
Remarks				

center of the track area or if there are too many obstructions the survey should be deferred. When the right of way is cleared, this is the best time for conducting the survey. The antenna shown in Figure 4.17 is shown in the horizontal position. Two different traces should be taken, one in a horizontal and the other in a vertical position; this represents horizontal and vertical polarization. This same technique should be used for all directional antennas including the log periodic antenna. The heavy

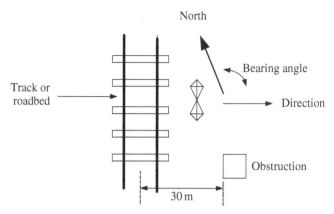

FIGURE 4.17 Survey antenna position; only biconical shown

gravel under the rails is commonly known as ballast; this will be discussed in detail in other case studies.

The obstruction shown in the diagram is a sample of what may be found along the right of way. This can be something as simple as a tool shed or as complex as a building but it should be noted in the remarks column. Occasionally in certain areas, trees may have a significance and may not be cut down; these in particular should be noted. In the case study in Section 4.4. several obstructions were apparent, such as the wrought iron fence, buildings and expressway. These all have an impact on radios operating near the right of way.

A sample of test equipment is provided below:

1. Tektronix-27120A EMI measurement system (far field measurements)
   Included in the measurement system:
   Calibrated biconical, log periodic, rod and loop antennas
   Spectrum analyzer
   RF pre-selector
   Software
   Instruction manuals
   Tripods
2. Other equipment
   Laptop computer with 488.2 bus card or USB port
   Miscellaneous cables
   Inverter
   GPS receiver with height measurements capability
   Four-wheel drive vehicle
   Digital camera
3. Spectrum analyzer 30 Hz to 30 GHz bandwidth (near field measurements)

4.5.1 Measurement Methodology

The equipment is set up as shown below in Figure 4.18.

The first measurement to be made is with the system in the 50 Ω terminated position, to give the noise floor of the system. All measurements will be made with vehicle engine, cell phones and radios off. The spectrum analyzer is adjusted in the wide band position. The next measurement is made with the antenna connected in the wide band position. An examination of the wide band result for large peaks is conducted. A hard copy of the survey is furnished as part of a report. The preamplifier may or may not be necessary depending on the sensitivity of the spectrum analyzer. It is internal and available if necessary for the spectrum analyzer presented in this example.

The frequency bands between 30 MHz and 1 GHz are generally done for licensed radio traffic and unlicensed equipment that radiates. The lower frequencies, in particular between 9 and 100 kHz, are in regions where signals implementation equipment can be affected. For example, audio frequency track circuits operate between 3 and 10 kHz. A different antenna scheme may be necessary for examining these frequencies. Radio communications are also of concern when conducting surveys along the railroad right of way because of the two-way radio communications used on rail services, both subways and freight.

A GPS which will provide longitude, latitude, elevation and bearing angles is a type that should be used to save time; otherwise calculations may be necessary to find the bearing angle using planar trigonometry. This data should be taken relative to the center of the right of way at between 10 and 30 m depending on the radiation observed in the sweep. The beam width of the directional antennas should be considered when taking measurements. For example, a beam width of 70° indicates at least six positions are necessary to measure the 360° required for the survey. The manufacturer of the antennas provides calibration curves with these devices so that corrections can be made to the spectral plots. The antennas will be used in vertical and horizontal positions respectively in the event that some of the radiation is horizontally polarized. This will be easy to observe when examining the spectral plots by comparing the magnitude of the particular frequencies in question and noting any difference based on horizontal or vertical polarization. An examination of Figure 4.19 (frequencies present across the whole bandwidth from 30 MHz to 1 GHz) shows a

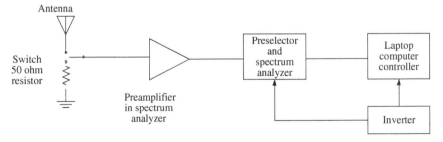

FIGURE 4.18 Simplified block diagram of survey test

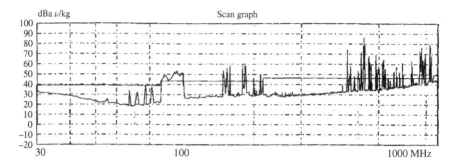

Title: FCC part 15, Class A—Radiated emissions, 3 meters
Data title #1: AMEC_OUT.FFA
Marker 1: 9962418 MHZ
Spec 1: FCC part 15, Class A—Radiated emission, 10 meters
Factor 1:em_6912a.rad
Factor 2:em_tpa30.rad

FIGURE 4.19   Spectral diagram of test across entire band (low accuracy)

horizontal line that represents FCC Part 15 radiation limits for unintentional or unlicensed radiation. It should be noted that a band of frequencies are shown between 90 and 108 MHz. This is the FM commercial broadcast radio band. The other frequencies above the limit line are open for investigation to determine whether they are licensed radio or unlicensed radio frequencies. This is only a sample to be used as a guide when analyzing the radio frequencies radiated that may result in EMC problems.

The next series of figures are the result of a survey taken near an airport, which is the harshest environment that may be encountered. The subway system right of way extends north and south parallel to an airport. The figures represent a sample of the total number of data taken at each point along the right of way. The total number for each position is about eight and the number of positions along the right of way is about 25 sites, with the total number of samples at about 200 per kilometer. This was all that authority requested. In some instances more samples might be necessary because of traffic and congestion in the area. The equipment to be installed is a series of cabinets for SCADA, audio, signals and video equipment and stations along the right of way.

While doing this survey it is prudent to analyze radio communications used by the authority requesting the survey. An analysis can be made to determine the fade in communication transmission along the right of way. If the radio can be keyed and the transmission observed during the survey, the receiver power can be calculated at a particular point where measurements are made. The distance from the transmitter to the position where the survey is taken can be found by using Google, aviation charts or some other available software. The received signal power at the location can be calculated using the plots of the survey. The calculation for free space loss, using the Friis transmission Equation 4.24, can be used to determine the fade at the location. The advantage to using this method for finding the fade is several samples are taken along the right of way at different times of day only 40 m apart. The graphic display in Figure 4.20 shows the lowest frequency band from 30 to 300 MHz. It is contracted

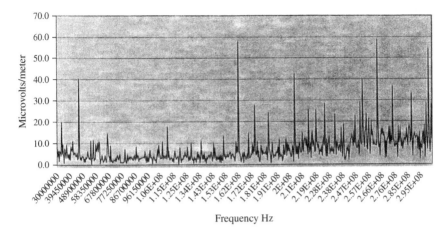

Frequency Hz

FIGURE 4.20   East vertical polarization

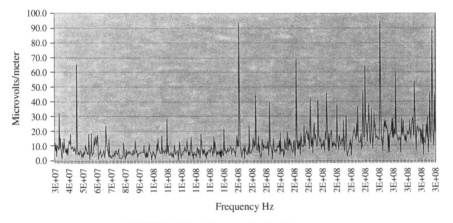

Frequency Hz

FIGURE 4.21   West vertical polarization

and is shown as a guide for the biconical antenna in a vertical polarized position. The antenna is facing West and the bearing angle is 270° for this particular plot. As can be observed several peaks are present; these are due to the mobile radios used at airports for ground control and various types of radar.

The bearing angle for the plot shown in Figure 4.21 is 270°. It should also be noted that many of the frequencies in this plot are quite large because the spectrum scans are taken toward the airport.

The spectral scans in Figures 4.22 and 4.23 are taken with the log periodic antenna of the frequency band 300 MHz to 1 GHz with horizontal polarization. The second set of scans (Figures 4.24 and 5.25) are taken with the log periodic antenna in vertical polarization.

The large peak in Figure 5.25 is most likely some form of radar. The spectral plots shown in Figures 4.19–4.25 are shown as examples. The spectral plots are stored in

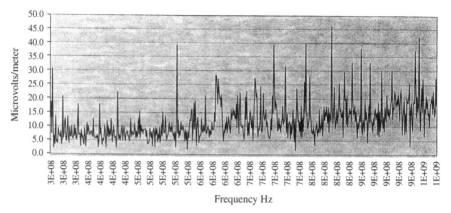

FIGURE 4.22   East horizontal polarization

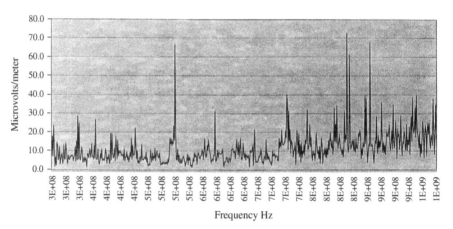

FIGURE 4.23   West horizontal polarization

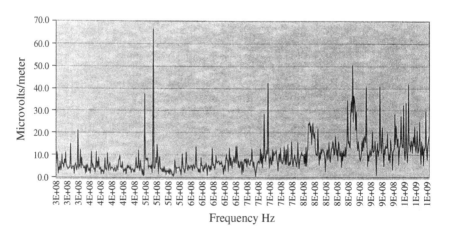

FIGURE 4.24   East vertical polarization

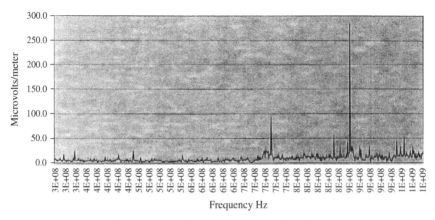

FIGURE 4.25   West vertical polarization

a computer so that they may be manipulated, that is cropped and expanded to improve resolution of the plot that is of interest. Also the correction factor for increments of frequencies and gain at that frequency are provided in the calibration Tables 4.27 4.29. These are to be used as a guide. The distance the measurements were made and the polarization are provided in the manufacturer's calibration table.

These calibration tables are provided as a guide to what typical values biconical, log periodic and active monopole antennas will exhibit.

Correction factors are added to Figures 4.20–4.25 using antenna and cable calibration data by the computer prior to displaying spectral plots. The antenna factor (loss) and cable loss at the particular band of frequencies must be entered into the computer and applied to the spectral plot as a correction. Some spectrum analyzers can compensate for antenna gain and cable loss automatically, thus relieving the operator from entering the information into the computer to make the necessary connection corrections. The gain is exhibited in dB and numeric form. To correct the cable and antenna factor loss from dBμV to the electric field intensity dBμV/m use SAE ARP-958, ANSI C 63.5 or IEEE 291 applied to the spectral plots. Some of the newer spectrum analyzers may have these features built into the software. The monopole antenna has only the antenna factor provided because the gain is adjustable for active antennas.

The calibrated cable described in Table 4.30 is RG-58, 16 m long with N-N connectors. When using this cable, conversion to a BNC type connector is necessary for use on the monopole antenna. The loss in this conversion is fairly small; however it should be considered to present improved accuracy. The measurements are made for all the spectral plots that may exhibit fade if measured at a different time or even given a small span of time with the equipment set up and remeasured. The objective is to determine if an EMC issue may occur on the right of way.

Conclusion   Since measurements are made from the center of the track, obstacle reflections will also be apparent. Later after the construction of tracks on the right of

TABLE 4.27   Biconical Antenna Calibration Sheet

Manufacturer_____   Polarization horizontal

| Model no. biconical _____ | Serial no._____ | Calibration date_____ | Calibration distance Meters_3_____ |

Frequency (MHz)	Antenna factor (dB/m)	Gain (dBi)	Gain (numeric)
20	14.0	–17.78	0.02
25	13.8	–15.56	0.03
30	14.0	–14.25	0.04
35	14.3	–13.18	0.05
40	13.7	–11.4	0.07
45	13.7	–10.42	0.09
50	12.7	–8.45	0.14
60	10.6	–4.75	0.34
70	10.2	–3.08	0.49
80	10.1	–1.77	0.67
90	10.4	–1.06	0.78
100	10.4	–0.17	0.96
110	10.5	0.56	1.14
120	11.1	0.69	1.17
130	11.5	1.01	1.26
140	12.0	1.12	1.29
150	12.4	1.36	1.37
160	12.9	1.44	1.39
170	13.7	1.17	1.31
180	14.6	0.75	1.19
190	15.5	0.33	1.08
200	15.9	0.36	1.09
210	16.8	–0.13	0.97
220	17.5	–0.45	0.90
230	18.5	–0.97	0.80
240	19.3	–1.46	0.72
250	19.9	–1.65	0.68
260	19.9	–1.39	0.73
270	19.9	–1.07	0.78
280	20.8	–1.57	0.70
290	20.4	–0.86	0.82
300	22.0	–2.20	0.60
310	22.6	–2.50	0.56
320	23.9	–3.56	0.44
330	25.3	–4.64	0.34

TABLE 4.28 Log-periodic Antenna Calibration Sheet

Manufacturer_____		Polarization	
Model no. log-periodic _____	Serial no._____	Calibration date_____	Calibration distance Meters__3____
Frequency (MHz)	Antenna factor (dB/m)	Gain (dBi)	Gain (numeric)
290	14.5	4.97	3.14
300	14.4	5.40	3.47
325	14.6	5.91	3.90
350	15.2	5.90	3.89
375	15.9	5.84	3.83
400	16.1	6.18	4.15
425	16.8	5.99	3.98
450	17.1	6.22	4.19
475	17.8	5.94	3.92
500	18.5	5.77	3.77
525	19.1	5.56	3.60
550	19.2	5.86	3.85
575	20.0	5.43	3.49
600	19.8	5.99	3.97
625	20.0	6.20	4.17
650	20.1	6.43	4.40
675	20.8	6.04	4.02
700	20.4	6.73	4.71
725	21.2	6.24	4.21
750	20.9	6.83	4.82
775	21.9	6.17	4.14
800	21.4	6.94	4.94
825	22.3	6.31	4.27
850	22.6	6.24	4.21
875	22.5	6.63	4.60
900	22.8	6.52	4.49
925	23.2	6.36	4.33
950	23.4	6.36	4.32
975	23.9	6.09	4.06
1000	23.3	6.99	5.00

way the communication and signal houses will most likely be installed. These houses are brought in on trucks and they are constructed of reinforced concrete; however some might be constructed of heavy gauge metals with protective coatings. Other obstructions may be supports whether catenary wire in open areas and tunnels or overpasses.

TABLE 4.29   Active Monopole Antenna Calibration Sheet

Standard calibration for active monopole antenna	
Frequency	Antenna factor
10 kHz	1.2
20 kHz	0.6
50 kHz	0.4
100 kHz	0.5
200 kHz	0.5
500 kHz	0.5
1 MHz	0.6
2 MHz	0.7
5 MHz	0.7
10 MHz	0.6
20 MHz	0.7
30 MHz	0.8
40 MHz	1.5
50 MHz	1.7
60 MHz	0.1

Since the electric field is measured and extrapolation can be made on either side of the right of way out to 30 m if necessary, as often requested in the authority specifications. This of course is another issue but in the same way a survey of radio coverage is also made along the right of way. This type of coverage is usually done using a signal generator and portable radios this will be discussed in Chapter 5 as a case study.

All of the data and other information in this case study are taken from actual measurements and manufacturer's specifications. One reason that actual manufacturer's names and addresses are not used in this study is to prevent book obsolescence. As antenna technology improves some manufacturers are no longer in business or they have improved versions that makes the previous one obsolete. Spectrum analyzers have improved a great deal since some of the surveys have been done, with improved software, higher resolution, lower noise and so on.

Many of the authorities constructing rail right of ways have changed requirements documents that may ask for surveys that include a multitude of wireless devices, that is Bluetooth or other types of spread spectrum technology. Also changing is some of the signals equipment that now uses spread spectrum technology. This can be an issue; it is studied in Chapter 5 of this book. The information provided is to be used as a guide to assist the EMC engineer in the preparation of reports that are required for either analysis, testing or both.

TABLE 4.30   Test Cable Loss Versus Frequency

Frequency (MHz)	Cable loss (dB)
10	0.8
20	1.1
30	1.4
40	1.6
50	1.8
60	2.0
70	2.1
80	2.3
90	2.4
100	2.6
200	3.7
300	4.7
400	5.5
500	6.3
600	7.1
700	7.8
800	8.5
900	9.1
1000	9.7

4.6   LEAKY RADIATING COAXIAL CABLE ANALYSIS

The material covered in this analysis will have a significant impact on total radio system design and implementation which will result in the reduction in cost to the system. This case study will only cover some of the details of this effort; it is ongoing for the continuous refinement of the simulation design and implementation of some of the earlier communication designs using a distributed antenna system to provide radio coverage within tunnels and other confined structures. Early uses of leaky radiating coaxial cable were used as a detection scheme, a sort of fence to detect intruders. Intruders near the radiating areas of the cable would cause reflections, thus detecting an intruder.

Several tunnel projects require radios as a means of communication with Fire Department, Police Department, Emergency Medical Services and facilities provider (Authority). Radiating cable (often called leaky coaxial cable) is the method preferred by most authorities for the implementation of communication design for tunnels. Some leaky cable has longitude slots that are aperture antennas along the length of the cable. Another type of leaky cable has lateral slots that are grouped to form an array of slot antennas. The latter type is the most prevalent at this time and it is discussed in this case study.

Most leaky cable manufacturers only provide the loss as a function of distance (coupling loss) without regard to the radiation patterns. In some areas extra cabling is used that may be unnecessary, due to the lack of coverage information. This requires more base stations and amplifiers to implement the design. The additional equipment will increase the cost of implementation and maintenance. If the pattern is known and can be modified using reflectors or other means to improve the electromagnetic fields, a savings in equipment cost can be achieved. The leaky radiating cable antenna will be analyzed to determine the radiation pattern for frequencies at 160, 460 and 870 MHz. These are the major frequencies for coverage of the various services.

The cable is characterized using Equation 4.31 for determining the characteristic impedance of the coaxial line evaluated using manufacturer's specifications. The model used to generate the E fields required the use of a 50 Ω end cap to prevent reflections and a source with 50 Ω impedance. The section of leaky radiating cable was 1 m long; if a longer section had been used the analysis would have taken weeks to produce E field patterns. Perhaps some of the newer computers with large memories and higher clock frequencies can produce multiple radiation patterns over the whole line that may be several meters long.

$$Z = \left[60 * \ln\left(D/d\right)\right]/\left(\varepsilon_r\right)^{1/2} \tag{4.31}$$

Where D = length, d = diameter, ε_r = relative permittivity.

Due to the limitation of computer power at the time that analysis is done, the apertures were assumed to be square rather than having rounded edges. For the length and width of each aperture (with some experimentation) the best case trade-off was a 17×4 mm that kept the runtime for the model down to 2.5 h. Some liberties were taken with the foil shell and jacketing; these were estimated at 4 mm each. The center section, a copper tube in the center of the cable, was made of solid copper instead of a tube which would make no difference in the analysis since the tube only acted as a waveguide. This change was necessary to decrease runtime of the computer because the airspace was also analyzed and contributed nothing to the analysis. The depth of penetration at each frequency is provided in Table 4.31 below.

Once the electric field is determined, there is a possibility of using a director to further focus the field in areas where it is necessary for communications, such as areas with stairwells and various portions of the station platform. Also in office areas leaky radiating cable may be used for wireless communications, such as Bluetooth and spread spectrum devices. At times the radiation must be focused to achieve better

TABLE 4.31 Depth of Penetration in Copper for VHF and UHF Frequencies

Frequency (MHz)	Depth of penetration (mm)
860	2.28×10^{-3}
460	3.102×10^{-3}
160	5.317×10^{-3}

communications; thus, by knowing the shape of the E field, a director may provide a necessary change in the field to improve the focus.

There are other methods of doing this same analysis. Method of moments (MoM) is another common one, alluded to in References [2] and [3]. Both FDTD and MoM produce excellent results that are close to the actual measurements. A discussion of these techniques will be presented in the appendix of this case study (it is available online at www.wiley.com/go/electromagneticcompatibility). Care must be taken not to take too many liberties with the dimensions of the model, that can destroy the accuracy of the analysis. However, to decrease runtime, some liberties must be taken.

The cable used in the study is characterized in Figure 4.26. The specifications for this cable are provided in the case study appendix. The cable comes in various sizes, from 5/8″ diameter to 1 5/8″ diameter. The latter is type was chosen for this study, that is 1 5/8″ diameter. Figure 4.26a is a cross-section of the cable structure. The inner corrugated copper tube that runs the length of the cable provides rigidity. Frequencies used on the cable depend on the skin effect of copper because it is in actuality a waveguide. Insulation material between the foil and the copper tube provides added rigidity to the cable. The cable foil is corrugated with slots laterally in the foil; these are the apertures that radiate the signal. The jacketing around the cable provides a protection from the environment, that is moisture, exhaust gases and other factors that may degrade cable performance. The cable has opposite the aperture array a guide for installing the cable. This allows the apertures to face away from the wall during installation. The minimum distance from the wall to the cable is approximately 3″.

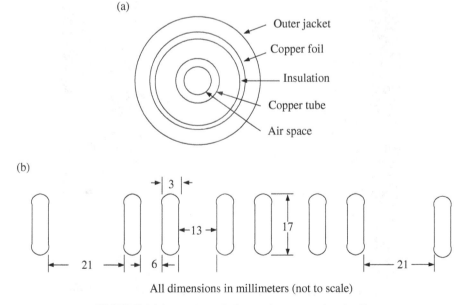

All dimensions in millimeters (not to scale)

FIGURE 4.26   Leaky radiating cable construction details

Figure 4.26b gives the layout and grouping of the aperture slots in the cable. An examination of the slot arrangements indicates three groups of two that can resemble a two-element array at very high frequencies, that is near 1 GHz; and the groups of two are arranged in a grouping of three. The total array of eight slots will support the lower frequencies in the VHF range. Figure 4.27 is a cross-section of the length of the leaky radiating cable showing the aperture radiation. This cross-section representation is generated with XFDTD software (finite-difference, time-domain; that X is added as part of the trade name). Note in the cross-section the yellow represents radiation emanating from the apertures through the jacketing; the pink area down the center is the airspace in the copper tube. The dark jacketing and the cross-section does not have holes or openings; it is shown to emphasize the radiation path. The analysis software requires a computer with a great deal of computing power. A three-processor computer with a 600 MHz clock took 8 h to run a simple program that generated Figures 4.27, 4.28, 4.29 and 6.30.

An examination of Figure 4.28 shows the electric field pointing outward (white arrow) from the axis going through the center of the field representation. All of the apertures and antenna factors will contribute to E field radiation.

Simulation results   A tabulation of the simulation results for Figures 4.28, 4.29 and 4.30 are provided in Table 4.32. The minimum runtime for any simulation is provided in the XFDTD manual. The runtime for this particular simulation took 8 h.

FIGURE 4.27   Cross-section of a length of leaky radiating cable, showing aperture radiation

FIGURE 4.28   E field simulation (160 MHz)

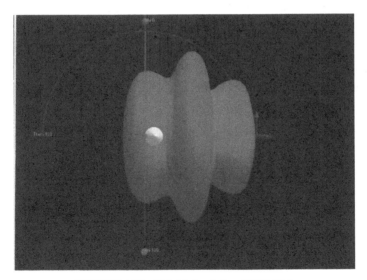

FIGURE 4.29   E field simulation (470 MHz)

A synopsis of the data provided in chart form for each run is shown in Table 4.32. It should be noted that the power level driving the cable is only in the milliwatt range. This is because the worst-case conditions occur at the end of cable. For a 1600-ft leaky radiating cable, the amplifier driving the cable is 1 W and the coupling loss,

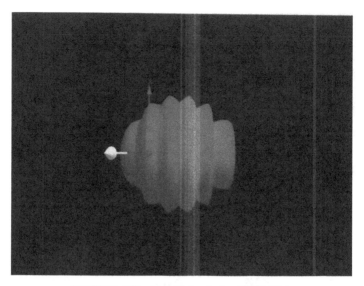

FIGURE 4.30   E field simulation (860 MHz)

85 dB, might be −55 dBm if the efficiency is low. When doing a simulation it is advisable to use worst-case conditions for simulating the end of the cable and assume the cable is terminated; and 50 Ω reduces reflections. During the simulation the power is provided as dBW but converted to dBm which are units of convenience for radio communications. It should be noted in Table 4.32 the efficiency is the best near the center of the frequency range.

The radiation resistance in Table 4.32 is calculated using Equations 4.32, 4.33 and 4.34, as shown below.

$$P_L = \frac{1}{2}\left(V_g\right)^2 \left(\frac{R_g}{\left[\left(R_L + R_g + R_r\right)^2 + \left(X_L + X_g\right)^2 \right]} \right) \tag{4.32}$$

Where R_L = load resistance, R_g = generator resistance, R_r = radiation resistance, X_L = load reactance impedance, X_g = generator reactants impedance, V_g = generator voltage, P_L = generator power and P_r = radiated power.

$$P_r = \frac{1}{2}\left(V_g\right)^2 \left(\frac{R_r}{\left[\left(R_L + R_g + R_r\right)^2 + \left(X_L + X_g\right)^2 \right]} \right) \tag{4.33}$$

$$R_r = \frac{P_r}{P_L} R_L \tag{4.34}$$

TABLE 4.32 Characteristics of Leaky Radiating Cable Using the Physical Properties of the Cable

	Tunnel leaky radiating cable results		
Frequencies (MHz)	160	460	860
Average input power (dBW)	−26.021	−26.021	−26.02
Average input power (dBm)	3.979	3.979	3.979
Radiated power (dBW)	−57.692	−55.84	−65.126
Radiated power (dBm)	−27.692	−25.84	−35.84
Efficiency (%)	0.07	0.11	0.01
VSWR (dimensionless)	1.27	1.44	2.74
Input port real (Ω)	50	50	50
Input port imaginary (Ω)	0.0447	0.129	0.24
Output port real (Ω)	56.91	72	35.732
Output port imaginary (Ω)	15.64	2.7	42.823
Radiation resistance (Ω)	0.034	0.0521	0.0061
Maximum gain (dB)	−51.48	−37.99	−33.3
Minimum gain (dB)	−55.67	−67.767	−58.83
Θ angle (degrees)	75	85	120
φ angle (degrees)	90	90	90

The effective apertures are calculated using Equation 4.35 and the gain used to make the calculations is the minimum gain under worst-case conditions. The calculation for directivity is Equation 4.36.

$$A_e = \lambda^2 \frac{G}{4\pi} \qquad (4.35)$$

$$D = \frac{4\pi}{\Omega} \qquad (4.36)$$

The coupling loss at 2 m is calculated at 54.48 dB at 160 MHz (i.e. 82.17 − 27.69 from Table 4.33) this an example of how to calculate coupling loss at 2 m. This same procedure can be used to calculate coupling loss at 7 m. The carrier to noise ratio (CNR) is calculated as 118 −100.3 dBm = 18 dB at 160 MHz with minimum gain at 7 m. A CNR greater than 15 dB is required for good clear audio communications. Other types of digital communications (such as P25) that use a combination of phase, amplitude and frequency for coding speech have a minimum sensitivity of −124dBm. Many of the transportation authorities required the new digital technology for speech. Instead of 25 kHz bandwidth for speech, it is reduced to 12.5 kHz and the newer one at 6.25 kHz. A discussion of this technology can be found on the Internet.

TABLE 4.33 Calculation of Signal Power Using the Physical Properties of a Leaky Radiating Cable

Calculations from simulation data

| Description | Frequencies | | | Units | Remarks |
	160	460	860	MHz	
Effective aperture	0.007	0.044	0.047	cm^2	
Power at 2 m	−86.36	−84.81	−87.67	dBm	at minimum gain G
Power at 7 m	−100.3	−97.64	−100.67	dBm	at minimum gain G
Power at 2 m	−82.17	−66.83	−62.14	dBm	at maximum gain G
Power at 7 m	−95.17	−79.83	−75.14	dBm	at maximum gain G
CNR minimum G at 7 m	14.74	17.36	14.33	dB	−115 dBm baseline
CNR maximum G at 7 m	19.83	35.17	39.86	dB	−115 dBm baseline
Wavelength λ	1.9375	0.4468	0.3563	m	
Directivity	98.42	2.84	4.68		
Ω solid angle	0.13	4.42	2.69	SR rad	
Solid angle omega	419	14 322	10 292	$Degrees^2$	

FDTD requirements for modeling the 1 m leaky radiating cable section are as follows:

1. Minimum cell spacing = λ/10 (36 mm) at 860 MHz since the slots in the cable are 4 mm wide; the distance between cells must not be more than 2 mm.
2. The z direction is 500 Δz.
3. Radius of the cable is 208 mm and the field is to extend from the surface of the cable for a total of 2208 mm or 1104 Δy and Δx.
4. Δr = (Δx + Δy + Δz)$^{1/2}$ = (3Δ)$^{1/2}$ then (Δ)$^{1/2}$ = Δr /1.73 = 1.15 mm. A 1 mm distance between cells was tried that resulted in excessive calculation time.
5. The total number of cells required = 1104*1104*500 = 609 million calculations.
6. The velocity v = 3.1*10^8 m/s for (ε$_r$)$^{1/2}$ = 1.96*10^8 m/s.
7. Sampling time Δt < Δr/v = 510 ps.
8. Leaky radiating cable placement must be greater than 3 inches from the tunnel wall and at a height of 14 feet above the rails, about 2 feet above the subway car roof.

As can be observed, single processor computer would have a lengthy calculation session for modeling a 1 m section of leaky radiating cable. The calculations are not quite as difficult as they may seem because each cell is updated and in some cases no calculations are necessary, for example at edges and corners where the fields are essentially have some components that are zero. The FDTD analyses are taken as excerpts from Reference [3].

An approximation of curl is shown in Figure 4.31 using Equation 4.37 for one cell; but this is equivalent to the curl at point P. Note that the value of P does not affect the curl.

$$\bar{\nabla}x\bar{E}_{at\_P} = \frac{E_y(1,0)-E_y(-1,0)}{2\Delta x} - \frac{E_x(0,1)-E_x(0,-1)}{2\Delta y} \tag{4.37}$$

To derive Equations 4.39 and 4.40 begin with Maxwell's equations with no sources or Faraday's law and Ampere's law for one-dimensional equations (Figure 4.32), as shown below.

$$\bar{\nabla}x\bar{E} = \frac{\partial E_x}{\partial x} = \frac{\partial B}{\partial t} = \mu\frac{\partial H_z}{\partial t} \quad \frac{\partial E_x}{\partial x} = \frac{\partial B}{\partial t} = \mu\frac{\partial H_z}{\partial t} \quad \text{or} \quad \frac{\partial H_z}{\partial t} = \frac{1}{\mu}\frac{\partial E_y}{\partial x}$$

$$\bar{\nabla}xH = \frac{\partial H_z}{\partial x} = \frac{\partial D}{\partial t} + J_c = \varepsilon\frac{\partial E_y}{\partial t} + \sigma E_y \quad \text{or} \quad \frac{\partial E_y}{\partial t} = \frac{1}{\varepsilon}\left(\frac{\partial H_z}{\partial x} + \sigma E_y\right)$$

$$\frac{\partial H_z}{\partial t} \approx \frac{H_z(t+\Delta t)-H_z(t)}{\Delta t}, \quad \frac{\partial E_y}{\partial x} \approx \frac{E_y(x+\Delta x)-E_y(x)}{\Delta x}$$

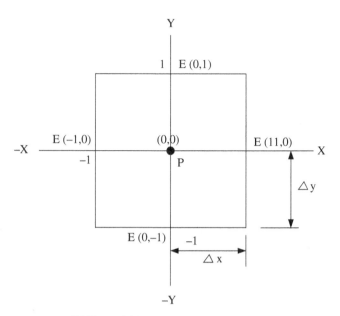

FIGURE 4.31   FDTD cell configuration

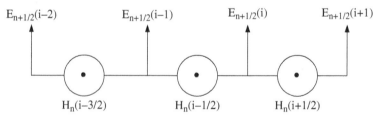

FIGURE 4.32   Block diagram of cell, one-dimensional layout

These two equations are representative of normal derivative functions; however for FDTD the derivatives must be taken from the center of each cell. This is the reason for the value of one-half in the equations that follow. The value n is the time step increment and i the distance increment.

$$\frac{E_{n+1/2}(i) - E_{n-1/2}(i)}{\Delta t} = \frac{1}{\varepsilon} \frac{H_n(i+1/2) - H_n(i-1/2)}{\Delta x} + \sigma E_{n+1/2}(i) \qquad (4.38)$$

$\dfrac{\partial E_y}{\partial t}$ = The left-hand term in Equation 4.38, $\dfrac{\partial H_z}{\partial x}$ = the right-hand term divided by Δx.

After rearranging Equation 4.38 several times, Equation 4.39 is the result. The objective here is not to present a tutorial in FDTD but to make the reader aware of how FDTD functions can be used to solve problems digitally. A more complete example of the derivation of various equations can be found in pages 560–563 of Reference [6] and online as a tutorial (www.wiley.com/go/electromagneticcompatibility).

$$E_{n+1/2}(i) = \frac{E_{n+1/2}(i)\varepsilon}{\varepsilon + \sigma \Delta t} - \frac{\Delta t}{\Delta x(\varepsilon + \sigma \Delta t)}\left[H_n(i+1/2) - H_n(i-1/2)\right] \quad (4.39)$$

$$\left[H_n(i+1/2) = H_{n-1}(i+1/2)\right] - \frac{\Delta t}{\mu \Delta x}\left[E_{n-1/2}(i+1) - E_{n-1/2}(i)\right] \quad (4.40)$$

The physical constants μ, σ and ε in Equations 4.39 and 4.40 can be a function of both time and position. These equations are taken from excerpts of lectures [3].

Current stability criterion   FDTD must have as a rule of thumb the distance step set to $\lambda_{min}/10$, where λ_{min} is the smallest wavelength or where the highest frequency expected within the computational space is also taken into account, where relative permittivity and permeability are greater than one. Sufficient time for the energy to property from one sample point to another is required. It is expressed as $\Delta t < \Delta r/v$. The sample points for FDTD are staggered in space and time because the derivative between x and x +Δx is most accurate for E and H fields at x +Δx/2. A lecture on modeling FDTD is provided by Dr. Kent Chamberlain at www.wiley.com/go/electro magneticcompatibility. The notes are provided in PowerPoint form and are complete with modeling and MathCad programs. Other programs for digital modeling may be available by searching the Internet.

4.7 EFFECT OF RAIL ON 26 PAIR CABLE BURIED ALONG RIGHT OF WAY

Occasionally along the subway right of way, instead of using the duct banks which are already present in new installations, the older ones may require buried cable in a conduit. A 26 pair cable is usually used for communications and buried along the right of way for telephone services. However with the advent of fiber optics the newer installations use optical fibers instead of copper cable. However, signals engineers still rely on 14 gauge twisted pair copper cable and sometimes duct banks are provided for the installation. In one instance the decision was made to bury a 26 pair 22 gauge telephone cable in a conduit, as shown in Figure 4.33. The objective is to determine if noise from the AC motor drives, transient current spikes or DC power supply have an effect on telephone communications between the communication and signal houses.

The layout of the rails shown in Figure 4.33 is not to scale. Distance d is 6 ft and depth shown as k is 5 ft from the rail. The distance from the rail is 7.81 ft (2.38 m). The conduit is buried below the frost line that can be 4 ft in the Northeast. The wall thickness is generally 1/8 inch and outside diameter is approximately 2 inches. The wire inside may or may not be shielded. If the conduit is 100% magnetic material $\mu_r = 1500$ hot rolled silicon steel. The soil resistance is approximately 1000 Ω/m (305 Ω/ft).

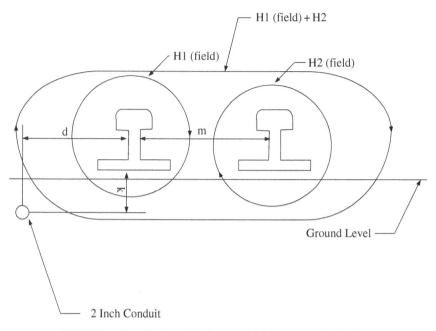

FIGURE 4.33 Track and buried conduit layout for 26 pair cable

The statistics for the 100 lb/yd rail used (1 lb = 0.37 kg, 1 yd = 0.91 m) is an impedance $720 + j\ 140\ \mu\Omega$/yd at a current of approximately 600 A at a frequency of 60 Hz and inductance of 0.41 μH/yd. Given the distance between the rails (4′ 11″, 1.502 m), the rails can be considered as a transmission line. The rail to ground level is filled with rock aggregate. This is called Balast and produces a capacitance. However, since the frequencies are from 3 to 1100 Hz the inductance and capacitance have an effect depending on these frequencies. The points of interest for this study are the harmonics of the traction motor drives in the DC current flowing through the rails during acceleration and the major harmonic of the DC power supply, that is the seventh harmonic of 60 Hz.

The older transit vehicles used DC motors that are more expensive than AC motors. The newer AC drives used in subway cars for 150 hp AC motors use pulse width modulated (PWM) drives. These PWM drives use a combination of frequency control and pulse width modulation. This produces a host of harmonics that must be filtered. There is also inrush current during acceleration. All of these anomalies produce waves that are rich in harmonics. The DC power supply produces 640 V at 800 A. The power is supplied to the transit system from two substations; one is the primary and the other a backup. At peak load a single traction motor will require 112 KW; a married pair car transit train (common in subway systems) requires four times that power or 448 KW at a current of approximately 650 A. An inrush current can be as high as 2000 A. The AC motor harmonics are discussed in Chapter 9. The harmonics due to the DC power supply are only odd harmonics (7th, 9th, 11th).

The even harmonics are canceled out in the power supply windings and the other odd harmonics are filtered. The 7th is the largest: 420 Hz at 18 A.

The most prevalent appear on the rails during speed control of the motor. The worst case is the 218 Hz harmonic: it is near the low end of the voice band for telephone. This transient inrush current is controlled during transit car acceleration.

Signaling data provided to the central control room via telephone cable to the communication house can be corrupted. This of course can become a very serious problem. Generally fiber optic cable is implemented but this is not always the case. Dispatchers must know at all times where trains are located within the system. This data is sent to the communication house and forwarded to the central control room, commonly called the operation control center (OCC). The data is sent digitally using a T-1 carrier (1.5 Mb) or RS 232, a voice band carrier.

The shielding effect of the steel conduit is calculated using Equation 4.41 with a relative permeability of 1500, wall thickness of a conduit is 1/8 inch with a radius of 1 inch.

$$
\begin{aligned}
A_{total} &= 20\log\left(1 + \mu_r * t/2r\right) \\
A_{conduit} &= 20\log\left(1 + 1500 * 0.125/2\right) = 39.53\,\text{dB}
\end{aligned}
\tag{4.41}
$$

The calculation for soil attenuation is as shown below with the following soil parameters: $\sigma = 5*10^{-4}$ and $\varepsilon_r = 3$.

$$
\left(\frac{\sigma}{\omega\varepsilon}\right)^2 = \left(\frac{5*10^{-3}}{2\pi * 218 * 8.854 * 10^{-12}}\right)^2 = 1.7*10^{11} \gg 1 \quad \text{Good conductor test.}
$$

The attenuation factor is calculated using Equation 4.42.

$$
\alpha = \sqrt{\pi f \mu \sigma} = 2\pi\sqrt{218 * 10^{-7} * 5*10^{-4}} = 6.559 * 10^{-4}
\tag{4.42}
$$

$$
A_{soil} = 20\log\left(e^{-6.56*10^{-4}*2.38m}\right) = 19.928\,\text{dB} \approx 20\,\text{dB}
$$

Total attenuation = 59.53 dB.

The number of subway cars used for this particular system is a married pair for harmonic content. As the subway cars start from rest the harmonic content increases for 10 s and reaches a peak of 80 A. After 30 s it reaches near 50% duty cycle, which is only 10 A, just part of the total current of all harmonics. The dominant frequency 80 A is the seventh harmonic of the inverter 1100 Hz. the DC power supply also has a continuous seventh harmonic at 420 Hz. The peak currents is 18 A. For d = 6 ft and k = 5, the rail is 7.8 ft (2.379 m) from the conduit. $H_\phi = H_1 + H_2 \cong 80/2\pi*2.379$ A/m = 5.34 A/m for the traction motor drive seventh harmonic and 1.2 A/m for the inverter power supply seventh harmonic.

$$
\left|\frac{H_r^2 xR}{2}\right| = P_P \quad \left(\text{Poynting vector W/m}^2 \text{ absolute magnitude}\right)
\tag{4.43}
$$

The Poynting vector for the seventh harmonic of the traction motor and the seventh harmonic of the DC power supply is 14.55 dBm/m^2 and 1.58 dBm/m^2 respectively, taking into consideration the attenuation of the soil and conduit –45 dBm/m^{2}*A$_{conduit\ surface}$ and –57.95 dBm/m^{2}*A$_{conduit\ surface}$.

For 1 m section of conduit surface, the area is 0.25 m^2 and the power delivered by the Poynting vector is {[–45 – 12 (area)] = –57 dBm} for the seventh harmonic of the traction motor drive and {[–57.95 – 12 (area)] = –69.95 dBm} for the seventh harmonic of the DC power supply. This is the calculated power delivered at the surface of the conduit. The traction power motor drive is –57 dBm (1.995 μW) and the seventh harmonic of the DC power supply is –69.95 dBm (0.257 μW). The noise level produced by the traction motor and the seventh harmonic of the DC power supply will have no appreciable effect on the telephone system and the buried cable communications.

Conclusion The noise level produced by the traction motor and the seventh harmonic of the DC power supply will have no appreciable effect on the telephone system and the buried cable communications.

4.8 RADIATION LEAKAGE FROM WAY SIDE COMMUNICATION HOUSES AND CABINETS

The construction of the communication houses is concrete with reinforcing bar embedded every four inches with doors made of stainless steel. The cabinets along the right of way are constructed of 10 gauge stainless steel with gasketed doors. The doors on the communication houses structures face away from the right of way at 2.5–5.0 m from the center of the track. Some of the cabinets at road crossing signals equipment are mounted with the door openings parallel to the track with both front and rear access doors. These cabinets are also 2.5–5.0 m from the center of the track. The differences between SCADA and signals equipment is signals equipment has both vital and non-vital logic. The vital logic has industrial hardened processors and other electronic devices and internal shielding. There are two industrial grade microprocessors, with one as the primary and the other as a secondary that shadows the primary processor. The configuration forms a failsafe feature necessary to prevent subway accidents.

The objective of the test is to determine if any of the clock frequencies from the SCADA, 10BaseT or 1000 BaseT systems radiate outside the cabinets or communication houses. All of the nine cabinets and two communication houses are located within 3 and 15 m from the right of way respectively. All of the radiation needs also be identified if it is licensed. The reason for this is to determine if any of the radiation has calls of a system malfunction of the communication, SCADA and signal (can also be located in the cabinets or communication houses) systems. Table 4.34 is a list of the harmonic emissions tests that were conducted. None of the harmonics failed the FCC part 15 standards.

The tests conducted on the communication houses and cabinets were done with a single monopole antenna and spectrum analyzer. The 1000 Base FL test was

TABLE 4.34   Communication Harmonics from SCADA Clock and Communication Cable Coding

System	Fund	2nd	3rd	4th	5th	6th	7th	Harmonic
SCADA	12	24	36	48	60	72	84	MHz
10 baseT	5	10	15	20	25	30	35	MHz
1000 base FL	500	N/A	N/A	N/A	N/A	N/A	N/A	MHz

unnecessary because this is the transmitter for the fiber optics that connected communications to the outside world. These consisted of a single module that had to pass FCC part 15 testing during the manufacturing phase of the particular part. The fiber optic receiver is a very high gain and high impedance device that is sensitive to electromagnetic E fields. These are generally a hardened industrial grade, but care must be taken when operating cell phones or two-way radios near these devices. Operation near these devices may result in corruption of the digital bit-stream and therefore produce erroneous data. Printed circuit boards can easily transfer transmission data to the receiver by having a standoff that may be used for an adjustment to the AGC circuit or another function that may cause radiation of the transmitter to impinge on the receiver. The high gain of the receiver will easily pick up such radiation corrupting the transmission. This is something that happened to the author during a design phase.

The next series of tests required is to test for radiation near the communication houses and cabinets with all doors. This is done to identify all radiation in the vicinity in the event that the extraneous noise is found to result in malfunctions of any of the microprocessors or other electronics equipment in the communication houses or cabinets. All wiring was in conduit or duct banks and prevented the radiation from these sources identified in Table 4.35 from entering the enclosures. This is equivalent to a survey taken previously in Section 4.5. There are only nine entries shown in Table 4.35. This is a summary of the largest signals encountered during the survey.

Since all these radios are licensed, they are legal to transmit at the various frequencies as identified. A column has been added to show the radios are operating above the FCC part 15 class A limits. The only reason this is shown is that it will identify any radiation that may be picked up by exposed wiring or door openings that present an aperture antenna to culprit frequencies. The length of the opening for an aperture is between 0.36 and 7.2 m. This leaves a wide variety of possible aperture lengths; for example a poorly fitted door or a hole cut in the enclosure for the purpose for installing a larger air-conditioner or other equipment on the outside of the cabinet with no gasket or seal for EMI protection. Another possibility is operating the equipment with the doors open, in particular cabinets along the right of way that open on both sides.

Often maintenance personnel use 2W two-way radios. They should be discouraged from using these radios in communication houses or near cabinets with the doors open. Cell phones used in communication houses or close to cabinets when the doors are open must also not be permitted. At 1 m a cell phone will radiate 5.4 V/m.

TABLE 4.35   The Nine Largest Survey Signals

Measure-ment no.	Frequency (MHz)	Electric field (μV/m)	Emission > standard	Remarks
1	662.83	271	71	Frequency of broadcast radio (TV) 614–698 MHz
2	940.50	680	480	Land mobile private personal communications frequency 940–941 MHz
3	571.83	371	171	Land mobile band 470–608 MHz
4	122.70	185	35	Aeronautical mobile transmit band 121.9375–123.5875 MHz
5	786.50	350	150	Broadcast radio (TV) 776–794 MHz
6	861.16	388	188	Land mobile private fixed frequency 854–861 MHz
7	120.90	161	11	Radio navigation 108.0–121.9375 MHz
8	43.05	107	7	Private land mobile band 42.0–43.69 MHz
9	830.83	1839	1639	Land mobile public fixed frequency 824–849 MHz

This type of radiation can be easily picked up by circuit board tracks or exposed wiring. It may or may not be detrimental to the equipment; however, as electronics components get smaller due to improved manufacturing techniques, the threat of malfunctions due to cell phones and two-way radios is greater.

4.9   LIGHTNING ROD GROUND EMC INSTALLATION

Often at stations lightning rods are installed to protect electronics equipment. The scope of this case study is to determine whether ground loops are present that can impact the performance of electronics equipment. In this particular case the racks are insulated from the electronics cabinet, as shown in Figure 4.34.

The construction data is as follows: insulation material area is approximately 1 ft² with a thickness of 0.5 inch, material used for insulation is a phenolic substance (physical parameters $\sigma = 10^{-14}$ S/m, $\varepsilon_r = 3$), the insulation on the communications bus bar has a capacitance of 10 pF for each standoff, the bus bars are all copper, the rack is constructed of cold rolled steel ($\sigma_r = 0.34$, $\mu_r = 2000$) and the enclosure is constructed with the same materials.

The wire sizes are listed in Table 4.36. There is a large variation in the wire. The communication ground grid and station ground grid in the diagram in Figure 4.34 are generally much larger than shown. As can be observed in the diagram, wire number 6 is used to prevent a shock hazard between the lightning rod pole and the cabinet

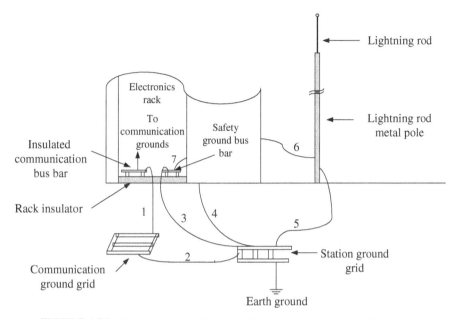

FIGURE 4.34   Lightning rod safety connection to cabinet and grounding layout

TABLE 4.36   Wire Resistance Calculation for Connection in Figure 4.34

Wire number	Gauge	Resistance (Ω/1000 ft)	Length of wire [feet (m)]	Resistance (mΩ)
1	2	0.162	10.0 (3.05)	1.62
2	2	2.59	16.0 (5.0)	0.91
3	2	0.162	22.0 (6.71)	3.56
4	10	1.039	19.5 (5.95)	2.03
5	2	0.162	5.2 (1.59)	0.842
6	14	2.626	4.0 (1.22)	10.50
7	14	2.626	0.5 (0.1525)	1.31
Mechanical connections				30.0

containing the electronics rack. In the case being examined the distance was about 2 feet. The wiring as shown in Figure 4.34 has a series of ground loops that are to be analyzed; the electronics rack within the cabinet contains audio equipment that may be affected by noise on the grounds. The communications ground grid and station ground grid are located below ground at greater than 4 feet. This is due to the frost line in the Northeast near where the subway system was installed. The station grid is connected to the rebar matrix in the concrete; these are all welded joints. Mechanically connected joints are on the bus bar racks and cabinet connections with a maximum resistance due to the connection of 30 mΩ, as shown in Table 4.36. These ground loops are listed in Table 4.37, with the particular area of each group estimated. The capacitance $C_{rack\ to\ cabinet}$ is calculated using Equation 4.44, using the parameters given previously. Since the two capacitors for the standoffs are parallel, they are equivalent to 5 pF and the rack to capacitance is 194 pF >> 5 pF. The total capacitance in series is approximately 5 pF. The length of the wire shown in Table 4.36 is not the actual perimeter of the loops because the communication ground and station ground may be tapped at different points on the grid. For example the station ground grid encircles the station and may be tapped at various points. The same holds true for the communication ground grid, but the communication ground does not encircle the communication house. The approximation for the loop areas were taken from drawings provided on the project. It could not be included in the book because it was unavailable.

$$C_{rack\_to\_cabinet} = \frac{\varepsilon_r \varepsilon_0 A}{s} = \frac{3*8.854*10^{-12}*9.3*10^{-2}}{1.271*10^{-2}} = 194 \text{ pF} \quad (4.44)$$

Test results for third rail (hot rail) current were obtained at the fundamental frequency 218 Hz and the seventh harmonic 1526 Hz. The maximum number of cars for this particular project was four cars or two married pairs, which is common practice in subway systems. When three cars are connected to form a train the fundamental current increases in additional cars cause the current to decrease.,Tthis is due to the increased inductance in the rails that begin to affect the rise time of the fundamental frequency current. This is also true for the seventh harmonic frequency. These are peak currents that appear on a hot rail when the train starts. The current increases until it peaks after 15 s, which is shown in Table 4.38. In the next 10 s, the current decreases to its normal running speed. These peaks occur after 244 m of movement along the rail.

The equation used to determine electromotive force (e.m.f.), due to the magnetic flux captured by the loop, is calculated using Faraday's Law, Equation 4.45. This equation is also considered one of Maxwell's equations in Chapter 3. The current envelope takes on a Gaussian shape with a spike of current occurring at 15 s. The spike is very short duration, in the millisecond range. The peaks given in Table 4.38

TABLE 4.37   Loop Areas and Impedance for Three Paths in Figure 4.34

Ground number	Ground path	Approximate area	Total resistance
1	1, 2, 4, $C_{com\ bus\ bar}$, $C_{rack\ to\ cabinet}$	7 m²	0.125Ω, $-jX_c$ > Meg Ω
2	1,2,3,7, $C_{com\ bus\ bar}$, $C_{rack\ to\ cabinet}$	9 m²	0.126Ω, $-jX_c$ > Meg Ω
3	1, 2, 5, 6, $C_{com\ bus\ bar}$, $C_{rack\ to\ cabinet}$	6 m²	0.134Ω, $-jX_c$ > Meg Ω

TABLE 4.38 Traction Motor Current at Fundamental and Seventh Harmonic versus Cars from Start to 244 m

	Fundamental, 218 Hz		Seventh harmonic, 1526 Hz	
Frequencies	0 Hz	244 Hz	0 Hz	244 Hz
Distance	0 m	244 m	0 m	244 m
Number of cars				
1	37 A	27 A	0.20 A	0.17 A
2	60 A	10 A	0.38 A	0.24 A
3	80 A	30 A	0.38 A	0.24 A
4	47 A	22 A	0.27 A	0.15 A

are not of the spike but of a gentle curve similar to the Gaussian shape alluded to previously. Equation 4.45 expresses the peak voltage that occurs. The peak values will be used to do the calculations.

$$\oint_C \overline{E} * d\overline{L} = -\frac{\partial}{\partial t} \iint_S \overline{B} * d\overline{S} \, \text{Volts} \quad \text{Where } \overline{B} = \mu \overline{H} = \mu \frac{I}{2\pi r} \hat{n} \quad (4.45)$$

The peak voltages are the ones of most concern. These can cause spikes to occur on telecommunications (T − 1). The result is bits are dropped or corrupted during data transmission; also audio can have annoying noise spikes. It must be kept in mind that noise on ground buses cannot be filtered easily. As the number of cars increases beyond three the hot rail inductance increases due to the extra length of track the cars occupy, so that longer strings of cars will act similar to an inductance and smooth out the current spikes.

$$\psi_0 = -\iint_S \overline{B} * d\overline{S} \quad (4.46)$$

The results of Equation 4.46 are in Table 4.39.

The intermediate calculation of peak magnetic flux density provided in Table 4.39 is calculated using Equation 4.46. This provides the reader with intermediate results to trace the calculations. Magnetic flux density is expressed in webers (Wb); the surface area is approximated for each loop. Assumptions are the following: (i) the loops are all in the same plane, which is a good approximation because the station ground and communication ground are generally fairly close to the cabinet and in about the same plane as the cabinet bus bars and (ii) the center of all the loops is about 2 m below grade from the hot rail. The geometry to calculate the area was done through the use of squares and triangles with dimensions provided in the project drawings. Often data is unobtainable and must be estimated using drawings provided by the authority or the general contractors. As can be observed in Table 4.39, the magnetic flux density peak occurs with two or three cars trains. The seventh harmonic exhibits the same trend a reduction in peak magnetic flux density due to train length occupying the rails.

TABLE 4.39   Loops of Magnetic Flux (in Webers; Wb) in Figure 4.34

	Fundamental, 218 Hz		Seventh harmonic, 1526 Hz	
Frequencies	0 Hz	244 Hz	0 Hz	244 Hz
Distance	0 m	244 m	0 m	244 m
Number of cars				
	Loop 1			
1	2.59E-05	1.89E-05	1.12E-08	1.19E-07
2	4.20E-05	7.00E-06	2.66E-07	1.68E-07
3	5.60E-05	2.10E-05	2.66E-07	1.68E-07
4	3.29E-05	1.54E-05	1.89E-07	1.05E-07
	Loop 2			
1	3.33E-05	2.43E-05	1.44E-08	1.53E-07
2	5.40E-05	9.00E-06	3.42E-07	2.16E-07
3	7.20E-05	2.70E-05	3.42E-07	2.16E-07
4	4.23E-05	1.98E-05	2.43E-07	1.35E-07
	Loop 3			
1	2.22E-05	1.62E-05	1.44E-08	1.02E-07
2	3.60E-05	6.00E-06	3.42E-07	1.44E-07
3	4.80E-05	1.80E-05	3.42E-07	1.44E-07
4	2.82E-05	1.32E-05	2.43E-07	9.00E-08

$$V = -\frac{\partial\left[\psi_0 \cos(\omega t)\right]}{\partial t} = \left[\psi_0\left(2\pi\omega\right)\right] * \sin\left[\psi_0 \cos(\omega t)\right] \qquad (4.47)$$

The most common train car configuration is a two car married pair, as previously alluded. However, as observed in Boston, train configurations are sometimes made up with five or six cars. In a single car application it is very difficult to find the cause of personnel efficiency, that is more crews are necessary to operate the equipment.

There are other harmonics besides the fundamental and the seventh, but these two are the dominating factor for calculating ground noise. This is not the same project that was discussed in Section 4.7. These results are from a transit project in California. A whole range of subway cars may found in this country, some of the older cars with DC traction motors and multiple AC traction motors by various manufacturers. Some cars are designed in Europe with parts for fabrication sent to the United States to be assembled. The traction motors are not standard but they must meet minimum safety criteria to be sold in the United States. Their electromagnetic emission requirements are fairly stringent. Most of the emissions have been tested by a car manufacturer. It is not usually done by EMC engineers unless they are employed by car manufacturers.

$$V_{Peak} = (\psi_0 \omega) \tag{4.48}$$

The final result is in Table 4.40. The equation for calculating the time-varying functions of voltage as a function of a time-varying magnetic field is Equation 4.47. A table of values is provided in Table 4.40 for the peak voltage only. As can be observed in the voltage trends, both the fundamental and seventh harmonics are dominated by two- and three-car trains. The resistances in the path for the three loops is very close to 0.1 Ω. The capacitance in the communication bus bar is very close to an open circuit at the frequencies involved in the cabinet.

A problem that can occur in the system is not from the traction motors but from the capacitance in the insulation under the rack in the communications bus bar standoffs. So far the traction motors are the only items analyzed. Since the lightning rod will short-circuit a lightning stroke to ground,. the stroke itself with the high potential may weaken the insulation in the capacitance parts of the loop. As the insulation becomes more leaky an impedance unknowingly drops and noise currents may flow that result in a short-circuited version of the loop. As an example, if the insulation is compromised and the impedance drops through the capacitance to a short-circuit or close to a short-circuit, 35 mA will flow through to ground. This is a substantial

TABLE 4.40 Loop Noise Voltages (mV) Coupled in Due to Magnetic Flux in Figure 4.34

	Fundamental, 218 Hz		Seventh harmonic, 1526 Hz	
Frequencies	0 Hz	244 Hz	0 Hz	244 Hz
Distance	0 m	244 m	0 m	244 m
Number of cars				
	Loop 1			
1	3.55	2.59	0.107	1.14
2	5.75	0.959	2.55	1.61
3	7.67	2.88	2.55	1.61
4	4.51	2.11	1.81	1.01
	Loop 2			
1	4.56	3.33	0.138	1.47
2	7.40	1.23	3.28	2.07
3	9.86	3.70	3.28	2.07
4	5.80	2.71	2.33	1.29
	Loop 3			
1	3.04	2.22	0.138	0.978
2	4.93	0.822	3.28	1.38
3	6.58	2.47	3.28	1.38
4	3.86	1.81	2.33	0.863

current. The worst-case example is when the short-circuit is not complete and the maintenance personnel only know that the equipment ground has been getting noisy and it is discovered as coming from the communication or station ground.

Several pieces of equipment have an inverter. When this equipment is used on more than one 120 AC 60 Hz branch circuits of the station power, the station ground may be connected to the neutral even though the three phases of the AC power are balanced. The phase with the inverter may have the frequency of the inverter on the neutral and it will appear on the station ground grid. This can be mistaken for a ground loop of the types presented here.

5

CASE STUDIES AND ANALYSIS OF LRT VEHICLE AND BUS TOP ANTENNA FARM EMISSIONS AND OTHER RADIO RELATED CASE STUDIES

5.1 INTRODUCTION

On several occasions transportation vehicles require several antennas mounted on the roof. The ground plane is stainless steel or cold rolled steel. These ground planes represent a challenge for most designs of antenna farms, that is multiple antennas located in one area. The transmission during normal operation of a single radio may interfere with the reception of the others. Each antenna presents an object that can reflect radio transmission back into the transmitter that is operational and noise on the ground plane because of the poor material electrical properties.

Some examples are as follows:

1. Trucks with a cab made of cold rolled steel will have paint or other protective coatings that will present poor grounding for the antenna, for example a maintenance vehicle.
2. Buses may be constructed of aluminum alloys that may be a good ground plane but will be painted with a fiberglass protective coating.
3. Railway locomotives often have large antenna farms mounted on stainless steel that is magnetic, therefore magnetic fields may circulate on the surface of the ground plane.

Electromagnetic Compatibility: Analysis and Case Studies in Transportation, First Edition.
Donald G. Baker.
© 2016 John Wiley & Sons, Inc. Published 2016 by John Wiley & Sons, Inc.
Companion website: www.wiley.com/go/electromagneticcompatibility

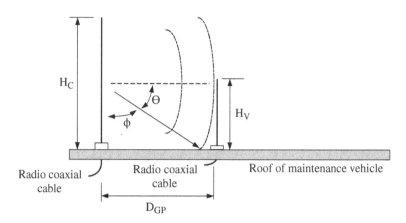

FIGURE 5.1   Ground plane reflection point

The number of radios implemented for transportation have various bandwidths and frequency bands for service. GPS antennas are often mounted on the surface of the vehicles and the noise induced by grounds from ground planes cannot be easily filtered. Radio such as used by police, fire and emergency medical services (EMS) have a band of frequencies from 150 to 800 MHz depending on the age of the equipment. Some forms of spread spectrum, short range, unlicensed radios are used along the right of way in subway and freight transportation systems for monitoring trains and communicating with signals equipment. With the advent of cell phones and other personal communication devices that are used near these antenna farms the ground noise may become intolerable. Therefore, the engineer working with this technology must constantly be aware of outside influences that may present noise on equipment or devices that may result in malfunctions.

Three examples of how radiation may affect ground plane noise are presented as part of this introduction. The first is the radio transmitting antenna with a receiving antenna nearby mounted on cold rolled steel ground plane; the second is a similar configuration with a stainless steel ground plane; and the last example is the same configuration with a copper ground plane. The relative permeability of copper, stainless steel and cold rolled steel are $\mu_{copper} = 1$, $\mu_{stainless\ steel} = 180$, *$\mu_{cold\ rolled\ steel} = 2000$, respectively; and the conductivity is $\sigma_{copper} = 5.7 \times 10^7$, $\sigma_{stainless\ steel} = 10^6$, $\sigma_{cold\ rolled\ steel} = 2 \times 10^6$. These different characteristics of the ground plane materials will give the reader a contrast between the noise components generated.

The radio used in all of the examples has a 600 mW monopole antenna (10 cm height, H_C) with an ERP 0.96 W, a beam width of 78°. The distance D_{GP} between the transmitting (culprit) and receiving (victim) antenna is 6 inches (15.25 cm). The transmitting frequency is 706 MHz; and the victim antenna height, H_V, is 6 inches. Figure 5.1 gives a description of the configuration for all examples. The electric field density for all the examples is provided in Table 2.1 in Chapter 2 for the magnitude, that is $E_0 = 35$ V/m and $H_0 = 0.928$ A/m.

---

*μ cold rolled steel is a function of frequency it is approximately 1 at 1 GHz

5.2 CIRCULATION CURRENTS IN THE GROUND PLANE

The ground plane in this example is cold rolled steel as installed on the roof of a maintenance truck that may have several antennas to various public authorities such as fire, police, EMS and transportation. The particular radio used in this application is for a subway radio 0.25 λ antenna mounted on a truck.

The electric field intensity is approximately 35 V/m and the magnetic field intensity is 0.928 A/m at the center of the beam from the transmitting antenna. This is considers at 0° since the beam width is 78°. The lower part of the beam below the center is 39°; this is the 3 dB point for the lower half of the beam, which is the point of interest for examining the magnetic field impinging on the ground plane.

Using trigonometry the calculation for the point where the lower edge of the beam intersects the ground plane is as shown below in Equation 5.1. The intersection of the electric field with the ground plane is at the base of the receiving antenna. The electric field E is 17.5 V/m; this also includes the roll off at the edge of the beam. The H field is normal to the xz plane, with the y direction into the page.

$$D_{GP} = 7.94 * \sin\left(51°\right) \approx 6 \text{ inches} \tag{5.1}$$

Where $\theta = 39°$, $\phi = 51°$ and $7.94 = 5/\sin 39°$.

The known values for the calculations are the angle of incidence ϕ at 51° (this is the angle between the vertical and where the wave is impinging on the ground plane) and the electric field intensity of 17.5 V/m. The assumption made is that $\Gamma_{RC} \approx -1$. The reflected wave is not shown in Figure 5.1 because the point of interest is the H field at the surface of the ground plane at the point where the electric field intensity of the wave impinges on the ground plane. In the calculations the reflected electric field intensity is used. The derivation of the equation for the magnetic field intensity H at the boundary between the ground plane and air is as follows:

In Equations 5.2 and 5.3 the electric field intensity at the y boundary between the air in the ground plane interface is equal to zero; this is as shown in Equation 5.4. Inserting Equation 5.2 into Maxwell's equation $\bar{\nabla}x\bar{E} = -j\omega\mu\bar{H}$, then taking the derivative of the electric field intensity E and using as some algebraic manipulation the resultant equation for the incident magnetic field intensity H_i is as described in Equation 5.5. This same procedure is used to derive Equation 5.6. These two equations are then used to find the sheet current at the surface z = 0 that is tangential to the ground plane surface, Equation 5.7.

$$\bar{E}_i = \hat{a}_y E_0 e^{-j\beta_0\left[x\sin(\phi_i)+z\cos(\phi_i)\right]} \tag{5.2}$$

$$\bar{E}_r = \hat{a}_y E_0 \Gamma_{RC} e^{-j\beta_0\left[x\sin(\phi_i)-z\cos(\phi_i)\right]} \tag{5.3}$$

$$\left(\bar{E}_i + \bar{E}_r\right)^{tan}_{z=0} = 0 = \bar{E}_i = \hat{a}_y E_0 \left(1+\Gamma_{RC}\right)e^{-j\beta_0 x\sin(\phi_i)}, \quad \text{Where} 1+\Lambda_{RC} \approx 0 \tag{5.4}$$

$$\bar{H}_i = \frac{E_0}{377}\left[-\hat{a}_x \cos\left(\phi_i\right)+\hat{a}_z \sin\left(\phi_i\right)\right]e^{-j\beta_0\left[x\sin(\phi_i)+z\cos(\phi_i)\right]} \tag{5.5}$$

$$\bar{H}_r = \frac{E_0}{377}\left[\hat{a}_x \cos(\phi_i) + \hat{a}_z \sin(\phi_i)\right]e^{-j\beta_0\left[x\sin(\phi_i)+z\cos(\phi_i)\right]} \tag{5.6}$$

$$J_S = \hat{n}x\left(H_i + H_r\right)_{z=0} = -\hat{a}_z x\left[\hat{a}_x\left(H_i + H_r\right) + \hat{a}_z\left(H_i + H_r\right)\right]_{z=0} \tag{5.7}$$

$$J_S = -\hat{a}_z x\hat{a}_x\left(H_i + H_r\right) = \hat{a}_y \frac{2E_0}{377}\cos(\phi_i)e^{-j\beta_0 x\sin(\phi_i)} \tag{5.8}$$

The final equation for the surface current density is Equation 5.8. Inserting numbers in Equation 5.8 is as shown below. The depth of the surface current density into conductor material is different because of the skin affect I/Δz.

$$J_S = \hat{a}_y \frac{17.5}{377}\cos\left(51°\right)e^{-j\frac{2\pi}{0.44}(0.152)\cos\left(51°\right)}$$

$$J_S = \hat{a}_y 29.21\underline{|78°}\ \text{A/m}$$

The next calculation to consider is the skin depth of the current in a copper ground plane; the numbers are inserted in Equation 5.9. Copper is calculated first because all other materials have only changes in the relative permeability and conductivity. As can be observed in all of the equations for skin depth, copper is the largest and cold rolled steel is the lowest.

$$\Delta z_{copper} = \sqrt{\frac{2}{\omega\mu\sigma}} = \sqrt{\frac{2}{2\pi * 705 * 10^6 * 4\pi * 10^{-7} * 5.7 * 10^7}} = 2.51 * 10^{-6}\ \text{m} \tag{5.9}$$

$$\Delta z_{copper} = 2.51\mu\text{m}\left(0.00251\text{mm}\right)$$

$$\Delta z_{cold\_rolled\_steel} = 2.51\mu\text{m}\left(0.00251\text{mm}\right)\sqrt{\frac{1}{2000 * 3}} = 32.4\ \text{nm} \tag{5.10}$$

$$\Delta z_{stainless\_steel} = 2.51\mu\text{m}\left(0.00251\text{mm}\right)\sqrt{\frac{1}{600 * 4}} = 51.2\ \text{nm} \tag{5.11}$$

The intrinsic wave impedance again begins with copper as shown in Equation 5.12 after a calculation is made for copper as can be observed in Equations 5.13 and 5.14; only the relative terms for the materials need to be considered. Note, the final result in all three equations have an imaginary term for equivalent inductive reactance. The magnetic intensity H field is very close to the connection by the victim antenna location. This antenna connection will have a capacitive reactance that may be small but it still may resonate with any inductance to produce radiation from the ground plane. The depth of skin effect penetration is very small for cold rolled steel and very large for copper, which means most of the current for cold rolled steel flows close to the surface as compared to copper.

Connectors and antenna mountings are very close to the surface of any ground plane. This means ground noise in cold rolled steel will be very high at the surface as compared to copper; the best trade-off is stainless steel. It is both non-corrosive and less likely to form a diode between the dissimilar metals of ground plane and connector. Most connectors are made with stainless steel, therefore the diode action may not be a problem.

Not considered in this study is the reflection from the victim antenna; this is considered elsewhere in this section. Reflections from the antenna are similar to radar cross-sections that present a scattering signature for the antenna. The reflections (scatter) may give phasing resultant peaks in the carrier of the transmitter or a reduction in the carrier magnitude. Another anomaly with rooftop installations is when antennas or other obstructions are too close or within the near field of the transmitter antenna. This results in an antenna or obstruction acting as a director for the electric field. Instead of the omnidirectional antenna the antenna may be directional or highly directional, depending where within the near field the culprit antenna or obstruction is located. Some of these anomalies are discussed in this section.

The permeability and permittivity of materials will have an imaginary part for frequencies above 1 GHz. These equations do not work well above that value. In certain instances in the case studies the imaginary permeability and permittivity will be investigated to determine their effect on radio transmission fields. One of the problems that occur that would also be investigated is radiation from the edge or corner of the ground plane. This in particular will have a tendency to distort the electric fields. If antennas are too close to the edge of the ground plane they will distort the radiated electric and magnetic fields. This is considered in some of the case studies in this section.

$$Z_{intrinsic\_copper} = \sqrt{\frac{\omega\mu}{2\sigma}}\left(1+j\right) = \sqrt{\frac{\pi\,705\,MHz * 4\pi * 10^{-7}}{5.7 * 10^{7}}}\left(1+j\right) \qquad (5.12)$$

$$Z_{intrinsic\_copper} = \left(6.99 + j6.99\right)m\Omega$$

$$Z_{intrinsic\_cold\_rolled\_steel} = 6.99\,m\Omega\sqrt{\frac{\mu_r}{\sigma_r}} = 6.99\,m\Omega\sqrt{\frac{75}{2}}\left(1+j\right) \qquad (5.13)$$

$$Z_{intrinsic\_cold\_rolled\_steel} = \left(0.428 + j0.428\right)m\Omega$$

$$Z_{intrinsic\_stainless\_steel} = 6.99\,m\Omega\sqrt{\frac{\mu_r}{\sigma_r}} = 6.99\,m\Omega\sqrt{\frac{600}{4}}\left(1+j\right) \qquad (5.14)$$

$$Z_{intrinsic\_stainless\_steel} = \left(85.61 + j85.61\right)m\Omega$$

5.3 ANTENNA INSTALLATION ON A RADIO MAST CASE STUDY

The objective of this study is to determine whether multiple UHF radios can be added to a radio mast without a sacrifice in performance due to the VHF radio. Several items that present a problem are as follows:

1. The UHF radios may saturate the VHF receiver and reduce the range of operation.
2. The VHF radio operational performance minimum is 7 miles with an adequate signal at a receiver sensitivity of −118 dBm.

---

*μ cold rolled steel = 75 at 705 MHz

3. The present radio operational performance is up to 30 miles with an adequate signal, also at −118 dBm.

4. The UHF radios may also suffer from degradation and quality of transmission due to the VHF radio in close proximity.

The VHF authority is willing to sacrifice some range of operation if the radios can all be as shown installed on the same mast, see Figure 5.2. If however this cannot be achieved the radios will be separated onto a separate mast away from VHF radio location.

The VHF radio antenna construction is a series of folded dipole antennas stacked in such a way as to provide an omnidirectional antenna suitable for the project. The VHF transmitter antenna is located at the bottom of the mast and the receiver is at the top of the mast. The UHF antennas are located near the center of the mast between the VHF transmit and receive antennas. All these antennas have dimensions accurately shown in Figure 5.2. This diagram however is not to scale but the dimensions are correct. Table 5.1 provides the relationship between antenna positions and wavelengths and where the electric field intensities are located. The Fire Department is one of the most critical for obvious reasons because of safety concerns. It has a high gain amplifier and antenna gain combination. As a first pass at the analysis all UHF antennas will be assumed to have an ERP of 5 W with the exception of the Fire Department. The VHF transmitter is 30 W minus all the losses, giving 28.1 W at the transmit antenna.

Table 5.1 has a great deal of information for use with case studies at these frequencies. Wavelength is in both meters and feet, as shown in column 2. Column 3 is the reactive near field; this is a radiating Frensel zone. When other antennas are within this zone they interact with the victim antenna. Column 5 is the far field; this is the one usually used in most calculations. Column 4 provides reactive radiating limits for the antenna. Table 5.2 gives all the distances from the various antennas that are used in the calculations.

The longest wavelength is from the 160 MHz radio transmitter. Using this as a guide none of the other radio antennas are within approximately 10 inches of each other. This is the approximate limit of the far field for the VHF 160 MHz. Since the other antennas are physically smaller they are all outside the far field limit. The next item to consider is the magnitude of the power at each receiver from each of the radio transmitters. For example, the 160 MHz transmitter imparts a certain amount of power to the antennas of all the other radios. The question to be answered is whether this power will saturate the receivers sufficiently to reduce the range of operation. Also examine the other radios to determine the effect on the 160 MHz VHF receiver. The DB 224 antenna has the same pattern as the 160 MHz in the antenna specifications sheet.

As can be observed in Figure 5.3 the beam width is fairly narrow, only about 20°. The lighter lines in the figure are a normal dipole antenna; however, the vertical pattern for the stacked folded dipole array antenna has lobes that are greater than a normal dipole. Since the two VHF antennas (i.e. transmit at the bottom and receive at the top) are the same, both have the same radiation pattern. A top view of the

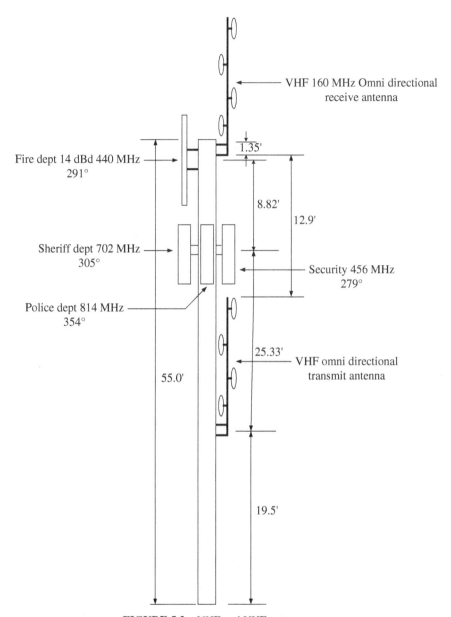

VHF 160 MHz Omni directional
receive antenna

1.35'

Fire dept 14 dBd 440 MHz
291°

8.82'

12.9'

Sheriff dept 702 MHz
305°

Security 456 MHz
279°

Police dept 814 MHz
354°

25.33'

VHF omni directional
transmit antenna

55.0'

19.5'

FIGURE 5.2   VHF and UHF antenna mast

radiation pattern is shown in Figure 5.4. The darker line is the antenna on the pole at present, the broken line is for a half wavelength 9 dB gain antenna that can be considered for use in the future.

Although the antenna structure for the VHF array appears to be a planar staggered array of folded dipoles, they are actually staggered in their angular positions on the

TABLE 5.1   Near Reactive and Radiating Near Fields and Far Fields for Antenna on Mast (Figure 5.2)

Frequency	Wavelength	Reactive near field	Radiating near field	Far field
Equations	$\lambda = 310/f_{MHz}$	$d_{react} = 0.62(D^3/\lambda)^{1/2}$ m		$d_{far\,field} = 2(D^2/\lambda)$ m
Conditions		$d_{react} > d > 0$	$d_{far\,field} > d > d_{react}$	Infinity $\geq d_{farfield}$
160 MHz	1.938 m (6.25 ft)	0.1502 m (0.49 ft)	0.242 > d > 0.1502 (0.79) > d > (0.49)	0.242 m (0.79 ft)
440 MHz	0.705 m (2.31 ft)	0.0546 m (0.18 ft)	0.088 > d > 0.0546 (0.29) > d > (0.18)	0.088 m (0.29 ft)
456 MHz	0.680 m (2.23 ft)	0.0527 m (0.17 ft)	0.085 > d > 0.0527 (0.28) > d > (0.17)	0.085 m (0.28 ft)
702 MHz	0.442 m (1.45 ft)	0.0343 m (0.11 ft)	0.055 > d > 0.0343 (0.18) > d > (0.11)	0.055 m (0.18 ft)
814 MHz	0.381 m (1.25 ft)	0.0295 m (0.097 ft)	0.048 > d > 0.0295 (0.16) > d > (0.097)	0.048 m (0.16 ft)

TABLE 5.2   Distances at Various Angles Between VHF and UHF Radio Antennas

Antennas	Radio antenna	Gain angle	Remarks
VHF Tx	19.5 feet (5.9 m)		From bottom of pole
VHF length Tx + Rx	21.25 feet (6.48 m)		Both same length
VHF Tx to Rx	12.9 feet (3.93 m)		
VHF Tx to UHF	4 feet (1.22 m)	$180° - 140° = 40°$	Max. toward UHF radio
UHF to VHF Rx	19.5 feet (5.95 m)	30°	Max. toward VHF radio Rx

mounting as well. The phasing is critical as one might surmise. The lobes shown in a vertical pattern are dependent on the spacing between folded dipoles in angular positions. Therefore, great care must be taken when installing these array antennas not to get objects close to any of the folded dipoles. This may modify or change the radiation pattern, both vertical and horizontal planes. In particular the installation definitely should not have obstructions in the reactive near field that will act as a director to one of the folded dipoles.

Shown in Figure 5.5 is a top view of the UHF radio radiation patterns. These radios are all directional where 0° maximum gain is present in the pattern shown in Figure 5.5 with a 6 dB gain in the direction they are facing on the pole. Since the transit authority that runs a subway system owns the pole the VHF radio is their only concern. The task is to determine if the VHF radio was affected by the installation of all the UHF radios, only one belonging to the authority. However it is prudent to investigate how all the radios will be affected by the mounting of the antennas on the pole (radio mast). Many of these other authorities, such as the Fire Department, will eventually ask for a similar analysis of the radios. If the analysis and testing indicates

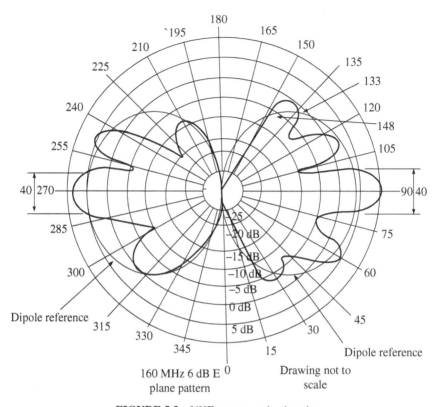

FIGURE 5.3 VHF antenna gain elevation

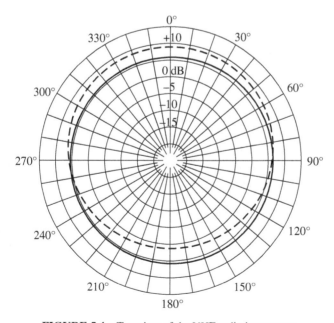

FIGURE 5.4 Top view of the VHF radiation pattern

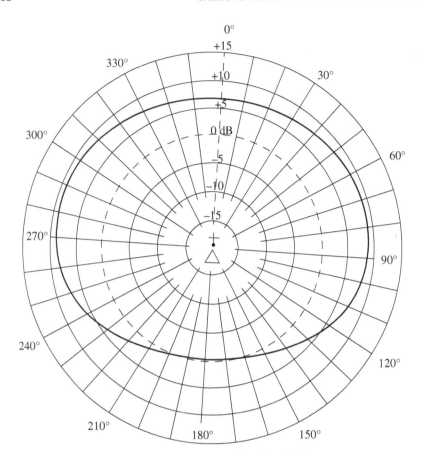

FIGURE 5.5 Top view of the UHF radio radiation pattern

serious problems, the various authorities must consider a separate place to mount their radios, perhaps on a pole or a building elsewhere. The FCC in Chapter 1 regards the VHF radios as a victim and any other radios installed afterwards as a culprit. The culprit must ensure that the victim is not affected by the installation, or if it is the victim must agree to allow the change.

The peak radiation of the VHF 160 MHz antenna closest to the UHF band radio antennas is approximately between 140° and 148° with a gain –6 dB and –9 dB respectively. Since these gains are dealing with the E field from the power point of view the peak at 140° is where the gain is one-half. Therefore the power is one-quarter of the peak radiated power of 28.1 W, or 7.02 W. Using this value, the calculation can be made of their proximate electric field at the UHF radio antennas. The distance the UHF radios are from the VHF antenna center is approximately 19 feet (5.7 m). The E field calculation at the center of the VHF array is as calculated below in Equation 5.15. This is the magnitude of the wave impinging on the UHF antenna, the vertical component in the plane of the E field of both radio antennas (center to center).

$$\bar{E}_{at\_UHF\_Ant.} = \frac{\sqrt{30 * ERP_{VHF\_Lobe}}}{R_{dist\_UHf\_ant}} = \frac{\sqrt{30 * 5.6}}{1.22} = 10.6 \text{ V/m} \qquad (5.15)$$

The steps required to find the power at the UHF antenna are described below. Step (1) is fairly straightforward: knowing the center to center distance and the angle of the lobe, the distance to the intersection with the UHF field is calculated. Step (2) with trigonometry the known data using the graphic data from Figure 5.3 a calculation can be made using Equation 5.15 to determine the E field at the intersection. The Poynting vector is calculated using the results from step (3). Step (4) provides the power density at the antenna; this is multiplied by the aperture calculated in step (6). Step (10) calculation provides a correction for the distance from the intersection to the UHF antenna; and step (11) provides a result of –34.32 dBm.

1. Distance to intersection = 1.22/cos(40°) = 1.59 m.
2. The distance from the intersection with the center of the UHF antenna E field using the law of cosines, $d_{intersection}$ = [(1.22^2 + 1.59 – 1.22*1.59*cos (40°)]$^{1/2}$ = 1.59 m.
3. Using the law of sines 1.59/sin(40°) = 1.22/sin (θ), θ = 56.88°.
4. Gain at 56.88° is –6 dB outside the beam width of the antenna.
5. The E field at the intersection using Equation 5.15 = 8.1 V/m.
6. Poynting vector = $E^2/(2*R_{air})$ = 8.1^2/(2*377) W/m^2 = 10.81 mW/m^2.
7. Power delivered to the UHF antenna = $P_{Poynting}$ * A_{UHF} (antenna aperture).
8. $A_{UHF} = \lambda^2/4\pi = 3.678 * 10^{-2}$ m^2 at 456 MHz.
9. $P_{UHF\,ant}$ = 10.81*10^{-3} *3.678*10^{-2} = 0.3976*10^{-3} W (–13.5 dBm) – distance correction – gain correction –2 dB UHF antenna.
10. Distance correction = 20 log (1.59) = –4.02 dB
11. $P_{UHF\,ant}$ = –28.3 – 4.02 – 2.0 = –34.32 dBm

The same method for calculating the radiated power delivered to the UHF antenna will be used to calculate the power at the VHF antenna as radiated from the UHF antenna. The steps are identical except for the dimensions. All of the dimensions used in these calculations are found in Table 5.2.

The UHF pattern peaks toward the VHF receiver antenna at 30°; the distance center to center from the UHF antenna to the VHF antenna is 5.95 m.

1. Distance to intersection 6.87 m.
2. $d_{intersection} = \left[(5.9^2 + 6.87^2 - 5.9 * 6.87 * \cos(30°)\right]^{1/2} = 6.84$ m.
3. Using the law of sines 6.84/sin(30°) = 5.9/sin(θ), θ = 25.55°.
4. Gain at 25.55° is –10 dB outside the beam width of the antenna.
5. E field at intersection 0.566 V/m.
6. Poynting vector = $E^2/(2*R_{air})$ = 0.566^2/(2*377) W/m^2 = 0.4248 mW/m^2. Power delivered to the VHF antenna equals $P_{Poynting}$ * A_{VHF} (antenna aperture).

7. $A_{UHF} = \lambda^2/4\pi = 0.2987\,m^2$ at 160 MHz.
8. $P_{UHF\,ant} = 0.4248*10^{-3}*0.2987 = 0.12688*10^{-3}$ W (−38.96 dBm) − distance correction − gain correction −3 dB VHF antenna.
9. Distance correction = 20 log(6.84) = −16.7 dB.
10. $P_{UHF\,ant} = -38.96 - 16.7 - 3.0 = -58.66$ dBm.

Conclusion   The calculation shows the power level on the antenna during VHF transmission with power radiated power onto the UHF receiving antenna. The second procedure shows the UHF transmission with power radiated onto the VHF receive antenna at the top of the radio mast. The UHF receivers have 85 dB rejection. The total signal level at the UHF receiver front end due to the VHF transmission is −107.01 dBm. The VHF receiver rejection is 80 dB with a total signal level of −136.66 dBm. As can be observed in Figures 5.4 and 5.5, both antennas are not completely omnidirectional; therefore there may be errors in these calculations because of the non-symmetry of the antennas. These calculations are only meant as an approximation until testing can prove the system is viable. Before the system is actually installed, the testing will include using a 5 W radio on the mast and a spectrum analyzer on the VHF antenna that are already installed to determine if any anomalies occur. Antennas in general can have anomalies due to simple installations by reflections for the antenna may have something in the radiating Frensel zone. After testing, refinements can be made to the modeling process. This will be helpful in the future since many installations are done after the project is completed. Generally most authorities will do this testing before the final batteries of tests are done for commissioning of the system. As mentioned previously, often EMI issues will be found during commissioning testing which becomes very costly. The engineer doing the design must always insist on solving EMI issues at the earliest possible time to prevent cost overruns due to fixing problems after the fact.

5.4   UNIQUE TESTING TECHNIQUE FOR EMI AND POLICE VEHICLES

This technique may be used for maintenance, emergency medical services (EMS), fire trucks, ferry operations and other vehicles for various authorities. With the advent of newer and higher powered computers, this technique may be expanded into other areas such as right of way surveys. These are of course very time-consuming and require a great deal of set-up time. If suitable antennas can be found or designed, this would be a big savings in both equipment and labor. The present-day spectrum analyzers and calibrated antennas for surveys cost US$ 30 000–40 000 and the testing is highly labor-intensive. To take measurements and record results for a 2 km survey required approximately 50 site set-ups, which took about three days (48 man hours, i.e. a two-man crew). The vehicles used for the survey had to be an SUV or similar vehicle with four wheel drive to operate in rough terrain and large enough to carry an inverter and other equipment necessary for the testing. The total cost at the time the test was being conducted was approximately US$ 38 000. The person doing the test had to be an EMC engineer or technician. The testing described in this document, if it can be extended for surveys, would be extremely helpful in lowering the cost of these right of way surveys. Generally all of the surveys are conducted during high periods of radio traffic during the day.

Dr. Chamberlin is considered an expert in electromagnetics. He has taught many courses on the subject and has also been the mentor of the author. This paper was presented at the 2004 International Symposium on Electromagnetic Compatibility, Sendai. It is presented as it was published, with no edits or changes to the document, with only changes to the reference numbers at the end of the document (source: Kent Chamberlin, *Measuring the impact of in-vehicle-generated EMI on VHF radio reception in an unshielded environment*, 2004 Internation Symposium on Electromagnetic Compatibility, Sendai. Reproduced by permission of Kent Chamberlin, University of New Hampshire).

MEASURING THE IMPACT OF IN-VEHICLE-GENERATED EMI ON VHF RADIO RECEPTION IN AN UNSHIELDED ENVIRONMENT

KENT AND MAXIM KHANKIN

Dept. of Electrical and Computer Engineering, University of New Hampshire
E-mail: Kent.Chamberlin@unh.edu

Abstract: Radiation generated by electronic equipment inside a vehicle can interfere with radio reception even though that equipment is in compliance with FCC standards. The result of that interference is an undesired reduction in radio coverage at frequen¬cies where the interference exists.

The objective of this paper is to present an approach for measuring electromagnetic interference (EMI) generated by in-vehicle electronic equipment without requiring the measurements to be made inside a shielded anechoic chamber. Such measurements are straightforward in a shielded chamber since interfer-ing signals from external radiation sources does not confound the measurement process. The contribution of the work presented here is to define a method for measuring EMI when external radiation is present. The basic approach is to identify regions in the spectrum where externally-generated signals exist and then to bypass those regions when measuring interference from in-vehicle-equipment. Because external interference can come from unlicensed as well as licensed sources, using the FCC database of licensed radiation sources to identify the regions to bypass will not achieve the desired goal. Rather, an analysis of the received spectrum is used to assess the presence of signals. Details regarding measurement equipment, procedures, accuracy and repeatability are given.

Keywords: EMI measurements, in-vehicle, radio coverage, FCC-compliant, external EMI.

1. INTRODUCTION

The motivation for the work presented in this paper stems from an effort to install networked computer equipment in police cruisers. That effort, named Project 54 [5.3.1], seeks to increase the efficiency and safety of police operations by networking equipment such as digital radios, computers, GPS units, radars, lights, sirens and other equipment. Networking equipment in this manner enables that equipment to be controlled by central computer, which affords the use of voice-recognition and touch-screen displays that are far more user-friendly than the original equipment.

The equipment used in these installations is off-the-shelf with the exception of the network itself, which was designed specifically for this application. Although the off-the-shelf equipment used is FCC compliant, it was observed that some of the equipment had to be turned off in order to maintain VHF radio reception in fringe-coverage areas because EMI from that equipment masked the desired signal.

To better understand EMI generation mechanisms, measurements of radiation were performed to determine explicitly the effect of in-vehicle EMI on VHF radio reception. Some of the initial measurements were performed in a shielded (not anechoic) enclosure to determine the spectral signature of EMI for specific equipment, while other measurements were made using the roof-mounted, VHF radio antenna for a completed installation. These latter measurements were performed in an unshielded environment, and hence radiation from external sources complicated the measurement process, making it difficult to quantitatively assess the EMI impact of a particular piece of equipment.

The significant and not surprising conclusions reached as a result of these initial measurements were that the EMI radiated was dependent upon installation practices, and that each vehicle installation would have to be measured to ensure that EMI was at acceptable levels.

One of the overall objectives of the project is to identify a system configuration that can be installed by the personnel who currently convert stock vehicles into police vehicles. These installers have experience with electronic equipment, although few if any have access to anechoic chambers or spectrum analyzers that would be needed for high-accuracy measurements of EMI. Consequently, one of the challenges for the project was to develop a method for measuring EMI that does not require expensive, specialized equipment or extensive expertise on the part of the user. The method developed is the topic of this paper.

To meet cost and ease-of-use objectives for the measurement procedure, a computer-controlled radio [5.3.2] is used. Once that radio is calibrated, the only equipment needed to perform the EMI measurements is a PC and the computer-controlled radio, as pictured in Figure 1. The cost for the equipment pictured is approximately $3500 US, and a calibrated signal source can be obtained for under $500 US.

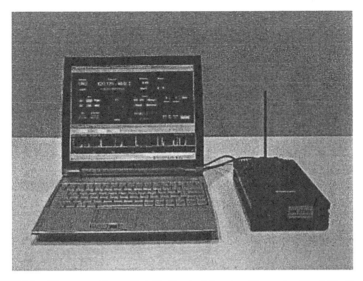

FIGURE 1   Computer-Controlled Radio And Laptop Computer Used To Perform EMI Measurements

When performing measurements, the radio's input is connected to the vehicle's antenna in place of the vehicle's VHP radio. Under computer-control, the radio scans across a predetermined range of frequencies and returns the signal level at those frequencies. The user is instructed to turn on or off the equipment under test (BUT) through a graphical user interface. After analysis, the software provides both tabular and graphical EMI data regarding interference generated by the EUT. These measurements typically require under 10 minutes for each device tested, and do not require in-depth knowledge about EMC on the part of the user. Further, the measurements show accuracy and repeatability within the requirements of this application, as is discussed in greater detail below.

2.   STRATEGY FOR BYPASSING EXTERNAL RADIATION SOURCES WHEN MEASURING IN-VEHICLE EMI

If the measurements described above were to be performed in a shielded anechoic chamber, the analysis of spectral data to determine EMI impact would be straightforward. However, performing these measurements without shielding necessitates bypassing frequencies where external radiation is present so that those sources do not contaminate the measurement process. Because the environment in which these measurements are to be performed can contain unlicensed radiation sources, such as radiation from nearby vehicles and equipment, the bypassing of local licensed radiation source frequencies determined from the FCC Frequency Assignment Database [5.3.3] would not be effective. The approach that has proven effective here is to scan through the band of interest to identify, and subsequently bypass, frequencies where signals are present when the EUT is turned.

Figure 3 plots the remaining bandwidth as a function of the number of sweeps, and it corresponds to the averaged signal level plot of Figure 2. It is worth noting that for all cases where the ambient noise threshold was reached, the remaining bandwidth was in excess of 90%, which is generally more than adequate to analyze the EMI characteristics.

Because the focus of the work presented here relates to interference in the VHP police band, the sweep range used is in the vicinity of that band (150-160

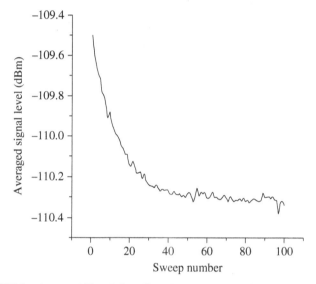

FIGURE 2   Averaged Signal Over Scan Range as a Function of Scan Number

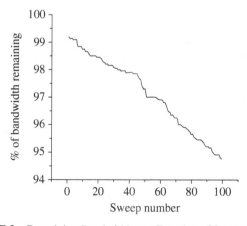

FIGURE 3   Remaining Bandwidth as a Function of Sweep Number

MHz). The frequency step size is 5 KHz, and the resolution bandwidth is 17 KHz. These parameters result in a sweep time of roughly 2 minutes for the 10 MHz frequency span. Convergence times in bands were signal transmission is not intermittent, such as the FM radio band, tend to be significantly less than the time reported here.

3. ANALYSIS OF EMI DATA

After the bypass list has been created, the BUT is turned on, and the frequency range is again swept, bypassing the frequencies on the list. The signal level for un-bypassed frequencies is averaged, and that average represents the EMI plus ambient noise in the frequency range. The same frequency range is swept and averaged a second time to ensure that no unaccounted-for external sources were radiated when the EUT measurements were being performed. If the two EUT measurements disagree by more than 2 dB, the user is notified.

An example of the result of an EMI measurement is shown in Figure 4, which plots the spectra of radiation from the onboard computer system with the vehicle engine running and ambient noise. These measurements were performed in an electromagnetically quiet setting, and hence the ambient noise threshold is lower than −110 dBm.

In addition to providing the Spectra of EMI, the measurement system also provides tabular data of the EMI averaged over the scanned frequencies. An example of tabular data is given in Table 1 for several in-vehicle components. As seen, some of these components would have a significant impact on radio reception.

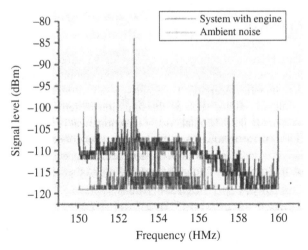

FIGURE 4   Spectra of Ambient Noise When Known Radiation Sources Are Bypassed and EMI from On-board Equipment and Engine

TABLE 1   Averaged EMI from In-Vehicle Devices

Device	EMI Level, dBm
Computer and Monitor	−106.23
Radio Control Head	−112.81
Light bar	−105.65
Entire System	−107.65
Entire System w. Engine Running	−106.08

4. ACCURACY AND REPEATABILITY

The concern about accuracy using this technique is that large spectral components of EMI might exist at frequencies bypassed in the measurement process. Were that to happen, the reported EMI averaged over the frequency range would be less than the actual EMI. However, the EMI spectra of all in-vehicle electronic devices investigated in this study tend to be distributed throughout the frequency range, similar to what is seen in the Figure 4. Consequently, errors resulting from the elimination of less than 10% of the frequency range are considered to be negligible for this application.

Even though external radiation sources and ambient noise can vary considerably with location, the repeatability observed using this technique is within 2 dB. Measurements of EMI on the same vehicle in significantly different electromagnetic environments provided very similar (i.e., within 2 dB) results. The explanation for this repeatability is that the EMI tends to be spread out over the frequency range and hence its average value is not changed when small segments of the spectrum are bypassed.

5. CONCLUSION

As stated, the objective of this work is to develop an affordable and easy to use tool that can be used by installers to determine if the equipment that they installed will interfere with radio reception. This approach has sufficient accuracy to detect common installation faults, such as the omission of ferrite beads or shielded covers and ground faults. The total cost of the equipment (under $4,000 US), the time required to perform measurements and the user-friendliness make this an attractive and affordable method for assessing EMI impact. This same approach should prove useful in other frequency bands and for other applications.

REFERENCES

[1]   The web site for Project 54 is: http://www.project54.unh.edu/about/
[2]   The computer-controlled radio selected for this project is called the WiNRADiO, which is described in detail at http://www.winradio.com/

[3] Information about the FCC Frequency Assignment Database is given at: http://www.fcc.gov/oet/info/database/

[4] "Man-Made Noise Power Measurements at VHP and UHF Frequencies", Robert Achaz and Roger Dalke, National Telecommunications and Information Administration, U.S. Dept. of Commerce, NTIA Report 02-390, December, 2001.

[5] Man-Made Noise in the 136-138 MHz. VHP Meteorological Satellite Band, Achatz, Lo, Papazian, Dalke, and Hufford, National Telecommunications and Information Administration, U.S. Dept. of Commerce, NTIA Report 98-355, September, 1998.

[5.3.1] The web site for Project 54 is: http://www.project54.unh.edu/about/.

[5.3.2] The computer-controlled radio selected for this project is called the WiNRADiO. which is described in detail at http://www.winradio.com/.

[5.3.3] Information about the FCC Frequency Assignment Database is given at: http://www.fcc.gov/oet/info/database/.

[5.3.4] "Man-Made Noise Power Measurements at VHP and UHF Frequencies", Robert Achaz and Roger Dalke, National Telecommunications and Information Administration, U.S. Dept. of Commerce, NTIA Report 02-390, December, 2001.

[5.3.5] Man-Made Noise in the 136–138 MHz. VHF Meteorological Satellite Band, Achatz, Lo, Papazian, Dalke, and Hufford, National Telecommunications and Information Administration, U.S. Dept. of Commerce, NTIA Report 98-355, September, 1998.

5.5 ANTENNA CLOSE TO THE EDGE OF THE GROUND PLANE

This is a study of how antennas close to the edge of the ground plane of a vertical polarized monopole antenna affect the field elevation of the antenna. A pictorial of the distortion is shown in Figure 5.6 (see Reference [2] for more details). This is generally a regular occurrence when antennas are mounted on vehicles such as buses, transit cars, railroad locomotive rooftops and maintenance vehicles where space is limited to the sites where installations may be positioned on the particular vehicle in question. This anomaly shown in the figure is due to diffraction at the edges of the ground plane. For a ground plane without diffraction, Equations 5.16 and 5.17 can be used to generate the elevation for power density and field normalized; as can be observed these equations present a smooth curve without the abnormalities of Figure 5.6.

$$W_{av} = \eta \frac{|I_0|^2}{8\pi^2 r^2} \left[\frac{\cos\left(\frac{\pi}{2}\cos\theta\right)}{\sin\theta} \right]^2 \tag{5.16}$$

$$U = r^2 W_{av} = \eta \frac{|I_0|^2}{8\pi^2} \left[\frac{\cos\left(\frac{\pi}{2}\cos\theta\right)}{\sin\theta} \right]^2 \cong \eta \frac{|I_0|^2}{8\pi^2} \sin^3\theta \tag{5.17}$$

Where η = impedance of air (377 Ω), θ = 0° to 180°, W_{av} = average power density, I_0 = antenna current and U equals power in watts. As can be seen in Equation 5.17 the polar plot is smooth with no abnormalities.

The polar plot shown in Figure 5.6 is for a 1 GHz antenna installation; however it should be noted that cell phone and some channels used by the authority in the 800 MHz region almost likely have the same abnormalities as shown. Variations and wavelengths, for example a 460 MHz radio channel will have a much more polar plot than illustrated in Figure 5.6. More information on this particular figure can be found in Reference [2]. It should be noted that the 1.22 m square of ground plane is a very large area to be occupied with antennas that have large spacing on the rooftops of vehicles. Most rooftop installations consist of a ground plane of no more than 2 m²; this implies that most radio antennas must be less than one-fourth wavelength, that is small antennas. By keeping the antenna small the range is very limited but repeaters can be used along the wayside to boost the signal for the antennas or may be made directional with the largest lobes in the direction of the longest ground plane. Edge reflection and diffraction are studied more in depth in Section 5.7.

It should be noted also that rooftop antennas will have several dead spots as radio operators approach closer to the rooftop, with distortions occurring as

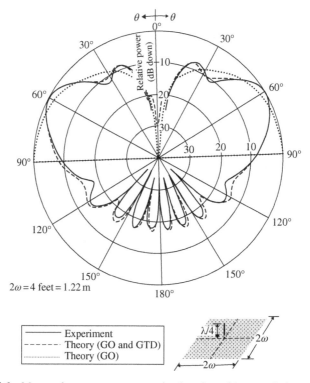

FIGURE 5.6   Monopole antenna antenna gain elevation with ground plane edge effects

shown in Figure 5.6. However, for tunnel operations where the radio is operated lower than the received portion of a leaky radiating cable installation, the abnormalities will not affect performance, that is they will be negligible. This includes maintenance vehicles, police, fire department and EMS vehicles where the ground planes are small and edge effects are present. Subway cars and buses in tunnels will have the same abnormalities at the lower region of electric field intensity as shown in Figure 5.6. Buses and subway cars however are generally made of aluminum, which of course is a very good ground plane and can present a very large area to install antennas. If stainless steel is used for any of these vehicles the ground plane will have an inductive reactance component that may aggravate the electric field intensity distortion, thus causing problems with the reception and transmission.

The center of the antenna beam or peak power occurs if an installation is shifted from 90° to 60° and the beam width is only 30°, as compared to a normal operation antenna that has a beam width of 78°. This will require that the AFC or AGC must have a wide dynamic range to accommodate the variation in signal levels. But the P 25 radio antenna uses spread spectrum techniques and it is an FM type radio that modulates angle, frequency and amplitude in the pattern to reduce the audio bandwidth to between 6 and 12 kHz instead of the commonly used analog bandwidth of 25 khz.

5.6 CASE STUDY: POSSIBLE FADE PROBLEM DUE TO ANTENNA REFLECTIONS ON THE ROOFTOP OF A LOCOMOTIVE

The scope of the work of this particular study is to analyze the rooftop radio installation of the locomotive. The particular radio was used to communicate to a control center location, sometimes using a two-way radio between personnel at grade level in the locomotive. Each of the locomotive radios is installed on a ground plane consisting of stainless steel 1/8 inch thick, width approximately 0.5 m and length 0.67 m. The radio antennas are encapsulated to protect them from the harsh elements of extreme heat and dusty conditions during a mining operation for iron ore. The dust may contain a great deal of ferrite particles. Figure 5.7 shows the two-way radio carried by maintenance personnel for two-way communication to the engineer and a similar type radio installation with a fixed antenna located along the right of way that is connected to a repeater that converts audio T-1 for transmission by the local microwave network connection to communicate with the operations control center (OCC).

Each radio antenna located on the roof of the locomotive will have reflections due to the close proximity that may impair radio communications. The request by an authority was only to study the possibility of the fade problem and provide some solutions if it did exist. The test that is required by the authority is to determine the overall performance of the communication system at commissioning. The reasoning behind it is that the solution to any problems that exist would be rather minor. However, the author at the time argued that they could be very expensive, for example changing radio configurations, frequencies and filtering. Getting approval to change

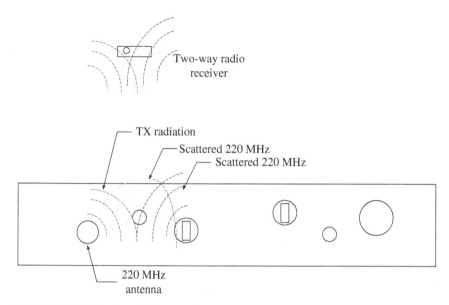

FIGURE 5.7   Relections off obstruction (other antennas) that affect handheld radios on the wayside

frequencies from regulatory authorities and other solutions would be very expensive unless done on only one or two locomotives and tested.

As shown at the top of Figure 5.7, radio waves appearing at the antenna from two different sources may impinge on the antenna of a two-way radio. One is due to the actual transmission from a radio on the roof of the locomotive and the other is from reflections from an obstacle which of course is the second antenna in close proximity. The phase component of the two waves will either add, creating a larger signal, or subtract, creating a fade. The worst case should of course always be considered, that is the waves are completely out of phase by 180°.

The lower portion of Figure 5.7 depicts reflections off the nearest antenna when transmitting. Note, there are other obstacles shown in the diagram. However this is the major one that is analyzed since the others are small in comparison to the antenna, shown in the diagram as reflected electromagnetic radio waves due to scattering. The antennas are mounted very close to the edge of the ground plane. This will result in a distortion of the electric field, as shown in Figure 5.6 in the previous study. The extent cannot be known until integration testing. The stainless steel ground plane is mounted on the cold rolled steel plate of the locomotive rooftop. This in itself presents a dislocation at the edge of the ground plane that can result in diffraction that can add to distortion of the electric field. These are some of the anomalies that may need to be addressed during integration testing; this should be kept in mind by the engineer doing EMC testing.

The VHF 220 MHz radio is transmitting with the RF electric field (E) shown in the diagram as curved waves radiating outward from the victim antenna. The other antennas, that is the culprits, will scatter the same signal with a lower magnitude curve than the radiated waves from its antenna. The distances between the antennas will determine the phase change, based on the wave number ($2*\pi/\lambda$; radians/meter). The impinging E on the VHF receiver will either cancel or re-enforce the transmitted signal. This can produce the fade alluded to previously. Only three antennas are shown, however they can all radiate the signal similar to an array antenna system. If this occurs the antenna will become directional, which is not desired for this installation. If a reflection is discovered, this will be the simpler problem to resolve. The out of phase waves will also appear at the transmitter and the wayside remote radio location. Either of these two situations can increase the noise level at the radio transmitter on the locomotive the remote location receiver along the right of way.

The power at the connectors, as measured in the report, is very large in some cases; this indicates the scattering antennas are fairly efficient radiators. The phase affects can be predicted at various distances from the antenna. If data is provided for where the VHF receiver will be (several distances may be used) and with the locomotive roof antenna location dimensions, perhaps a graphic representation can be derived using MetLab, EXCEL or another type of software to generate curves.

Calculations   The general equations are provided. They indicate how the dimensions affect the behavior of the antennas, when radiation from an external source is scattered (Figure 5.8). The equations are similar to those used in the radar range equation.

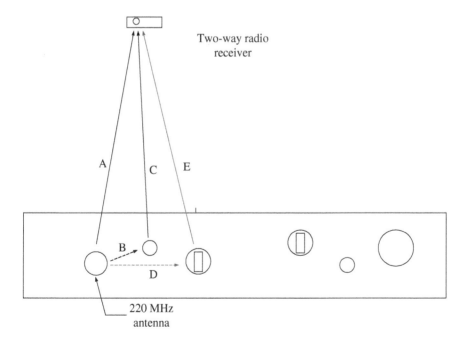

FIGURE 5.8   Distances of various antennas to the handheld radio

Let B distance = 6 inches, D = 12 inches, λ = 3.1/2.2 = 1.409 m

$$B = 0.5\,\mathrm{ft} * 0.305\,\mathrm{m}/\mathrm{ft} = 0.1525\,\mathrm{m}$$
$$D = 1\,\mathrm{ft} * 0.305\,\mathrm{m}/\mathrm{ft} = 0.305\,\mathrm{m}$$
$$\beta = 2\pi/\lambda = 4.459\,\mathrm{rad}/\mathrm{m}$$
$$\theta_B = 0.680\,\mathrm{rad}\left(38.96^\circ\right)$$
$$\theta_D = 1.359\,\mathrm{rad}\left(77.92^\circ\right)$$

A ≈ C ≈ E for distances greater than 3 m for all antennas. Transmitted power from the 220 MHz radio is Tx_{220} = 220 MHz power = 9.08 dBm (8.09 mW).
The height of the 220 MHz antenna is 6 inches (0.151 m), λ = 1.409 m (wavelength). The radiation resistance (R_R) of the 220MHz is as calculated using Equation 5.18.

$$R_R = 20\pi^2 \left(\text{antenna length}/\lambda\right)^2 \tag{5.18}$$

R_R = 2.26 Ω VHF antenna (length of antenna ≈ 1/10 λ).
All of the calculations are based on antenna lengths of less than 1/10 wavelength.
The calculation for radiated power is assuming the antenna is matched and only the radiation resistance and transmission line resistance are considered, that is $P_R = PT * R_R/(R_R + R_{Tline})$.
Power radiated = 8.09 mW [(2.26)/52.26] = 0.3498 mW (–4.56 dBm).
The calculation of the radiated power of cell A antenna is as shown in Equation 5.21.
The Poynting vector power (power density) is calculated where cell A is 6 inches from the 220 MHz antenna.

$$\text{Poynting vector power} = 0.3498/4\pi * \mathrm{r}^2 \tag{5.19}$$

$$P_p = 1.277\,\mathrm{mW}/\mathrm{m}^2$$

$$A_{eff} = 3\lambda^2/8\pi = 0.2369\,\mathrm{m}^2 \text{cell A aperture} \tag{5.20}$$

$$\text{Received power at cell A}, P_{ant} = Pp * A_{eff} = 0.302\,\mathrm{mW}\left(-5.2\,\mathrm{dBm}\right) \tag{5.21}$$

The difference in power measured at the connector and receive power is dissipated in heat.
Power radiated by cell A to the remote receiver, that is the radio along the right of way = 0.302 mW (–5.19 dBm).

Radiated power calculation of VHF power radiated by cell A   Radiation resistance at cell A for a 6-inch antenna is 2.26 Ω. Radiated power 2.26/52.26* 0.302 = 0.01297 mW (220 MHz from cell A), that is the Poynting vector at the antenna of the remote receiver from cell A (at 3 m).

$$P_{pfrom} = P_{CellA} / 4\pi * r^2 = (0.01265/4\pi * 9) \, mW/m^2$$
$$P_{pfrom} = 1.118 * 10^{-4} \, mW/m^2$$

(5.22)

For the power received at the VHF antenna from cell A due to scattering and 220 MHz Tx for a wayside radio with a $\lambda/4$ monopole antenna with gain G = 1.6 and wavelength λ = 1.409, the effective aperture is calculated using Equation 5.23.

$$A_{eff} = G * \lambda^2 / 4\pi \quad \text{Aperture for } \lambda/4 \text{ antenna}$$
$$A_{eff} = G * \lambda^2 / 4\pi = 0.2527 \, m^2$$

(5.23)

The calculated power at the antenna of the wayside radio is accomplished using Equation 5.24

$$P_{Ant} = P_{Poynt} * A_{eff}$$

(5.24)

After calculating the Poynting vector at 3 m and using Equation 5.24, the results are as follows:

$$P_{pfrom220} \, 220 \, MHz \, T_x = 0.366 / 4\pi * r^2 = 0.00297 \, mW/m^2$$

P_{Rx} = 0.00297 mW/m^2 * 0.2369 m^2 = 6.87*10^{-4} mW (−31.63 dBm). This is the power at the wayside receiver from the 220 MHz transmitter.

P_{PfromA} = 1.118*10^{-4} mW/m^2 * 0.2527 m^2 = 2.825*10^{-5} mW (−45.48 dBm). This is the power at the wayside receiver due to scattering at cell A.

The power at the 220 MHz transmitter from radio cell A is calculated using a scattering reflected power at 6 inches (0.1525 m) from the transmitter, $P_{CellA}/4\pi*r^2$ = 0.302 mW/4π*(0.1525 m)^2 = P_{poyntA} = 1.033 mW/m^2. The power at the antenna due to the scattering.

P_{220MHz} = P_{poyntA}*0.2369 = 1.033 mW/m^2 * 0.2369 = 0.244 mW (−0.61 dBm) at cell A. The Poynting vector at the 220 MHz antenna is at 0.83 W/m^2 due to scattering from cell A.

The power at the transmitting antenna 220 MHz on the locomotive roof is 0.83*0.2369 = 0.1966 mW (−7.1 dBm) the CNR at the 220 MHz transmitter is [−4.56 − (−7.1)] = 2.54 dB.

Carrier to noise ratio (CNR) = −31.29 − (−45. 48) = 14.19 dB due to the signal radiated by cell A as the VHF transmitter does not include other noise sources. The transmitter itself may present a problem because of the noise impinging on its antenna from the scattering noise of cell A. This noise cannot be filtered out because it is the same frequency as the 220 MHz carrier. The solution is to find a 220 MHz antenna that is directional away from the cell A antenna.

Since both signals appear at the antenna the ratio of the two is approximately the CNR. The phase can vary, due to atmospheric conditions, and this will add to the fade margin required. The variations to be considered are the distance from the two antennas, the fade margin and the noise floor.

Assumptions

Length of the 220 MHz antenna is 6 inches.

Cell A antenna is 6 inches high.

Distance between cell A and the VHF antenna is 6 inches.

The input resistance is 50 Ω and the line is matched.

Distances are planar.

All of the antennas are $\lambda/50 <$ antenna length $< \lambda/10$.

The radius of the antenna can be neglected.

Only RMS values are considered.

All distances are in the far field. Distances $\geq 2*$ (antenna length)^2 /λ.

Distance ≥ 0.0331 m (1.302 inches) for the 220 MHz 6 inch antenna. Cell A is not in the near field of the VHF antenna.

Cell A antenna at 925 MHz has a wavelength of 0.335 m. The near field for a 6-inch antenna at this frequency is 0.1388 m (5.46 inches).

The noise floor is approximately –124 dBm. This was not included in the calculations due to the large value of the interference. At longer distances where the receiver is operating near its maximum sensitivity the noise floor must be considered.

As personnel with handheld two-way radios get closer to the engine (within 1 m) there are diffraction losses due to the edge of the engine roof ground plane. A shadowing effect can be expected or communication is blocked. The calculations for these anomalies are performed using Equations 5.25, 5.26 and 5.27.

Diffraction loss

$$\Delta d = d_{Act} - d_{S\_line} \qquad (5.25)$$

Where $d_S = 2$ m straight line distance, the 220 MHz antenna from handheld remote radio held by personnel is 1 m from the rails at the track right of way and $d_{Act} = 2.75$ m.

$$\Delta d = 0.75 \, m$$

$$v = 2\sqrt{\frac{\Delta d}{\lambda}} = 1.459 \qquad (5.26)$$

$$L(v) = 6.9 + 20\log\left(\sqrt{1+v^2} + v\right) \, dB = 17.08 \, dB \text{ total diffraction loss} \qquad (5.27)$$

The Friis transmission equation is provided with diffraction included, with only an estimate of the fade provided for the 2.75 m distance power at the 220 MHz receiver of the locomotive. The antenna at the tower is in the same plane as a rooftop antenna on the locomotive.

$$L_T = 32.4 + 20\log\left(f_{MHz}\right) + 20\log\left(d_{km}\right) + L(v)_{dB} + \text{fade}_{dB} \qquad (5.28)$$

$$L_T = 32.4 + 46.8 - 50.46 + 17.08 - 20 = 65.82 \, dB.$$

This is the loss including diffraction for a person standing 1 m from the right of way. Received power at the antenna portable radio is

$$Rx = -4.56\,dBm - 65.82\,dB = -70.47\,dBm$$

The power due to scattering waveform from cell A is equal to $Rx = -45.48 - 65.82 - 20 = -131.3$ dBm. Extra fade has to be added because there is a fade between the transmitted 220 MHz waveform and the scattering object cell A that retransmits the waveform.

Cell A operates at approximately 857 MHz. A 6 inch height antenna structure indicates the antenna is approximately 0.5λ, with radiation resistance of 36.5 Ω and radiated power at 680 mW (−1.675 dBm).

Poynting vector power $= P_{at220} = 680\,mW/4*\pi*r^2$

$P_{at220} = 2.32$ W/m². This is the power density at the 6 inch antenna for the 220 MHz radio.

The power radiating from the 6 inch antenna back toward cell A radio is as follows:

$A_{eff} = G*\lambda^2/4\pi = 0.0167$ m², where $G = 1.6$ and $\lambda = 0.3617$ m.

$P_{object} = P_{at220}*A_{eff} = 38$ mW. This is the reflected power from the 6 inch 220 MHz antenna.

Poynting vector power $= P_{atCellA} = 38/4*\pi*r^2$, where $r = 0.1525$ m and $\lambda = 0.3617$ m.

$P_{atCellA} = 130$ mW/m². This is the power density at the antenna of cell A.

The reflected power on the antenna of cell A due to scattering is $P_{reflectA} = A_{eff}*P_{atCellA}$.

$P_{reflectA} = 0.634$ mW (−32 dBm). This is the reflected power on the cell A antenna. Therefore, the signal to noise ratio at the antenna when it is transmitting is 30.32 dB [−1.675 − (−32)].

Conclusion The calculations in this case study are a good starting point. However it should be kept in mind that rooftop radios as alluded to before have a stainless steel ground plane mounted to a cold rolled steel rooftop. These are generally less than 0.5 m from the edge of the rooftop. The electric field intensity close to edges will resemble those shown in Figure 5.9. This is for illustration purposes only to show the audience that, as two-way mobiles approach a locomotive, dead spots may occur in transmission and reception because of the anomalies shown here. What generally is needed is test results using the actual components involved in this study.

The radio installation along the wayside must be elevation above the cab of the locomotive to get the full gain of the antenna. As shown in Figure 5.9 it must be at least 30° above the horizontal where the main lobe is at 60° as shown. Due to this short range communications as shown the distance as calculated using Equation 5.29. The height of the antenna must be at least 9 m above grade level.

$T_x - R_x = 92.02$ dB. This is the maximum allowable loss for the radio communication from locomotive to the wayside radio receiver

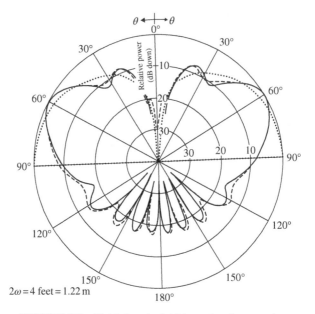

FIGURE 5.9   Field electric field intensity close to edges

$$= 32.4 + 20\log\left(f_{MHz}\right) + 20\log\left(d_{km}\right) + Fade$$
$$9.08\,dBm - 102\,dBm = 32.4 + 46.8 - 50.46 + 17.08 = 20 = 85.82\,dB \quad (5.29)$$

$T_x - R_x = 92.02$. This is the maximum allowable loss for the radio communications from the locomotive wayside radio receiver.

$$\left[92.02 - 32.4 - 20\log\left(f_{MHz}\right) - fade\right]/20 = \log\left(d_{km}\right) \quad (5.30)$$

The maximum distance to the receiver using Equation 5.30 is approximately 0.5 km with a loss margin of approximately 16 dB (with the receiver sensitivity as prescribed by the authority at −102 dBm) with the maximum radio sensitivity at −118 dBm, as in the radio specifications. If a P 25 radio is implemented instead of a conventional radio the maximum sensitivity is increased to −124 dBm. The maximum distance would be increased by 6 dB or double to 1 km. The P 25 radio would use a different band of frequencies.

For transit systems, the short-range radio links are probably okay; however when dealing with remote areas, such as may be encountered with freight services and hundreds of kilometers, this type of communication must be modified for longer-range communications. Most radio communications along the right of way of subway systems are short-range. In many cases the communication houses are close enough that subway trains can communicate directly with radios that are connected to the OCC directly with fiber optic equipment. However, bus systems and other types of vehicles that travel over hilly or mountainous terrain may need several of these short-range links connected to a microwave system backbone to maintain communications. The case study in Section 5.8 will deal with such a system.

This same type of antenna can be used on buses and maintenance vehicles, where multiple antennas may be installed, to prevent reflections due to scattering. This will result in noise above the noise floor, which may be as low as −124 dBm. It is always prudent to use printed circuit or patch antennas whenever possible. The EMC engineer correcting these antenna problems should be aware that there are always new designs that may improve not only the antenna characteristics but also their radar cross-sections which result in scattering and can cause increase in noise problems. Also the use of ground planes with a low magnetic permeability should always be used when possible. At high frequencies highly conductive paints, flame spraying and other methods may be available since the skin effect reduces the depth of penetration of current in the ground plane.

As a suggested radio installation that does not reflect as much signal back into the transmitter from an antenna resembling an obstacle, a patch antenna should be considered. This particular patch antenna is for a spread spectrum unlicensed radiator and is only meant as an example of what is available. Patch antennas have a hemispherical radiation pattern that can be used for transmission to a cell phone tower. The general frequency band is very close to 857 MHz. Other patch antennas may have a smaller footprint than the one depicted in Figure 5.10. Patch antennas are constructed with a local amplifier built in as part of the antenna structure. They require a power source for the amplifier and a matching network that are all incased in the structure similar to the one shown. These types of antennas are generally not good for use on a locomotive or for communication close to the right of way because of the hemispherical electromagnetic intensity E field. There are many more applications where the communication is between the locomotive and a tower. These will present a compact and low profile antenna. Many of the GPS antennas available are very similar to the one depicted in Figure 5.10. Most are approximately 0.75 inch high, with a diameter of about 2 inches. These types of dimensions are much less prone to scattering radiation that is detrimental to other radio antennas that are transmitting data or voice.

If cell A radio is transmitting at frequencies of 902–928 in the spread-spectrum region or cell phone frequencies in the 800 MHz region with reflections from the 220 MHz antenna, it is prudent to search the web for these low-profile antennas. Directional antennas are also available in these low-profile packages; however, for this analysis only cell A need be considered because the many twists and turns of these vehicles along the right away may require omnidirectional antennas.

This document provides a rough estimate of how the VHF antenna may affect cell A and the scattering affect by the cell A antenna on a remote VHF receiver. The CNR is adequate for good reception: 15 dB is usually required for voice and 10–12 dB for data. Decreasing the length of cell A, making either antenna slightly directional or increasing the spacing will improve the CNR. The fade can be calculated

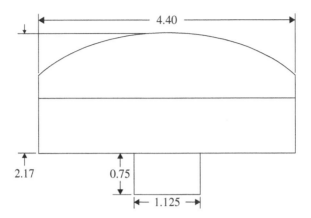

Dimensions (D × H):	4.4 × 2.2″ (11.2 × 5.6 cm)
Weight:	1.5 lb, (0.675 kg)
Input voltage	4.5–18 VDC
Nominal input current:	0.4 A
Max. rated input wattage:	680 mW
Frequency range:	902–928 MHz

FIGURE 5.10   Patch antenna

using phase changes with distance; however, this is difficult with any high degree of accuracy due to terrain, ground reflections and obstacles that are present. A rough approximation can be made. The field pattern should be mapped using a field strength meter or spectrum analyzer with directional antenna to determine the interference due to scattering.

A map should be constructed with a series of equi-potential lines around the antenna system with the VHF antenna transmitting. The conversion to the Poynting vector power is easily accomplished using $E^2/120\pi$, where 120π is the resistance of air. Once the Poynting vector power is known, the power at the remote radio receiver can be calculated using the antenna aperture to calculate the captured signal power. If possible, make measurements around the antenna system, then remove the cell A antenna and make the same measurements again. If the cell A antenna cannot be removed, shield it with conductive foam rubber or a similar material that is conductive and does not reflect electromagnetic fields.

Given test results, a model can be developed for analysis that can save time in the future when changes are made to the structure. The test data is analyzed to extract as much detail as possible. If an anechoic chamber is available, testing should be conducted on the antenna system to provide an unobstructed measurement of the behavior of the system. Open air testing (OATS) can also be conducted, provided an electromagnetic quiet area can be found. MIL-STD-462E has some information that can be useful when testing.

5.7 CASE STUDY: ANTENNA REFLECTION AND DIFFRACTION AT THE EDGE OF THE GROUND PLANE

A case study is required because several high-frequency radios were mounted on a copper ground plane and installed as a unit mounted close to other antennas. The radios in question are short-range, spread-spectrum, unlicensed band 902–928 MHz radios, with the longest wavelength at 0.344 m. At four wavelengths and a distance of 1.376 m from the edge of the copper plate, the power level is approximately 600 mW (from Table 2.1 in Chapter 2). This would produce a field strength of 3.5 V/m.

A plane wave at angle θ impinging on the edge of a ground plane and the diffraction due to the plane wave are depicted in Figure 5.11. Point P in the figure is a point of observation and d is the distance at point of observation from the edge of the ground plane.

The diffraction may be rather weak and insignificant if the source such as a radio antenna is far from the edge of the ground plane. However, as can be seen in previous examples, the diffraction may result in a large amount of electric E field intensity distortion. Equation 5.31 is for the geometrical optics electric intensity incident field, as described in Reference [1].

$$E_i = E_0 e^{j\beta d \cos(\phi-\theta)} \tag{5.31}$$

Inserting all the variables into Equation 5.31, the result is Equation 5.32. The magnitude of the electric field is not shown in the result because a table of values is calculated to observe the result for various field strengths.

$$E_i = E_0 e^{j\frac{2\pi}{\lambda}4\lambda\cos(180°-60°)} = E_0 e^{j8\pi\cos(120°)} = E_0\left(-0.998 + j0.0\right) \tag{5.32}$$

$E_r = 0$ at $\phi = 180°$ this field does not exist.
$E_t = E_0(-0.998)$

The diffraction calculations are made using (Keller's) Equation 5.33 with their results shown as Equation 5.34 for the incident field. The result of the calculation is Equation 5.35. The intermediate step is provided in Equation 5.35; this allows the reader to trace the calculations.

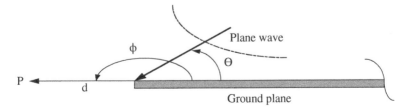

FIGURE 5.11 Diffraction at the edge of a ground plane

The reflected field is calculated using Equation 5.36 with the result as Equation 5.37. The incident diffraction with all variables inserted in performing the calculations is Equation 5.37.

$$E_D^i = E_0 \frac{e^{-j(\beta d + \theta)}}{\sqrt{2\pi\beta d}} \frac{1}{n} \sin\left(\frac{\pi}{n}\right) \left[\frac{1}{\cos\left(\dfrac{\pi}{n}\right) - \cos\left(\dfrac{\phi-\theta}{n}\right)} \right] \tag{5.33}$$

$$E_D^i = E_0 \frac{e^{-j\left(\frac{2\pi}{\lambda} 4\lambda + \frac{\pi}{3}\right)}}{\sqrt{2\pi \dfrac{2\pi}{\lambda} 4\lambda}} \frac{1}{2} \sin\left(\frac{\pi}{2}\right) \left[\frac{1}{\cos\left(\dfrac{\pi}{2}\right) - \cos\left(\dfrac{180^\circ - 60^\circ}{2}\right)} \right] \tag{5.34}$$

$$E_D^i = E_0 \frac{e^{-j\left(8\frac{1}{3}\pi\right)}}{4\pi} \frac{1}{2} \sin\left(\frac{\pi}{2}\right) \left[\frac{1}{0 - \cos 60^\circ} \right] = E_0\left(-0.06892 + j0.03979\right) \tag{5.35}$$

The total diffraction is calculated using Equations 5.35 and 5.39 for the incident and the reflected waves. The reflected shadow boundary (RSB) and incident shadow boundary (ISB) are both far away from the observation point at four wavelengths; this is the reason for using Keller's equation, because the calculations are asymptotic due to the long-distance observation point.

$$E_D^r = E_0 \frac{e^{-j(\beta d + \theta)}}{\sqrt{2\pi\beta d}} \frac{1}{n} \sin\left(\frac{\pi}{n}\right) \left[\frac{1}{\cos\left(\dfrac{\pi}{n}\right) - \cos\left(\dfrac{\phi+\theta}{n}\right)} \right] \tag{5.36}$$

$$E_D^r = E_0 \frac{e^{-j\left(8\frac{1}{3}\pi\right)}}{4\pi} \frac{1}{2} \sin\left(\frac{\pi}{2}\right) \left[\frac{1}{0 - \cos 120^\circ} \right] = E_0\left(0.06891 - j0.03979\right) \tag{5.37}$$

$$E_D^t = E_D^i - E_D^r \tag{5.38}$$

$$E_D^t = 2E_D^i = \left(-0.1378 + j0.07958\right) E_0 \tag{5.39}$$

$$E^t = E_i + E_D^t = \left(-1.1378 + j0.07958\right) E_0 \tag{5.40}$$

All of the calculations are done as a planar representation including diffraction. The data converges near the observation point of the magnitude and phase angle in Equation 5.41.

$$E^t = -3.992 \underline{|-4^\circ} \tag{5.41}$$

The result presented in Equation 5.41 indicates that distortion will be present due to diffraction. The use of Keller's equation makes the calculations rather simple as compared to making calculations using other techniques. In most cases using asymptotic type functions relieved the person doing the calculations of a great deal of mathematical manipulation. The downside of asymptotic calculations is that many people will use them without realizing the limitations of the equations involved. Engineers and scientists are generally vigilant in determining if the mathematical tool, that is the equation is beginning to break down and providing erroneous calculations based on experience or estimates of the answer. To find out more about diffraction principles read Chapter 13 in Reference [1] which has a very good analysis of diffraction.

The magnitude of the wave impinging on the edge of the ground plane shown in Figure 5.11 is on a per unit basis using 1 V/m. Tables 5.3 and 5.4 are generated using this rationale. Table 5.3 is estimating the reflection of the wave on a copper ground plane. In all cases the observation point has been taken directly off the edge of the ground plane; however Example 5.1, in analyzing the diffraction, is taking a

TABLE 5.3   Incident and Reflected Waves at the Edge of the Ground Plane

Freq. (MHz)	Nxλ	λ (m)	Incident θ (degrees)	Obser. Nxλ	Obser. φ (degrees)	E_i field incid. (V/m)	E_r field refl. (V/m)	Total E field = $E_i + E_r$
902	5λ	0.3437	30°	1.718	180°	−0.9434 + j0.0648	0	−0.9434 −j0.0648
902	5λ	0.3437	45°	1.718	180°	−0.9752 + j0.2214	0	−0.9752 + j0.2214
902	3λ	0.3437	30°	1.031	180°	−0.8618 + j0.506	0	−0.8618 + j0.506
902	3λ	0.3437	45°	1.031	180°	−0.861 + j0.507	0	−0.861 + j0.507

TABLE 5.4   Incident and Reflected Diffracted Waves at the Edge of the Ground Plane

Freq. (MHz)	Nxλ	λ (m)	Incident θ	Obser. Nxλ	Obser. φ	E_i field diffracted incid. (V/m)	E_r field diffracted refl. (V/m)	Total E field = $E_i + E_r$
902	5λ	0.3437	30°	1.718	180°	0.101− j0.0925	0.1089− j0.0997	0.2094− j0.1922
902	5λ	0.3437	45°	1.718	180°	0.0657− j0.0657	0.0657− j0.0657	0.1314− j0.1314
902	3λ	0.3437	30°	1.031	180°	0.176− j0.0207	0.7528− j0.654	0.9288− j0.6747
902	3λ	0.3437	45°	1.031	180°	0.0841− j0.0841	0.0841− j0.0841	0.1682− j0.1682

more general case. Table 5.4 is just an extension of the analysis examining the diffraction, as shown in Figure 5.11. The tables are useful to generate modeling equations for use with Mathcad or MatLab to generate curves for electric field intensity, that is E field elevations for a particular antenna. Some manufacturers have self-contained antennas that will provide a graphical illustration, such as shown in Figure 5.9.

An analysis of diffraction may only be required at certain points to determine if an EMC issue will occur. To do the modeling for each EMC diffraction issue could be quite time-consuming. However, in many situations where the same radio is to be used continually on various projects, it may be necessary to generate a model for the particular antenna being implemented.

Example 5.1

This is both an example and a problem to be analyzed.

The observation point is taken at $120°$ instead of $180°$, as shown in Tables 5.3 and 5.4, with the observation point that 5λ has a reflection at $45°$. The objective is to find the reflected and diffracted electric intensity E field at the observation point with given parameters as follows: $\phi = 120°$, $d = N\lambda = 5\lambda$ and $\theta = 45°$.

Use Equation 5.31 to calculate the incident field $E_i = -0.273 + j0.962$ and Equation 5.42 to calculate the reflected electric field intensity $E_r = -0.273 + j0.962$.

$$E_r = E_0 e^{j\beta d \cos(\phi+\theta)} \tag{5.42}$$

The diffraction electric E field intensity is calculated with Equation 5.33 for the incident E electric field intensity with $E_{i\_D} = 0.0311 + j0.0311$ as a result and the reflected diffraction electric field intensity calculated using Equation 5.36 with the result $E_{r\_D} = 0.02605 - j0.0256$.

Equation 5.43 provides the total field present at the observation point with the result as calculated. As can be observed by the result at the observation point in Example 5.1 very little distortion is apparent, only about 5%, which is to be expected. If the observation point is selected at $165°$, the distortion would be much higher. This is reserved for a problem at the end of this chapter.

$$E_t = \left(E_i - E_r\right) + \left(E_{i\_D} - E_{r\_D}\right) \tag{5.43}$$

$E_t = 0.0311 + j0.0311 - 0.02605 + j0.0256 = 0.00505 + j0.0567\,\text{V/m}$. The answer is also provided in polar form $\left(0.05692\underline{|5.09°}\right)\text{V/m}$.

As the transportation vehicle travels through the tunnel the reflection from the wall is calculated to determine if the transmitter is affected by wall reflections. The reflection coefficient is calculated below using Equation 5.44 and the transmission coefficient calculated using Equation 5.45. A diagram of this is not shown but this is considered a one-dimensional situation because the signal will travel in a straight line and reflect back from the wall because the approximation for the cosine is (5.09°). The next case is a bit more complex because the reflections are oblique.

$$\Gamma_R = \frac{\eta_2 - \eta_1}{\eta_2 + \eta_1} \tag{5.44}$$

$$T_{TX} = \frac{2 * \eta_2}{\eta_2 + \eta_1} \tag{5.45}$$

Where η_1 and η_2 are the impedance of air and walls, respectively.
    The wall has a permittivity of 3.5 (i.e. glazed concrete).

$$\eta_1 = 377\ \Omega\ \text{air},\ \eta_2 \frac{377}{\sqrt{3.5}} \cos 5° \approx 217.7\ \Omega$$

$$\Gamma_R = \frac{217.7 - 377}{217.7 + 377} = -0.268 \text{ and } T_{TX} = \frac{2 * 217.7}{217.7 + 377} = 0.732$$

The radiated power from the vehicle (Poynting vector) is as follows:

$$P_p = (E_t)^2 / (4\pi * 377) = 6.83878 \times 10^{-7}\ \text{W/m}^2 \left(6.8387 \times 10^{-4}\ \text{mW/m}^2\right).$$

The power absorbed by the wall and reflected at the wall are shown in Equations 5.46 and 5.47, respectively.

$$\text{Power absorbed by the wall} = P_p |T_{TX}|^2 = 3.664 \times 10^{-4}\ \text{mW/m}^2 \left(-34.36\ \text{dBm/m}^2\right) \tag{5.46}$$

$$\text{Power reflected by the wall} = P_p |\Gamma_R|^2 = 4.912 \times 10^{-5}\ \text{mW/m}^2 \left(-43.09\ \text{dBm/m}^2\right) \tag{5.47}$$

$$A_{eff} = \frac{G\lambda^2}{4\pi} = \frac{1.67 * (0.674)^2}{4\pi} = 6.037 * 10^{-2}\ \text{m}^2$$

$$\text{Power at the transmitting antenna} = A_{eff} P_{R\,from\,wall} = 2.965 \times 10^{-6}\ \text{mW} \left(-55.279\ \text{dBm}\right)$$

As can be observed from the graphical presentation shown in Figure 5.12, it would be difficult to show general trends in fields because of the non-linearities exhibited in this figure. At this point only the asymptotic approach of Keller's functions are used. These are rather simple equations as compared to other approaches when these functions breakdown. It is beyond the scope of this book to present all the implications; however Reference [1] alluded to previously has methods that complement the asymptotic behavior of Keller's function.
    When a subway car or train passes through a tunnel or underpass the wall reflections as calculated are reflected back into the transmitter, assuming the tunnel walls are constructed of glazed concrete. A certain amount of absorption will occur and

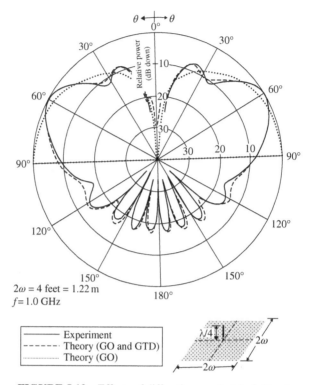

FIGURE 5.12   Effects of diffraction on electric field gain

the reflections are rather small. However, if the walls are coated with conductive material, such as paint or conductive panels, the reflections will be greater since the wave impedance into the wall will be much lower, as can be observed in Equation 5.44.

The −55.279 dBm for a 600 mW (−2.22 dBm) transmitter represents a transmit power carrier to noise ratio of approximately 53 dB. This may or may not be acceptable to the radio engineer designing the radio system, but this is the answer asked for by the authority. In a section near the end of this chapter a discussion will be presented with this example used for describing scattering. Scattering will most likely have less impact on radio system behavior as compared to this section.

5.8   ANTENNA APPLICATION WITH REFLECTION ALSO AT THE EDGE OF THE GROUND PLANE

The analysis is radiation by diffraction, as shown in Figure 5.13, where a ground wave travels along the ground plane and that is diffracted at the end of the ground plane. The diagram shows a one-quarter wavelength monopole antenna on a ground plane with the image shown below. This is the situation most encountered in transportation because the antenna may be located on the center of the rooftop of a bus or at the center location on the locomotive with a copper ground plane or stainless steel.

The application to be considered is a 2 W (33 dBm) 460 MHz radio mounted on the aluminum roof of a transit car with 90° bends at the edges of the roof. The antenna is a one-quarter wavelength monopole. The objective is to estimate the reflected radiation from the transit car and the diffraction at the corner of the transit car. Again this task is requested by an authority to determine if the radiation has detrimental effects to the performance of equipment along the right of way. If the radiation from the equipment results in a reduced performance of equipment already installed, the authority must either find another frequency or modify the equipment so that it does not interfere with other communications equipment that has previously been installed under another contract. Much of this testing is done at the final integration tests before commissioning of the system.

The total incident and reflected electric intensity fields are calculated using Equation 5.48. This is presented on p. 811 in Reference [1] and developed in Section 7.5. For r the distance from the antenna at 3 m and θ at 75° E_0 is calculated at 1.8698 V/m. All of the necessary parameters are available to calculate the total emission of incident and reflected fields.

$$E_\theta\left(r,\theta\right) = E_0\left[\frac{\cos\left(\dfrac{\pi}{2}\cos\theta\right)}{\sin\theta}\right]\frac{e^{-j\beta r}}{r} \tag{5.48}$$

Where $0 \le \theta \le \pi/2$.

To calculation the electric field intensity use Table 2.1 in Chapter 2 or use the calculation for generating a table as described in Chapter 2. This provides the magnitude scalar for the electric field intensity E_0. The phase term is provided as $e^{-j\beta r}$ where r is equal to 3 m, the observation point.

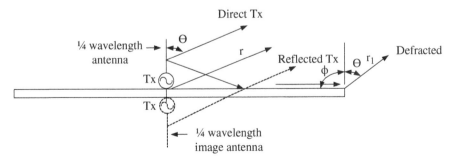

FIGURE 5.13   Effect of reflection and ground waves at the edge of a ground plane

$$E_0 = \frac{\sqrt{30 * ERP}}{R} = \frac{\sqrt{30 * 1}}{3} = 1.8698 \text{ V/m}$$

$$E_\theta (r,\theta) = E_0 \left[\frac{\cos\left(\frac{\pi}{2}\cos 75^0\right)}{\sin 75^0} \right] \frac{e^{-j\frac{2\pi}{0.674}3}}{3} = -0.31696e^{-j4.451\pi}$$

$E_\theta(\theta,r) = -0.04862 + j0.31639$ polar form $(0.32\underline{-81^\circ})$

Since the point of interest is limited to one edge of the ground plane, the diffraction is only considered on one edge. The general formula in Equation 5.49 is implemented to calculate the incident and reflected diffraction.

$$E_{\theta 1}^d = E_0 \left[\frac{e^{-j\beta w/2}}{\sqrt{w/2}} D^{i,r}\left(\frac{w}{2}, \theta + \frac{\pi}{2}, n = 2\right) \right] \frac{e^{-j\beta r_1}}{r_1} \tag{5.49}$$

Where $D^{i,r}\left(\dfrac{w}{2}, \theta + \dfrac{\pi}{2}, n = 2\right)$

Evaluating the trigonometric terms in Equation 5.50 will reduce them to Equation 5.51.

$$D^i(\rho, \phi - \phi', n) = \frac{e^{-j\frac{\pi}{4}}}{\sqrt{2\pi\beta}} \left[\frac{\frac{1}{n}\sin\left(\frac{\pi}{n}\right)}{\cos(\pi/n) - \cos(\{\phi - \phi'\}/n)} \right] \tag{5.50}$$

$$D^i(\rho, \phi - \phi', n) = \frac{e^{-j\frac{\pi}{4}}}{\sqrt{2\pi\beta}} [-0.5928] \tag{5.51}$$

Where $(\phi - \phi') = (140^\circ - 75^\circ)$.

$$D^r(\rho, \phi + \phi', n) = \frac{e^{-j\frac{\pi}{4}}}{\sqrt{2\pi\beta}} \left[\frac{\frac{1}{n}\sin\left(\frac{\pi}{n}\right)}{\cos(\pi/n) - \cos(\{\phi + \phi'\}/n)} \right] \tag{5.52}$$

$$D^r(\rho, \phi + \phi', n) = \frac{e^{-j\frac{\pi}{4}}}{\sqrt{2\pi\beta}} [1.6627] \tag{5.53}$$

Where $(\phi + \phi') = (140° + 75°)$.

The total diffraction is expressed in Equation 5.54. Inserting Equation 5.54 into Equation 5.49. The result is Equation 5.55.

$$D^s = D^i - D^r = \frac{e^{-j\frac{\pi}{4}}}{\sqrt{2\pi\beta}}[-2.255] \tag{5.54}$$

$$E_{\theta 1}^d = [-2.255]E_0 \frac{e^{-j\left(\frac{\pi w}{\lambda} + \frac{\pi}{4} + \frac{\pi r_1}{\lambda}\right)}}{2\pi r_1\sqrt{(w/2\lambda)}} \tag{5.55}$$

Evaluating Equation 5.55, w = 2 m, λ = 0.6739 m and r_1 = 3 m.

$$E_{\theta 1}^d = -0.0982E_0 * e^{-j7.6687\pi} = 1.8698 * (-0.0496 - j0.08472)$$
$$= -0.09274 - j0.1584 \, V/m.$$

$E_{\theta 1}^d = -0.18355\lfloor 59.65°$ The answer in polar form.

$$\sin^{-1}\left(\sqrt{\frac{\varepsilon_0}{3.5\varepsilon_0}}\right) = 32.3° = \theta_{critical} \tag{5.56}$$

Since the angle of incidence 60° is greater than the critical angle of 32° the waveform will be assumed to be normal to the wall.

$$\left(\frac{\sigma_2}{\omega\varepsilon_2}\right)^2 = \left(\frac{10^{-2}}{2\pi 460 * 10^6 * 3.5 * 8.854 * 10^{-12}}\right)^2 = 1.246 > 1 \, Test \tag{5.57}$$

This equation indicates that the wall is a conductor but not a very good one; if the Test >> 1, the wall could be considered a good conductor. However, the exact Equation 5.58 is used for the attenuation constant (see Table 3.1 in Chapter 3). It is reproduced here.

$$\alpha = \omega\sqrt{\mu\varepsilon}\left\{\frac{1}{2}\left[\sqrt{1 + \left(\frac{\sigma_2}{\omega\varepsilon_2}\right)^2} - 1\right]\right\}^{1/2} \tag{5.58}$$

Where f = 460 MHz, $\mu = 4\pi \times 10^{-7}$, $\sigma_2 = 10^{-2}$ mhos/m and $\varepsilon_2 = 3.5*8.854 \times 10^{-12}$ F/m.

To evaluate the equation and inserting all the variables, the intermediate result is shown below:

$$\alpha = 2\pi * 456 * 10^6 \sqrt{4\pi * 10^{-7} * 3.5 * 8.854 * 10^{-12}}\left\{\frac{1}{2}\left[\sqrt{1 + 1.246} - 1\right]\right\}^{1/2}$$

$\alpha = 9.612$. This attenuation constant is used to calculate the skin effect, as shown below in Equation 5.59:

$$\delta = \frac{1}{\alpha} = 0.104 \text{ m} \qquad (5.59)$$

The calculation of this particular construction is necessary to determine the surface current in the glazed concrete. To calculate the intrinsic impedance of the concrete, the exact equation for the wave impedance is used. This is shown below with the result as Equation 5.60. Consider a strip of concrete in the y direction at ± 0.5 m in each direction (i.e. 0.5 m looking into the page and 0.5 m out of the page).

$$\eta_2 = \sqrt{\frac{j\omega\mu}{\sigma + j\omega\varepsilon}} = \frac{j2.89*10^9*4\pi*10^{-7}}{10^{-2} + j2.89*10^9*3.5*8.854*10^{-12}} \approx 201.5\Omega \qquad (5.60)$$

$$\eta_1 = 377\Omega$$

$$\Gamma_R = \frac{\eta_2 - \eta_1}{\eta_2 + \eta_1} = \frac{201.5 - 377}{201.5 + 377} = -0.303$$

The total electric field at the surface is calculated using the results in polar form, Equation 5.55 (see Figure 5.14 for the details). The resulting E field radiation from the diffraction (0.18355 V/m) impinges on the wall with a surface current, Equation 5.61.

$$E_{total} = 0.18355(1 - \Gamma_R) = 0.18355(1 - 0.303) = 1.2793 \times 10^{-1} \text{V/m}$$

$$J_C = \hat{a}_x J_0 = \hat{a}_x * \sigma_2 E_{z=0,wall\_surface}^{Total} \qquad (5.61)$$

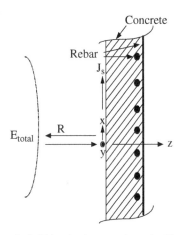

FIGURE 5.14   Electric field impinging on a tunnel with glazed concrete wall

Inserting all of numerical data variables into Equation 5.61 the result is as shown below in A/m² J_c. This is ignoring skin effect; inserting the skin effect into Equation 5.62 provides the J_s current.

$$J_c = 1.2793 \times 10^{-3} \, A/m^2 \text{ is a vector in the x direction.}$$
$$J_s = \delta * J_c = 0.104 \times 1.2793 \times 10^{-3} \, A/m = 1.2793 \, mA/m \tag{5.62}$$

The results shown is in the x direction in the xy plane with y the width. Therefore the total current flowing on the surface of the wall and not for just 1 m strip of the wall in the y direction would be 1.2793 mA of current. This could become a problem because the hangers for leaky radiating cable are made of metal; so the wall leakage current in the vicinity of the hangers could appear on the cable itself. Since the frequency of this current is at 460 MHz, coupling to other areas affecting radio reception may become a problem. The solutions are to make the wall more conductive with conductive paints, or to use non-conductive hangers for installing the leaky radiating coaxial cable, or automatic reduction of radio signal power when entering a tunnel or underpass.

5.9   ANTENNA APPLICATION WITH REFLECTION BETWEEN ANTENNAS IN A ROOFTOP ANTENNA FARM

Often in the transportation industry many radio and wireless devices are installed on the rooftop of vehicles. In this study several examples will be analyzed to determine if reflections from the various antennas affect the performance of the radio or device. The isolation issue is of great importance to maintain omnidirectional transmission or reception when required by the radio or device. The placement of these antennas is crucial; often very little consideration is given when installing antennas because they may be installed as an afterthought when a project is completed. Interaction between the antennas may result in one or more of the antennas acting as a director that provides extra gain in one direction and blind spots in E and H field patterns. This case study provides the EMC engineer with some tools to analyze these anomalies.

All radio and wireless device antennas must be presented at the onset of any project that requires rooftop installation on vehicles or other areas where space may be at a premium. This same type of analysis is required when installing radios on top of buildings that may have several antennas already installed. A careful study must be made with all dimensions and obstructions shown on a plan layout. It is also necessary to provide height dimensioning for any device, pole or structure (no matter how small and insignificant) that may distort the E and H fields.

The calculations shown are a step by step progression to determine far field and near field power attenuation between the various antennas for a vehicle rooftop installation. Tables 5.1–5.4 provide an estimate that can be used in the future if the antennas are repositioned. Several assumptions are made, as follows:

1. Antenna gain is not included in the calculations. Antenna gain must be subtracted from the original estimates provided in Table 5.5. If ERP is used the gain is included. Physical dimensions are required for height and separation between antennas for near field calculations.
2. The antenna lengths are all less than 0.25 λ (wavelength), omni-directional.
3. The elevation electric field pattern is similar to a monopole.
4. The attenuation does include the dielectric dome for protection from the elements.

Near field $= 2\dfrac{(k\lambda)^2}{\lambda}$. Where k is in wavelengths or parts of a wavelength, such as k = 0.25 λ, the near field is 0.125λ.

Figure 5.15 gives an example of antenna installations on a flat plate of stainless steel. These are for a short range radio system communications between a railroad locomotive and an antenna tower or pole along the right of way.

The use of the Friis equation for free space is used for signal attenuation between all the nearest obstacles, that is antennas. These are shown in column 6 of Table 5.5.

$$P_{Poy} = \frac{P_R}{4\pi r^2} \tag{5.63}$$

Where P_R is in watts and the area is in m²

$$Loss_{RCS} = P_{Poy} * A_{eff} \tag{5.64}$$

Where P_{Poy} is from the transmitting antenna and A_{eff} (RCS equivalent) is the nearest antenna which is considered an obstacle. Since all of the antennas are extremely small compared to the wavelength of the transmitter, Equation 5.65 can be used to calculate RCS.

$$A_{eff} = \frac{3\lambda^2}{8\pi} \tag{5.65}$$

Where λ is the obstacle antenna with aperture is calculated based on the receiver of that particular radio. Most of the antennas that are very short for transmitting to a tower or pole above the level of the cab of the locomotive can use patch antennas or active antennas with amplifiers included. Patch antennas have very low profiles, as shown previously in Figure 5.10. In all cases in Table 5.5 the longest wavelength is considered when calculating the RCS of a particular antenna that is a dual frequency type, such as the dual band 860–960 MHz and 1710–2500 MHz antenna. The worst case is a small high-frequency antenna that radiates towards a larger antenna. The larger antenna will resemble a large obstruction. Such is the case for the dual band cell radiating toward the 220 MHz voice band radio, as shown in row 2 of Table 5.5.

TABLE 5.5 RCS of Obstructions and Attuation

Frequency (MHz)	Type	Height (inches)	Near field (inches)	From–to	Atten. (dB)	RCS	Remarks
220	Voice	4.09	1.3 inches	220–860 (3.5 inches)	18.86	0.234 m² (8.1 dB)	$\lambda/50 < L_{ant} < \lambda/10$ ($T_{att} = 45.82$ dB)
2500	Cell	2.75	0.06776	2500–220 (3.5 inches)	19.33	0.632 m² 6.25 dB	
860–960, 1710–2500	Dual band cell	2.75	0.252	860–2430 (3.62 inches)	25.5	0.0155 m² 16.8 dB	$\lambda/10 < L_{ant}$ ($T_{att} = 67.8$ dB)
2430	WAN	1.75	1.22	2430–860 (7.85 inches)	26	0.00194 m² 9.1 dB	$\lambda/10 >> L_{ant}$ ($T_{att} = 61.1$ dB)
860–960, 1710–2500	Dual band cell	2.75"	0.252	860–460 (3.62 inches)	25.5	0.0155 m² 13.2 dB	$\lambda/10 < L_{ant}$ ($T_{att} = 64.2$ dB)
460	Voice	3.0	0.521	460–1500 (3.5 inches)	25.2	0.0542 m² (12.9 dB)	$\lambda/50 < L_{ant} < \lambda/10$ ($T_{att} = 63.3$ dB)
1500	GPS	5/8	0.0469			0.00509	$\lambda/10 >> L_{ant}$

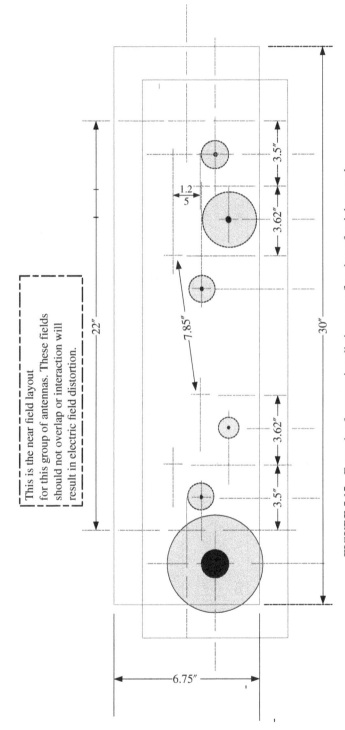

This is the near field layout for this group of antennas. These fields should not overlap or interaction will result in electric field distortion.

22″

1.2 5

7.85″

3.5″

3.62″

3.62″

3.5″

30″

6.75″

FIGURE 5.15 Example of antenna installations on a flat plate of stainless steel

Since patch antennas are used in a host of applications in transportation, a discussion of these devices is necessary to understand their importance in this industry. Patch antennas have several well-known advantages over common wire antennas. They are lightweight, easily manufactured (which makes them lower cost, low profile as alluded to previously), mechanically robust and they can easily be hardened for harsh environments such as railroad and subway transit conditions. They can be easily implemented into a circuit board with other components that will make them very compact. However, patch antennas have some disadvantages. They have narrow bandwidths, generally between 4 and 5%, and the radiation pattern is heavily dependent on the shape and size of the ground plane.

The metal patch is grounded on a substrate patch which radiates from the fringing fields. At the edges it resembles a resonant cavity which is ideal for modeling. When resonance occurs radiation resistance dominates the input impedance. The feed line must be positioned in the right place to achieve peak efficiency. This would be similar to using a wire (cat's whisker) in a crystal set for tuning. The length of the patch for the non-radiating edges is 0.5λ (wavelength inside substrate). If the physical parameters are varied, such as the patch is made smaller, the frequency of operation will be increased or the thickness of the substrate increased bandwidth will also increase (this can also cause higher-order modes to appear). This will degrade performance.

Patch antennas are constructed by etching or metal deposition on a low-loss dielectric or ceramic substrate. The thickness of the substrate is between 0.01 and 0.05 of the free space wavelength, as mounted between the patch and ground plane that will provide mechanical integrity to the patch antenna. The permittivity of the substrate material has an impact on its size. Higher permittivity decreases the volume of the antenna but also reduces the bandwidth. This is ideal for GPS units since bandwidth is not as critical for this application.

Ground plane size has an effect on the patch antenna resonant frequency. Where optimal matching occurs, gain at the matching frequency and center frequency occurs. Also, with gain and patch orientation close to the edge of the ground plane E field intensity distortion will occur. When an antenna is installation, from the EMC point of view, the engineer should check with the manufacturer specifications sheet for the particular antenna and determine how the measurements were made with a given ground plane size. This should then be compared to the installation to observe whether EMC principles are violated that may result in poor performance of the antenna. Since most antennas are packaged with a coaxial connector and power to a low noise amplifier, the only parameter that will need to be thoroughly analyzed is how close to the edge of the ground plane is the antenna mounted. The package itself will have a ground plane included for the patch antenna so that the unit is self-contained, but edge diffraction can still distort the E field.

The antenna farm shown in Figure 5.15 is enclosed in a cover to protect the circuitry from the elements. The ray dome cover must have certain attributes, such as it must be slippery and not allow dust to settle on its surface easily, particularly in any area where ferrite dust is present (such as mining operations for iron ore), and the permittivity must be as close to air as possible to prevent loss from reflections. Teflon is the best choice but it lacks mechanical strength, so it must be reinforced with glass fiber. The relative permittivity is approximately 2.3. The area where the mining of iron ore takes place may have high temperatures, along with the ferrite dust; and all of this could be a hindrance to transmission.

Analysis of the ray dome cover will have two possible scenarios, as shown in Figure 5.16. The first scenario shows the electric field impinging on the cover from the inside, normal to the surface; and the second scenario is an oblique electric field at the surface of the cover. The shape of the ray dome cover is designed with no sharp interior corners but gentle rounded ones that must be close to the top of the cover. The Cartesian coordinate system is used with the z direction normal to the xy plane, with the x parallel to the wall as shown and the y direction out of the page.

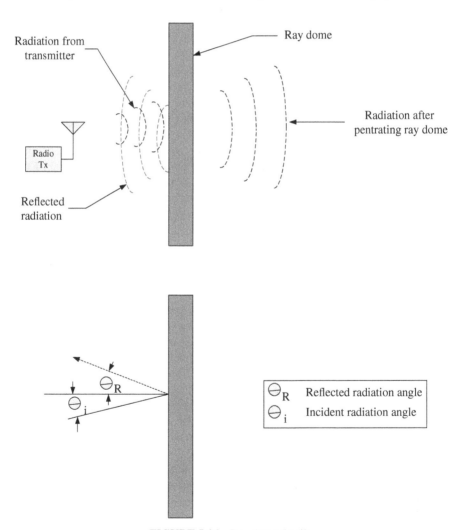

FIGURE 5.16   Ray dome details

Ray dome cover parameters:

1. Thickness 1/8 inch ($3.21*10^{-3}$) m.
2. Constructed of glass re-enforced Teflon.
3. ε_r is 2.3 for this material.
4. σ is 1.45 S/m.
5. μ_r is approximately 1.0.
6. Dust $\varepsilon_r = 9$.
7. Fe_2O_3 $\varepsilon_r = 12$ to 16.

The first scenario analysis: The electric field intensity E_1 and magnetic field intensity H_1 are found using Equations 5.66 and 5.67, respectively. The first term at the right of the equal sign is an incident wave; the second term is a reflected wave.

$$E_1 = E_1^+ e^{-j\beta_0 z} + E_1^- e^{+j\beta_0 z} \quad z \leq 0 \tag{5.66}$$

$$H_1 = \frac{E_1^+}{\eta_0} e^{-j\beta_0 z} + \frac{H_1^-}{\eta_0} e^{+j\beta_0 z} \quad z \leq 0 \tag{5.67}$$

$\eta_0 = 377\ \Omega$, air

$$\eta_c = \sqrt{\frac{\mu_0}{\varepsilon_r \varepsilon_0}} = \frac{377}{\sqrt{\varepsilon_r}} = 248.6\ \Omega \text{ dielectric cover impedance.}$$

The reflection coefficient and transmission coefficients are calculated below at the inside surface of the Ray dome cover. The antenna closest to the ray dome is the 220 MHz at 2 inches. Equations 5.68 and 5.69 give the reader an example of the reflections coefficient and transmission coefficient respectively if the surface were infinite.

$$\Gamma_a = \frac{\eta_c - \eta_0}{\eta_c + \eta_0} = -0.205 \tag{5.68}$$

$$T = \frac{2*\eta_c}{\eta_c + \eta_0} = 0.795 \tag{5.69}$$

Equation 5.70 is a reflection coefficient corrected for the internal reflections from the cover itself. The approximation is due to ignoring the phase terms. The complete analysis to find all effects due to internal reflections in the cover including phase terms can be found in Reference [1]. An entire section of this reference is devoted to reflections in various materials.

For an electric field at 100 V/m, the approximation is for 1 W of power at 2 inches from the 220 MHz antenna. This reflection results in a 21 V/m field reflecting back to the transmitter. The pointing vector is 93 mW/m^2 (P = E^2/4π*377). The aperture for the antenna is 0.156 m^2. The antenna will receive a reflected power of 14 mW. This could cause some concern as a noise received at the transmitter output.

$$\Gamma_t \approx \frac{(\eta_c + \eta_0)(\eta_c - \eta_0)}{(\eta_c + \eta_0)^2 - (\eta_c - \eta_0)^2} = \frac{-626 * 128}{3.75 * 10^5} = -0.213 \qquad (5.70)$$

$$T_t \approx 1 - 0.213 = 0.787 \qquad (5.71)$$

The transmitted electric field intensity E at the outside surface of the cover is 78.7 V/m. The pointing vector is 1.3 W/m^2; only about 7.3% of the power is lost passing through the cover. The ratio of (reflection coefficient/transmission coefficient)^2 × 100 is a good estimate of the power loss due to the cover. It must be kept in mind that multiple reflections continue inside the cover itself. They can be neglected because of the small thickness of the cover. However, for larger structures these intrinsic reflections may not be negligible.

The critical angle is θ_{crit} = 41.25°, using Equation 5.72. This angle is important. Some of waves incident on the ray dome cover are greater than θ_{crit}. These will become surface waves and will die out unless the surface becomes conductive due to paint or dust. Surface waves can permeate to the corners or edges and radiate on conductive surfaces. If other surface waves from radios are also present some mixing may occur, forming intermodulation products.

$$\theta_{crit} = \sin^{-1}\sqrt{\frac{\varepsilon_0}{\varepsilon_r \varepsilon_0}} = \sin^{-1}\sqrt{\frac{1}{\varepsilon_r}} = \sin^{-1}\sqrt{\frac{1}{2.3}} = 41.25° \qquad (5.72)$$

The phase constant is evaluated below using the permittivity of the cover.

$$2\beta_c t = 2\omega\sqrt{\mu\varepsilon} * t = 2 * 2 * \pi * 2.2 * 10^8 * 5.05 * 10^{-9} * 3.21 * 10^{-3} = 0.045 \text{ rad } (2.256°)$$

$$\theta_{cover} = \sin^{-1}\left(\sqrt{\frac{1}{2.3}}\sin\theta_i\right) = \sin^{-1}(0.329) = 19.23° \qquad (5.73)$$

At the cover the angle of the waves leaving is 19.23°, as shown calculated in Equation 5.73. The angle impinging on the inside of the cover is 30°. This is effectively narrowing the beam width of the antenna radiation pattern of the E field. The intrinsic impedance is altered by the phase of the wave impinging on the cover surface, as indicated in Equations 5.74 and 5.75.

$$Z_{air} = \eta_0 \cos\theta_i = 377\cos 30° = 326\Omega \qquad (5.74)$$

$$Z_{cover} = \eta_c \cos\theta_c = 249\cos 19.23° = 235\Omega \qquad (5.75)$$

The calculations for the reflection coefficient and transmission coefficient are shown in Equations 5.76 and 5.77.

$$\Gamma_t \approx \frac{\left(\eta_c + \eta_0\right)\left(\eta_c - \eta_0\right)\left(1 - e^{-j2\beta_1 t}\right)}{\left(\eta_c + \eta_0\right)^2 - \left(\eta_c - \eta_0\right)^2 e^{-j2\beta_1 t}} = \frac{-561 * 91}{3.1 * 10^5} = -0.168 \tag{5.76}$$

$$T = 1 - \Gamma = 0.832 \tag{5.77}$$

During the course of the study, several suggestions were made to protect the antennas from the harsh elements. The first was to eliminate the ray dome cover and to coat the antennas with Teflon or some other substance for protection; since the ground plane was made of stainless steel it required no protection. However, this could result in maintenance problems and the total assembly could not be removed easily for replacement. Using a ray dome cover which fastens to the ground plane made the complete assembly easy to remove and replace when necessary. The final analysis could not be done with ferrite (F_2O_4) dust due to mining and high-intensity AC magnetic fields from other radios and traction motors. These all affect the ferrite dust on the ray dome. When no magnetic fields are present it is an insulator, $\varepsilon_r = 12$–16, which changes the backscatter considerably. If the dust becomes baked-on due to the harsh temperature conditions near the mining operation, a different scenario must be considered. Note: $(\sigma/\varepsilon\omega)^2 \ll 1$. Criteria for a good insulator $= 1.05 * 10^{-2} \ll 1$. As the frequency gets above 1 GHz, the permittivity will become complex, resulting in anomalies in the ray dome cover material. The dust at the mine may be a combination of earth ($\varepsilon_r = 9$; Fe_2O_3 $\varepsilon_r = 12$–16). Ferrites can have $\mu_r > 1.0$. This gives them a magnetic component (inductance) when exposed to a DC or AC magnetic field. Normally they are an insulator. Traction motors have strong magnetic fields that may affect ferrites in the dust. It would be prudent to examine the dust for the consistency of mixture described. If a coating can be softened by the sun and the dust can get baked-on, this could become a maintenance problem. As a precaution, keep the ray dome free of dust to prevent this from occurring.

5.10 ANTENNA FARM APPLICATION WITH PATCH ANTENNAS

In this application, four patch antennas are clustered in an area and sheltered by a ray dome cover, as shown in Figure 5.17. The assembly consists of four antennas two 455 MHz, one 457 MHz and one 161 MHz antenna. The patch antennas have a hemispherical field; they are used for short range communications only. They all communicate through a tower or an antenna mast. The antennas are all located on the roof of a locomotive and have a shadow effect below the rooftop of the cab. They are constructed of the patch and insulation material with a circular ground plane, as shown in Figure 5.17. The outer circle as shown in the figure is the near field. All of these antennas are physically small and they are used for very short-range communications along the right of way only.

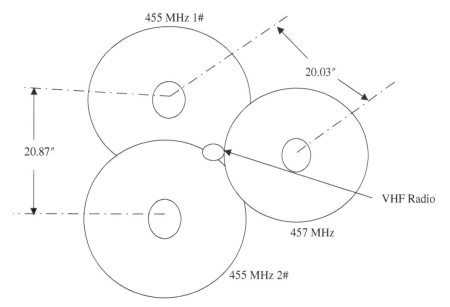

FIGURE 5.17   A four-antenna farm with near and far field distances

The antennas are mounted on a stainless steel ground plane but have a ray dome cover (not shown in the diagram) to protect the antennas from the environmental elements. In some cases antennas have been used that are completely enclosed in a case and required no special mounting ground planes. The assemblies are hardened for harsh environments. This is all built into the antenna itself with the necessary DC power for the low noise amplifiers (LNA). Occasionally no manufacturer can be found that has exact frequencies required for a particular project. The antennas may need to be manufactured as a specialty item, which becomes very expensive.

The losses between the various antennas in Figure 5.17 are reflected in Table 5.6. These losses are due to signal attenuation between antennas. These must be doubled because a loss from one antenna is reflected back to the transmitter. The losses shown in this table are for only one direction.

A summary of the total loss of the reflected transmission from the nearest antenna is provided in Table 5.7. These tables are provided as a guide for the EMC engineer when determining loss between patch antennas that are closely spaced. The ray dome cover analysis is done the same as provided in Section 5.9. The materials are all the same, that is including the thickness.

Table 5.6 provides the isolation between all the center section antennas, based on the distance to the nearest antenna. The distances are very short for some of the antennas such as the VHF and the EOT/HOT UHF antennas. However the short height presents a very small RCS cross-section; this removes the reflection problems to some extent. The ray dome will have reflections from inside while transmitting and from outside when receiving signals. These are very difficult to predict due to the geometry. The rounded corners and angular surfaces give it a small RCS for outside

TABLE 5.6 **Attenuation of Signal Level Between Various Antennas Due to Distance**

Audio start point	Audio to DP1/2	Audio to 457 MHz	Distance = $k\lambda$
455 MHz DP 1	−19.85 dB	−19.49 dB	0.779 = k
455 MHz DP 2	−19.85 dB	−19.49 dB	0.779 = k
VHF start point	Audio to DP 1/DP2 k_1	Audio to 457 MHz	
VHF 161 MHz	−6.845 dB	−3.122 dB	0.175 = k_1, 0.114 = k_2
Radio start point	Audio to VHF 161 MHz		Distance in λ
455 MHz DP1/DP2	−16.13 dB		Approx. 0.508 = k
457 MHz EOT/HOT	−12.21 dB		0.325 = k

TABLE 5.7 **Attenuation of Signal Level Between Various Antennas Due to RCS as an Obstruction**

Radio start point	Radio to DP1/ 2 (RCS)	Radio to EOT/ HOT(RCS)	Total =RCS + Dist.
455 MHz DP 1	−23.99 dB	−23.95 dB	−63.6 dB
455 MHz DP 2	−23.99 dB	−23.95 dB	−63.6 dB
Radio VHF start point	Radio to DP 1/DP2 (RCS)	Radio to EOT/HOT k_2	
VHF 161 MHz	−33 dB	−33 dB	−56 DPs / −56 EOT HOT
Radio start point	Radio to VHF 161 MHz	Radio to VHF 172 MHz	
455 MHz DP1/DP2	−25.98 dB	−25.98 dB	−58.42 dB
457 MHz EOT/HOT	−25.98 dB	−25.98 dB	−50.4 dB

signals inward (similar to stealth aircraft). The reflections from transmitters outward have reflections from walls and corners; these could present a problem. Corners can produce gain (corner cube reflectors).

Table 5.7 provides the RADAR cross-section (RCS) for each antenna. This is the same as the scattering of the signal due to retransmission of the signal from the adjacent antennas. Scattering, due to a high frequency (short λ) impinging on a lower frequency antenna, will result in a greater scattered signal strength than the reverse situation. The lower frequency antenna is a more efficient radiator to the high frequency signal. The total return signal to the transmitting antenna is the sum of the scattered signal (RCS) plus the isolation.

This section of the locomotive rooftop is inundated with multiple short antennas for communications. These are four assemblies each with four or five short-range antennas, the lengths of which are less than one-tenth of a wavelength. These are both licensed and unlicensed types of radios; some are spread-spectrum, others are not. This is one reason that it is coined as an antenna farm. This is not the author's expression; other people coined it first. All of the antennas are not within each other's near field; however, each antenna is viewed as an obstacle to an operational antenna. Some of the construction details of the antenna farm are as follows:

1. The center section VHF antenna is 4.06 inches high at a frequency of 161 MHz.
2. The wavelength (λ) is 75.76 inches.
3. The adjacent UHF antennas at 455 MHz are 13.66 inches from the VHF antenna.
4. The largest VHF dimension is 2.375 inches.
5. The near field is 0.149 inches.
6. The distance from the adjacent antenna is approximately 13.66 inches or 0.18 λ.
7. The UHF antenna is not in the near field of the VHF antenna. This will most likely not distort the E field of the VHF antenna.
8. The exact value for the isolation is -16.33 dB. As can be observed by the calculations, the isolation is fairly poor.

If the VHF and UHF antennas are transmitting at the same time, even with sharp filtering, saturation can be a problem. The UHF antenna near field is very close to the antenna and will present no problem when the isolation is -16 to -12 dB. This isolation will most likely not result in saturation of the receiver of the VHF radio. Reflection due to scattering at the VHF antenna can result in noise at the UHF transmitter. The scatter due to a 2.375 obstacle VHF antenna has a cross-section of -25.98 dB. The total attenuation is -41.98 to -37.98 dB. A 1 W transmission (30 dBm) results in a reflection of -7.98 dBm back to the transmitter. Most radios have a 1–2 MHz bandwidth at the transmitters, with a thermal noise floor of -114 dBm.

The uses for railroad and subway are for wireless communications at 900 MHz that are unlicensed and radio communications that are licensed but are made for short-range use. They can be arranged in an array that can be controlled and made directional with increases in gain and power handling capability. There are some excellent design papers, one in particular is Reference [7] that has all the manual calculations for the design of circular patch antennas. Reference [1] has more in-depth design aspects of patch antennas.

The antenna electric E field strongly resembles the electric field, as shown in Figure 5.12. As can be observed, because radios with patch antennas have a hemispherical E field, the radio communications is with a tower higher than the cab of the locomotive. It should be noted that the GPS antennas are very short, approximately 5/8 inch, because of this hemispherical shape. Patch antennas are ideal for GPS reception since they communicate via satellite. Due to the complexity of calculating the diffraction due to the edge radiation of the ground, most of the calculations require the use of MathCad or other software.

Microstrip circular patch antenna construction consists of a dielectric substrate sandwiched between a ground plane and a circular patch made of conductive material, such as copper or silver. The patch is made very thin, that is $t \ll \lambda_0$ (the wavelength at the radiating frequency in free space). The height of the substrate is within the range of 0.03 λ_0 ≤ thickness h ≤ 0.06 λ_0. h is the thickness above the ground plane. The circular patch is positioned and tuned so that the maximum field is normal to the patch. The mode excitation beneath the patch, that is the frequency of resonance, is determined by the physical dimensions.

The physical material constants of the substrates are very important as well as the physical thickness. The relative permittivity is in a range from 2.2 to about 10 or 12 for best-case performance. The substrate thickness is the largest possible with a permittivity as low as possible. This will provide the best performance, bandwidth and efficiency. Most microstrip antennas are integrated along with other electronic circuits. However, most of the patch antennas used for rail communications are standalone type with everything included, such as amplification and matching circuitry. Patch antennas are of many forms: rectangles, triangles and circular (as shown in Figure 5.18).

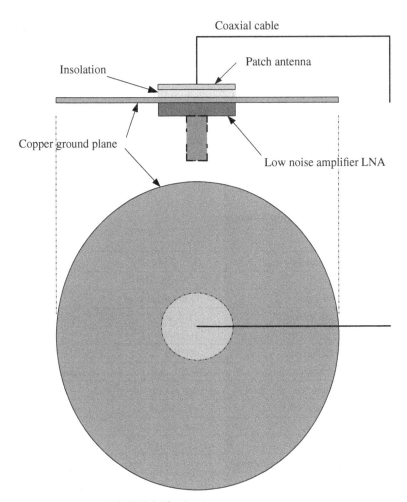

FIGURE 5.18 Patch antenna components

Since the material presented in this section is of a tutorial nature, all the equations and design information are not included. For more information on patch antennas, use References [3] and [7]. For data necessary to make judgments about a particular antenna, a table should be constructed similar to Table 5.8. The frequency band being used, whether short- or long-range, is determined by the antenna length. For transportation applications, many wireless devices are used along the right of way. Some must be considered as obstacles that reflect radiation. This is of particular importance when installing rooftop antenna farms on vehicles. As one may have observed previously, even short antennas present reflections to a transmitter that will inject noise. Since the frequency of the transmitter is the same as the reflection, with only difference being the phase, this cannot be filtered out as would be the case if it were the receiver.

The radiated power on some licensed radios is of such sufficiently low power that it will present no appreciable noise to the other antennas in the antenna farm. The patch antennas alluded to previously have certain characteristics due to length and width or diameter that provide the reader with sufficient information to judge whether it will be a hazard to communications for the other radios being used along the right of way. As a first pass to any analysis, try to get the cut sheets for all the antennas to be used on antenna farms. These will provide such information as height, gain, radio sensitivity and E field patterns. Table 5.8 is a minimal list of information necessary to do an adequate analysis to determine whether the antennas will interact. Not all antennas use ray dome covers but many of those that do must have a permittivity as close to one as possible yet be protective to the antennas enclosed.

Some ray domes protect the antennas from ferrite, silver, gold and other mining dust where metal ore is mined, particularly iron mines where ferrite dust is plentiful.

TABLE 5.8   Data for Rooftop Antennae

Item	Description	Data	Remarks
1	Transmit frequency band	455–457 MHz	Short range
2	Radiated power	2 W (33 dBm)	
3	Tx antenna type	Omnidirectional	
4	Antenna cut sheet	TLA 400-71	
5	TX antenna height	100mm	
6	Antenna factor	dBV/m	
7	Line and connector loss		Cable length
8	Ground plane stainless steel	$\mu_r = 600$, $\sigma = 2*10^7$	
9	Receiver sensitivity	−118 dBm	−96 dBm for margin
10	Rx antenna type	Omnidirectional	
11	Antenna cut sheet	TLA 400-71	
12	RX antenna filter characteristics		
13	Line and connector loss		Cable length
14	Ray dome cover	$\varepsilon_r = 2.2$	

Ferrite dust in particular is hazardous for communications because its permittivity will change with magnetic fields in the local area. There are a host of magnetic fields on railroad locomotives, traction motors and other electric devices that produce large magnetic H fields. Some of these anomalies will be presented in Chapter 8 of this book.

Most of the antenna farms that have an encapsulation of fiberglass or some type of variation have a stainless steel (and in some cases copper) ground plane coated with an anticorrosion substance. This whole assembly will be affixed to the rooftop of the locomotive or other vehicle in question. If the antennas are mounted directly on cold-rolled steel with no protection except for a small encapsulation around the antenna itself, this all must be included in the analysis when calculating reflections. Some of this type of problems will be provided in Chapter 8 of this book.

5.11 REVIEW PROBLEMS

Problem 5.1
A 220 MHz radio is transmitting. The far field for this radio is $d_{farfld} = 2*d^2/\lambda$, where d is the antenna's longest dimension is 0.131 meters. Calculate this value.

Problem 5.2
Is cell A within the near field of the 220 MHz radio antenna (see Figure 5.8)?

Problem 5.3
What is the scattering signature (equivalent to a radar cross-section) of cell A if the transmitting antenna is at 440 MHz instead of 220 MHz?

Problem 5.4
What is the power density at cell A if the radiated power from the 440 MHz antenna is 1 W (30 dBm)?

Problem 5.5
What is the power impinging on the cell A antenna from the 440 MHz radio due to scattering (where the cell A antenna is considered an obstruction)?

Problem 5.6
Find the total edge distortion of the ground plane due to diffraction and reflection, using the following parameters with the observation distance at 5 wavelengths, angle of 165° and angle of incidence 45°. Assume the observation point is far enough from the edge of the ground plane to use Keller's functions.

Problem 5.7
A monopole antenna is mounted on a bus with a sharp corner at the edge of the ground plane, because it is modular and mounted on a copper ground plane. The incidence angle is 45°, the frequency is 460 MHz, the radio power to the antenna is 2W at 5 m, w at 2 m from the edge, $\phi = 120°$, $\phi' = 45°$ and the point of observation is 5λ.

Problem 5.8
Calculate Eθ (θ, r) for a frequency of 310 MHz, 0.25λ monopole antenna at incident waveform at 35°, gain 1.67 and radiated power of 1 W. Assume the source of radiation is 3 m from the antenna. Hint: the necessary information is available to easily calculate E_0.

Problem 5.9
Find the diffraction n = 1, ϕ = 140, w = 5 m and ϕ' = 30°. Evaluate the trigonometric terms first before applying them to the final equation.

Problem 5.10
Figure 5.19 is an example of a communication cabinet made of stainless steel with insulation as shown in the figure. This particular cabinet was only 2 m from the right of way and the magnetic field from the traction motors and other devices emitted sufficient radiation to affect the video devices.

The objective here is to determine the electromagnetic radiation at the conductive surface. The radio transmission that results in diffraction from the transit car operates at 460 MHz and radiation from the edge of the car is 3×10^{-2} V/m. The field is normal to the surface of the wall or cabinet. Find the following:(a) skin effect at the stainless steel surface (μ_r = 600 stainless steel, σ = 1.45×10^7 mho); (b) surface current density; (c) incident H field intensity; (d) reflected H field intensity; (e) incident E electric field intensity; (f) reflected E field intensity. Other frequencies that may be detrimental to electronics equipment and wayside cabinets are analyzed later in Chapter 8. This particular problem should be studied as more a tutorial than a problem.

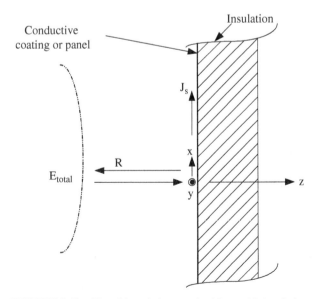

FIGURE 5.19   Wayside stainless steel cabinet with insulation

Problem 5.11

Two radio antennas are located on top of a transit car. One is a short-range unlicensed 902 MHz spread-spectrum radio antenna for use as a signal communication device along the right of way. The other is a 460 MHz radio voice communication antenna. The 460 MHz radio delivers 2 W of radiated power with a 0.25λ monopole antenna; and the 902 MHz antenna delivers 600 mW of radiated power with a 0.1λ monopole. The two antennas are 1 m apart.

Find the following: (a) the Poynting vector for each antenna; (b) the effective aperture A_{eff}; (c) the isolation due to distance between the antennas; (d) the effect due to reflections from each antenna.

Problem 5.12

A fiber glass enclosure covers several radios. The one nearest the cover is to be analyzed. It is located 6 inches from the cover. The construction of the cover of the material is as follows: relative permittivity $\varepsilon_r = 2.8$ and conductivity $\sigma = 7.8 \times 10^{-4}$ S/m. Problems: (a) is the cover a good insulator; (b) find the impedance of the cover; (c) find the reflection coefficient of inside air to the cover; (d) find the reflection coefficient of cover to outside air using Equation 5.78; (e) find the transmission coefficient. Assume all electromagnetic radiation is normal to the cover. What is the power loss due to the cover percentage, neglecting other parameters that may cause power loss, such as internal reflections in the cover due to absorption?

Problem 5.13

A patch antenna with a substrate that is 10 mm thick is most likely in the band of frequencies installed on a vehicle rooftop. Should the radio operating with this antenna be: (a) a 902 MHz spread-spectrum radio or (b) a 470 MHz licensed radio?

Problem 5.14

The height of an antenna is three-quarters of an inch. Because it is an encapsulated patch antenna, should this be considered as a reflection object when making calculations to determine if noises generated from it can be neglected, if the transmitting antenna is a one-quarter length monopole operating at 450 MHz.

Problem 5.3

If radial antennas are installed on cold rolled steel instead of stainless steel, what are some of the problems that may occur due to this decision made by the installer?

5.12   ANSWERS TO REVIEW PROBLEMS

Problem 5.1

$d_{far\ fld} = 0.033$ m

Problem 5.2

No

Problem 5.3

$S = A_{eff} = \lambda^2 / 4\pi = 0.0395$ m^2

Problem 5.4

$P_p = 1 / 4\pi r^2 = 3.42 \text{ W}/ \text{m}^2$

Problem 5.5

$P = A_{eff} * P_p = 0.0395 * 3.42 = 13.34 \text{ mW} (-18.7 \text{ dBm})$

Problem 5.6

Use Equations 5.31, 5.42, 5.33, 5.35 and 5.43.

$$E_i = E_0 e^{j\frac{2\pi}{\lambda}5\lambda\cos(165°-45°)} = E_0 e^{j10\pi\cos(120°)}$$
$$= -1.0 + j7.96 * 10^{-3} \text{ V}/\text{m} \quad \text{Incident calculation}$$

$$E_r = E_0 e^{j\frac{2\pi}{\lambda}5\lambda\cos(165°+45°)} = E_0 e^{j10\pi\cos(120°)}$$
$$= -0.482 + j0.4083 \text{ V}/\text{m} \quad \text{Reflection calculation}$$

These equations are the results of incident waves. The reflected waves on the surface may or may not be used to complete the analysis. This depends on how close these incident and reflected waves appear near the edge of the ground plane.

Diffraction incident calculation

$$E_D^i = E_0 \frac{e^{-j\left(\frac{2\pi}{\lambda}5\lambda+\frac{\pi}{4}\right)}}{\sqrt{2\pi(2\pi/\lambda)5\lambda}} \frac{1}{2}\sin\left(\frac{\pi}{2}\right)\left[\frac{1}{\cos(\pi/2)-\cos(\{165°-45°\}/2)}\right]$$
$$= -1.414 + j1.403$$

Diffraction reflection calculation

$$E_D^r = E_0 \frac{e^{-j\left(\frac{2\pi}{\lambda}5\lambda+\frac{\pi}{4}\right)}}{\sqrt{2\pi(2\pi/\lambda)5\lambda}} \frac{1}{2}\sin\left(\frac{\pi}{2}\right)\left[\frac{1}{\cos(\pi/2)-\cos(\{165°+45°\}/2)}\right]$$
$$= 2.732 - j2.725$$

Total distortion calculation

$$E_t = (E_i - E_r) + (E_{i\_D} - E_{r\_D}) = (-0.518 - j0.4003)$$
$$+ (-4.1464 + j4.128) = -4.664 + j3.7277$$
$$E_t = 5.971 \lfloor -38.63° \quad (\text{Polar})$$

Problem 5.7

Calculation of the electric field at 5 m:

$$E_0 = \frac{\sqrt{30 * ERP}}{R} = \frac{\sqrt{30 * 2}}{5} = 1.549 \text{ V}/\text{m}$$

Total emission for reflection and incident fields as shown below:

$$E_\theta(r,\theta) = E_0 \left[\frac{\cos\left(\frac{\pi}{2}\cos 45°\right)}{\sin 45°} \right] \frac{e^{-j\frac{2\pi}{0.674}5}}{5} = -0.1877 * e^{-j4.451\pi}$$

The result is shown in italics below:

E_θ (θ, r) = *-0.02938 + j0.18539* V/m polar form (0.1877|-81°)

This is the calculation for the incident diffusion field.

$$D^i(\rho,\phi-\phi',n) = \frac{e^{-j\frac{\pi}{4}}}{2\pi\sqrt{1/\lambda}} \left[\frac{\frac{1}{2}\sin\left(\frac{\pi}{2}\right)}{\cos(\pi/2)-\cos(75°/2)} \right]$$

The result is shown with only the trigonometric values evaluated.

$$D^i(\rho,\phi-\phi',n) = \frac{e^{-j\frac{\pi}{4}}}{2\pi\sqrt{1/\lambda}}[-0.63]$$

This is the calculation for the reflected diffusion field. The variable ρ is replaced by r for these particular calculations. The details of these equations and how they were derived can be found in Reference [1]. In many cases the results are only shown to relieve the reader of the drudgery of going through the analysis derivations.

$$D^r(\rho,\phi+\phi',n) = \frac{e^{-j\frac{\pi}{4}}}{2\pi\sqrt{1/\lambda}} \left[\frac{\frac{1}{2}\sin\left(\frac{\pi}{2}\right)}{\cos(\pi/2)-\cos(165°/2)} \right]$$

The result is shown with only the trigonometric values evaluated.

$$D^r(\rho,\phi+\phi',n) = \frac{e^{-j\frac{\pi}{4}}}{2\pi\sqrt{1/\lambda}}[-3.82]$$

This is the sum of the incident and reflected diffusion terms, again with only the trigonometric values evaluated.

$$D^s = D^i - D^r = \frac{e^{-j\frac{\pi}{4}}}{\sqrt{2\pi\beta}}[3.19]$$

This is the total field at 5λ with the polar form included.

$$E_{\theta 1}^{d} = -0.0534 + j0.1139$$

$E_{\theta 1}^{d} = 0.1258\underline{|-64.88°}$ Polar form

Problem 5.8
The calculation of the electric field at 3 m:

$$E_0 = \frac{3\sqrt{30*1.67*1}}{3} = 2.36 \ \text{V/m}$$

$$E_\theta(r,\theta) = E_0 \left[\frac{\cos\left(\frac{\pi}{2}\cos\theta\right)}{\sin\theta} \right] \frac{e^{-j\beta r}}{r} = 2.36 \left[\frac{0.209}{0.5} \right] \frac{e^{-j2\frac{\pi}{1}*3}}{3} = 0.3288 - j0.01$$

Problem 5.9
The result is shown with only the trigonometric values evaluated for both the incident and reflected diffraction, using the two equations below.

$$D^i\left(\rho,\phi-\phi',n\right) = \frac{e^{-j\frac{\pi}{4}}}{\sqrt{2\pi\beta}} \left[\frac{\frac{1}{n}\sin\left(\frac{\pi}{n}\right)}{\cos(\pi/n)-\cos(\{\phi-\phi'\}/n)} \right] = \frac{e^{-j\frac{\pi}{4}}}{\sqrt{2\pi\beta}}[-0.872]$$

$$D^r\left(\rho,\phi+\phi',n\right) = \frac{e^{-j\frac{\pi}{4}}}{\sqrt{2\pi\beta}} \left[\frac{\frac{1}{n}\sin\left(\frac{\pi}{n}\right)}{\cos(\pi/n)-\cos(\{\phi+\phi'\}/n)} \right] = \frac{e^{-j\frac{\pi}{4}}}{\sqrt{2\pi\beta}}[-5.737]$$

The sum of the incident and reflected diffractions is calculated with the equation shown below, with the final result shown in both Cartesian and polar coordinates.

$$D^s = D^i - D^r = \frac{e^{-j\frac{\pi}{4}}}{\sqrt{2\pi\beta}}[-6.609]$$

$$E_{\theta 1}^d = E_0 \left[\frac{e^{-j\beta w/2}}{\sqrt{w/2}} D^{i,r}\left(\frac{w}{2},\theta+\frac{\pi}{2},n=2\right) \right] \frac{e^{-j\beta r_1}}{r_1}$$

$$E_{\theta 1}^d = [-6.609]E_0 \frac{e^{-j\left(\frac{\pi w}{\lambda}+\frac{\pi}{4}+\frac{\pi r_1}{\lambda}\right)}}{2\pi r_1\sqrt{w/2\lambda}}$$

$$E_{\theta 1}^{d} = -0.303e^{-j12.19\pi} = -0.303(0.826 - j0.562) = -0.25 + j0.1703)$$

$$E_{\theta 1}^{d} = 0.302\underline{|34.3°}$$

Problem 5.10

Skin effect calculation

This test is necessary to determine how good a conductor or insulator is. This need not always be done, for example when dealing with metals such as steel or copper. Most are good conductors and need not be tested to determine if they are good conductor or insulator. The difficult ones that are borderline conductors or insulators are semiconductors such as concrete, depending on its consistency.

$$\left(\frac{\sigma_2}{\omega\varepsilon_2}\right)^2 = \left(\frac{4.1*10^7}{2\pi460*10^6*8.854*10^{-12}}\right)^2 = 1.6*10^5 \gg 1$$

$$\alpha = \sqrt{\frac{\omega\mu\sigma}{2}} = \sqrt{\frac{2\pi*460*10^6*600*4\pi*10^{-7}*1.45*10^7}{2}} = 3.975*10^6$$

$$\delta = \frac{1}{\alpha} = 1.581\mu m \; (0.001581mm) \qquad (a)$$

This is the skin depth where 63% signal attenuation occurs.

$$\eta_0 = 377 \; \text{Air impedance;} \; \eta_{ss} = \sqrt{\frac{\omega\mu}{2\sigma}}\underline{|45°} = 0.274\underline{|45°} \quad \text{Stainless steel impedance.}$$

$$\Gamma_R = \frac{\eta_{ss} - \eta_0}{\eta_{ss} + \eta_0} = \frac{0.274\underline{|45°} - 377}{0.274\underline{|45°} + 377} \approx -1 \quad \text{Reflection coefficient}$$

The E electric field intensity wave is normal to the surface of the stainless steel cabinets or wall. Since the field of interest for the analysis is at the surface of the stainless steel, z = 0 which reflects the phase term in equation. If the angle had been oblique rather than normal to the surface, an exponential E would need to be considered.

$$E_x^i = \hat{a}_x E_0 e^{-j\beta z} = \hat{a}_x 3*10^{-2} e^{-j\frac{2\pi}{\lambda}z} = \hat{a}_x \left[3*10^{-2}\right]_{z=0} \quad \text{Incident E field} \qquad (e)$$

$$E_x^r = \hat{a}_x \Gamma_R E_0 e^{-j\beta z} = -\hat{a}_x 3*10^{-2} e^{-j\frac{2\pi}{\lambda}z} = -\hat{a}_x \left[3*10^{-2}\right]_{z=0} \quad \text{Reflected E field} \quad (f)$$

The incident H field is normal in the y direction to the E field impinging on the surface.

$$H_y^i = \hat{a}_y \frac{E_0}{\eta_0} e^{-j\beta z} = \hat{a}_y \frac{3*10^{-2}}{377} e^{-j\frac{2\pi}{\lambda}z} = \hat{a}_y \left[7.957*10^{-5}\right]_{z=0} = \hat{a}_y 79.57 \, \mu A/m$$

$$(c)$$

$$H_y^r = \hat{a}_y \Gamma_R \frac{E_0}{\eta_0} e^{-j\beta z} = -\hat{a}_y \frac{3*10^{-2}}{377} e^{-j\frac{2\pi}{\lambda}z} = -\hat{a}_y \left[7.957*10^{-5}\right]_{z=0} = -\hat{a}_y 79.57\,\mu A/m$$

(d)

$$H_y^t = H_y^i - H_y^r = 2H_y^i = \hat{a}_y 159.14\,\mu A/m$$

The surface current density J_s is normal to the E and H fields, as shown in the equation below. Note, this is a vector cross product equation $-\hat{a}_z x \hat{a}_y = \hat{a}_x$.

$$\bar{J}_s = \hat{\eta} x \bar{H}_y^t = -\hat{a}_z x \hat{a}_y 159.14\,\mu A/m = \hat{a}_x 159.14\,\mu A/m$$

(b)

The results from calculations indicate that the signal is very weak and should not present a problem within the cabinet itself or within the structure of the wall. The radio system will have no detrimental effect to the electronics inside.

Problem 5.11

$$RCS = \frac{Le^{j-\frac{\pi}{4}}}{\sqrt{\lambda r}}$$

$$RCS = \frac{(0.1525)e^{j-\frac{\pi}{4}}}{\sqrt{0.674*1.0}} = 0.185e^{j-\frac{\pi}{4}} = 0.131 + j0.131$$

$$RCS = -17.7\ dB$$

Friis equation for free space loss

$$Loss = 32.4 + 20Log(460)_{MHz} + 20Log(0.001)_{km} = 25.65\,dB$$
$$Total\ round\ trip\ loss + RCS = 51.3\,dB + 17.7 = 69\,dB$$

Aperture for 0.25λ antenna

$$Length = \lambda/\pi = 0.318\lambda$$
$$A_{eff} = (\lambda/\pi)^2/4\pi = 3.65*10^{-3}\,(48.7\,dB)$$

Total reflection at the transmitting antenna is 33 dBm − 69 − 48.7 = −84.7 dBm

The RCS reflection and free space loss will attenuate the reflection sufficiently to not present an EMI interference issue at the transmitter.

Problem 5.12

(a) Evaluation using the equation below whether the cover is a good insulator. As can be observed it is a fairly good insulator.

$$\left(\frac{\sigma}{\varepsilon_r \omega}\right)^2 = \left(\frac{7.8*10^{-4}}{2.48*10^{-11}*2.89*10^9}\right)^2 = 1.18*10^{-4} \ll 1 \quad Dielectric\ test$$

The wave impedance of the cover is calculated using the equation below 225 ohms. The equation for well-known errors has a value of 377 ohms.

(b)

$$Z_c = \sqrt{\frac{\mu}{\varepsilon_r}} = \sqrt{\frac{4\pi * 10^{-7}}{2.48 * 10^{-11}}} = 225\Omega, \quad Z_{air} = 377\Omega$$

(c) and (d)

Using Equation 5.68 the calculations for the reflection coefficients are:

$$\Gamma_{12} = \frac{225 - 377}{225 + 377} = -0.252, \quad \Gamma_{32} = \frac{377 - 255}{225 + 377} = 0.252$$

The total reflection coefficient for the cover is as follows:

$$\Gamma_{in} = \frac{\Gamma_{12} + \Gamma_{23}e^{-j2\beta z}}{1 + \Gamma_{12}\Gamma_{23}e^{-j2\beta z}} = \frac{0.503}{1.063} = 0.473 \tag{5.78}$$

The total transmission coefficient is calculated using Equation 5.75.

$$T = 1 - 0.473 = 0.527$$

The percentage of power lost due to the cover = $0.473^2 \times 100 = 22.4\%$

Problem 5.13

The correct answer is (a): $0.03\lambda = 10.3$ mm

Problem 5.14

Yes, all obstructions will have some effect on radios. It is best to analyze them to ensure that the effect has not reduced radio performance.

Problem 5.15

One of the problems with cold rolled steel is that it is highly magnetic and can resonate with other objects on the ground plane to produce radiation affects. It is advisable to use a highly conductive paint or some other type of covering to reduce these effects.

6

CASE STUDIES AND ANALYSIS OF COMMUNICATIONS EQUIPMENT AND CABLE SHIELDING AND GROUNDING FOR BUS AND FERRY OPERATIONS

6.1 INTRODUCTION

This chapter of the book is devoted to the analysis of a bus and ferry system. The reason this is unique as compared to other communications systems is that the bus grounds and ferry grounds are independent of other grounding systems. Buses for example can be made of non-conductive materials, such as composites, or limited conductive materials, such as steel. However, the majority of new buses are usually made of aluminum alloys which are good conductors. Another major problem is that even aluminum when bonded to copper or other highly conductive materials can have corrosion at the joints and present a rusty bolt problem. The rusty bolt problem was coined from towers that had joints that were bolted and not bonded that produced a diode effect, causing electromagnetic radiation that disturbed communications equipment.

Buses, for example, have a host of onboard controls for engines, ticket collection, passenger counting and egress purposes. This is only the onboard part; it also requires radio antenna farms on the roof such as GPS, licensed radio and wayside spread-spectrum communications antennas. Buses usually upgrade their location using GPS; however the master GPS must provide location coordinates when the bus is in tunnels or in areas with limited access due to heavy iron deposits in the local vicinity that block GPS signals. The roof of the bus will contain as many as four radio antennas, each with a specific purpose. They sometimes are self-contained

Electromagnetic Compatibility: Analysis and Case Studies in Transportation, First Edition.
Donald G. Baker.
© 2016 John Wiley & Sons, Inc. Published 2016 by John Wiley & Sons, Inc.
Companion website: www.wiley.com/go/electromagneticcompatibility

along with an amplifier, that is such as patch antennas. The licensed radio antenna is generally a one-quarter wavelength monopole. These are generally used for communications between tower or poles at strategic points along the right of way to interrogate and communicate with the bus communication system. A short-range unlicensed spread-spectrum radio is used to interrogate the bus system as it enters the yard parking area.

The ferry communication system has similar problems with grounding. Because it is over water it has an additional issue: it must communicate with land-based licensed radio. This type of communication is difficult because of atmospheric conditions, reflections from the water and other anomalies that could affect the performance of communications. It also has controls and monitors for engines, passenger comforts and other features similar to land-based bus systems. However the GPS is always functioning correctly and need not be corrected due to tunnels as buses do.

The backbone structure of the communication system for this application is generally a series of microwave towers because fiber-optic installations are difficult due to terrain. The communication with the ferry is generally done through a two-way licensed radio or multiple radios that communicate between the ferry and land base station and between base station and microwave towers. Separate radios for navigation and other maritime communication are required. The layout of the backbone towers for this particular project was done previously. This is an upgrade to existing equipment as presented in this chapter. This part of the study pertains to EMC/EMI grounding issues of various components within the bus or ferry. Communications can be impacted by signal mixing from various sources such as radios and communications with frequencies sufficient that, when mixed, produce intermodulation problems as an end result.

In some cases, the shadow effect of the roadway will be explored to determine if the necessary license communications can be maintained. The designer that installs radios generally resorts to topographical maps such as EDX and software to determine the terrain typography. This can also be done when examining EMI/EMC issues using Google maps. They are not as accurate as the software; however they give one a first pass at determining all the obstructions in the area that may hinder licensed radio communications, such as buildings, canyons in the case of buses and other anomalies that will result in poor performance of radio systems. This section of the book is not meant for the design of a particular network; it is only a proposed design encountered during the author's employment over the past 20 years. The data presented here is meant to bring out the subtleties of designing radio systems that have EMC/EMI impact.

6.2 COMMUNICATION SYSTEM OVERVIEW

As a first pass to the analysis and overview, this system will be examined as shown in Figures 6.1 and 6.2a, 6.2b. The network shown in Figure 6.1 is a counter-rotating ring using a backhaul system of microwave radios. It can just as easily

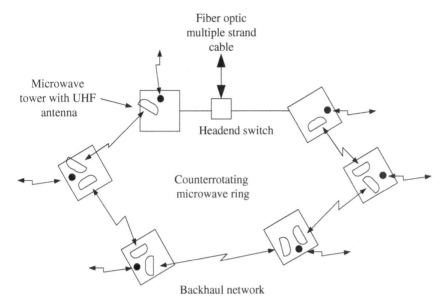

FIGURE 6.1 Block diagram and details of a microwave backhaul radio system

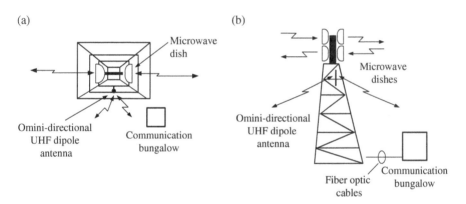

FIGURE 6.2 Description of bus communication system. (a) Tower detail 1. (b) Tower detail 2

be designed with a star configuration or some other type of configuration that is a combination of the two types of topology. If fiber optics were implemented, replacing the backhaul microwave system, it would be implemented with SONET nodes. However, in mountainous terrain where roads may not be readily available or other means for routing fiber-optic cable such as rail systems have microwave it is generally the choice for a design. For the topology shown in the figure only one radio antenna is shown on the tower; however, multiple antennas may be necessary in order to produce the coverage desired over the terrain. EDX software or other types of terrain analysis software are always available to the radio and EMC engineer to determine what obstacles might cause a variation in coverage. Each tower may be equipped with one or more antennas used for various other purposes.

For multiple antenna installations, care must be taken to prevent various radios saturating other radios in the system. Also, the antenna must be installed in such a way as not to affect stress on tower components. Generally, a mechanical or structural engineer must analyze the antenna integrity when adding antenna masts to the structure. The main concern of the EMC engineer is the interaction between the frequencies emanating from the antennas and impacting on tower components that are not bonded. This may produce intermodulation products due to mixing. The intermodulation products can develope frequencies that can interfere with radio performance. The backhaul microwave network shown has a failsafe advantage: if the microwave node is damaged or completely destroyed, the whole network does not go down. One of the only drawbacks of such a design is that maintenance is increased by adding the additional components to the network; but reliability is increased by a factor of about 50%. As an example, increasing redundancy in a network will increase reliability by approximately 3/2. This is where the 50% factor was generated.

An important factor that is not shown in Figure 6.1 but is shown in Figure 6.2a, 6.2b is the communications bungalow, because it would only clutter up the drawing. The communication bungalow is connected to the tower using fiber-optic cable. The tower itself has other communications such as DS3 (28 DS1 or T-1) or SONET terminals that are housed in an enclosure near the microwave transmitter and receivers. This provides the add-drop capability to the network. The communication bungalow is connected to this and the drop terminal to provide the necessary processing of the data, video, dispatcher and audio from the UHF radios. As mentioned previously, multiple radios may be installed on the tower, but only one is shown in the diagram of Figure 6.2a (tower detail 1). The tower detail in Figure 6.2b is a side view of the tower, showing the dual microwave dishes and the dipole antenna mounted on the side of the tower. All communications between the bungalow and tower are through fiber-optic cable that will provide isolation between the equipment on the tower and the processing equipment which is highly sensitive to radiation from external sources (and EMC issues). The power provided to the radios, microwave electronics and bungalow is either from local substations, photo optical arrays with battery backup, fuel cells or other state of the art power supplies.

The communication bungalows will have equipment that require EMI/EMC attention during the design phase due to the diversity of the data coming in from the local radios. The data will consist of video information due to cameras mounted in the buses, audio messages from the dispatcher, audio messages from the bus driver, positional data from the GPS, system-timing functions, people-moving functions such as passenger count, ticket information and bus maintenance information that may be necessary in an emergency. For ferry operations, which have other functions required to maintain a smooth and functional schedule, other necessary data may be necessary such as weather, sea conditions and so on.

When processing all this data, it is imperative to have good EMC capability because small glitches in the system can cause large problems in the field. Examples of some problems that may crop up are:

1. The ring communications has dropped bits or corrupted bits in digital communications that can cause large-scale errors in data tabulation.
2. Errors in digital video cause corrupted frames. This may be especially important when an investigation is underway for accidents or emergencies.
3. Data sent for signage in buses may be corrupted or hacked in some cases to produce undesirable statements in the signage itself.

This last has happened on several occasions for highway signage. It is not as serious in audio processing because many times the ear will integrate the loss of data due to excessive noise. All this information must be processed locally in the bungalows and sent to the main control room for analysis and action by the dispatching group on duty in the control room. Some of this information represented is for buses and maintenance vehicles.

There are multiple type vehicles to be analyzed for EMC/EMI issues; the networking is not quite complete at this point. There are several areas that have a small mesh of wireless devices that extract data from buses, ferry and maintenance equipment. Again, this is not a complete design but an overview of where wireless communications radios and other types of radiation may impinge on certain areas that cause a reduction in the performance of equipment. The main system control room has the usual EMC/EMI issues when stacking equipment in racks or enclosures near workstations or underneath workstations. The large racks and cabinets of equipment in control rooms are generally studied on a case by case basis depending on whether an EMC/EMI issue are present. The problems usually occur when external radiation impinges on ground loops, communication lines or reflections, producing unintentional radiation that must be shielded or filtered out.

Figure 6.3 depicts a layout of the bus communication system and wayside equipment communications bungalow/house. The bus has continuous positioning using

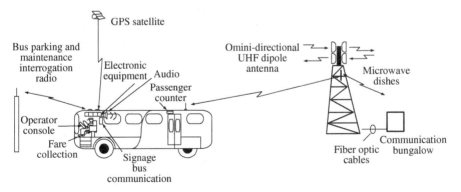

FIGURE 6.3   Bus communication system

a GPS receiver on top of the bus, shown in the figure. A short-range unlicensed 900 MHz radio transceiver is used for communications when pulling into the bus yard, the parking area. The communication with the bus is an interrogation for all the data accumulated during the trip and this is stored and transferred to the main system control center. The data accumulated is used for statistical purposes for bus maintenance, scheduling, fare collection, passenger statistics and other types of data. A rack of equipment is located above the operator. It includes the following: the unlicensed 900 MHz short range radio equipment, GPS receiver, various radio filters, audio amplifier for both voice communications and PA system, passenger counting equipment, signage drivers and interfaces to the control module located in the driver's console. Not shown is a gooseneck microphone used for both PA and communications. All the engine monitoring equipment and controls are located in the back of the bus. It is not necessary to analyze these for EMC/EMI issues. The reason is that engine controls, sensors and controls in the engine compartment must pass EMC/EMI emission and performance test prior to delivery of the equipment. The only EMC/EMI testing necessary is when the installation of the communication system is complete. The cabling from engine components must not radiate sufficient EMI to hamper performance of the equipment. This means an examination of all cabling to determine if shielding is necessary.

The bus driver operator console is an industrial grade computer that provides the communication and monitoring functions on the bus and communications with the main system control. The console interfaces with all functions on the bus and also provides a storage medium for messages sent from the main system control room. These messages may be in the form of audio or signage messages that can be displayed by the operator or used by the operator, thus relieving the operator of the more mundane tasks. Not shown in the diagram, some buses are equipped with cameras that may record events in real time as they occur. The data is relayed back to the main system control room, that is for emergency situations and accidents the information is stored in the control room data base. This will provide this dispatcher with information that may impact the bus routing.

Each bus parking area generally has a maintenance terminal so that maintenance personnel may observe abnormalities in the bus and engine functions. Thus buses may either be taken out of service, or be added to service or taken out of service for routine maintenance, which generally occurs on a monthly or bimonthly basis, depending on the needs of the authority. As can be observed in the diagram, the radio on the bus with the monopole antenna in the back of the rooftop is used to communicate with the local towers. These radios are P 25 type. P 25 radios have a combination of modulations of amplitude, phase and frequency shift keying (FSK) to shrink the bandwidth to about 6.5 kHz. These radios are 700 MHz band.

The EMC/EMI issues that may require analysis are as follows:

1. Antenna placement on the roof top of the bus that will not interfere with communications,

2. Cabling installed between the rack mounted equipment above the operator,

3. Interface cabling between the operator's console and of various components controlled and monitored,

4. Grounding and bonding issues because the bus is a self-contained unit similar to an aircraft that may be made of composites or other non-conductive components,

5. External radio dead spots that may be caused by hilly terrain or tunnels that may hamper the performance of radio equipment.

The first item in this list is of paramount importance because buses must maintain communications with the control room at all times. When adding a rooftop radio to the bus it is not only reflections from that radio to the other antennas but also ground currents that may circulate around the rooftop that can add noise to the grounding system. For example, a rooftop antenna using a connector that is composed of material other than that of the top may have a diode reaction as corrosion sets in between the two materials. Once ground circulation currents are present they cannot be removed by filtering. But they will provide noise components in the receiver. This may or may not corrupt the receiving radio communications; however, it can produce intermodulation noise by possibly mixing components from other frequencies that can produce undesirable effects on communications. Therefore, it should always be kept in mind to check this situation to ensure that the materials are compatible. Some of the more rugged radio antennas are self-contained as a module and are isolated to prevent this type of situation. It should always be kept in mind that positioning may cause diffraction radiation problems that will affect the E field patterns when placed too close to the edge of the ground plane, as alluded to previously in Chapters 4 and 5.

The next item in the list is cabling between the rack-mounted equipment. This is generally analyzed prior to installing equipment on the bus itself. The cabling input/output (I/O) to the rack is most likely the only EMC/EMI issue that may come up. However, much of the equipment is connected to fiber-optic I/Os which eliminate emissions problems. This has been done in modern aircraft. Today most of the cabling is connected via fiber-optic I/O interfaces. This same treatment should be observed for the I/O interfaces to the operator's console.

Grounding in the bonding of materials to provide ground paths for all equipment to a central location is of paramount importance because the bus has a self-contained inverter that produces the voltages necessary to supply communications equipment. Bus equipment operates at 48 V DC. This is a fairly common voltage found in the industry for everything from computers to sensors. When dealing with grounding, using frame components that may be made of steel and a range of materials other than the aluminum of the bus skin, the best solution is to provide ground return wires to equipment.

Finally dead spots in radio communications may occur from time to time due to fades caused by temperature inversions or other anomalies when equipment must operate over a short term. When a dead spot is encountered, for example going through an underpass, the GPS and radio may be blocked for a short period. However, when operating in tunnels, radio signals are simulcast both above ground and through leaky radiating cable type installations. At the mouth of the tunnel is where the most critical communications occur because the simulcast signals may be out of phase with the tunnel E field, thus reducing the magnitude of

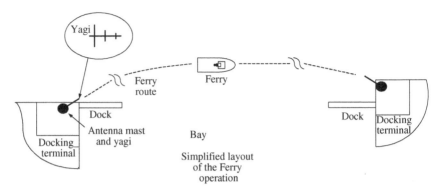

FIGURE 6.4   Ferry communication layout

a communication signal, or may be in phase, thus providing a larger signal level. This was alluded to in Chapter 4 with the subway system. Subway systems have a gradual grade where this occurrence is more likely to occur for a much longer period of time, as compared to a bus traveling through a tunnel which generally has a very sharp cut off at the mouth and at the exit. Generally the radio engineer designing a leaky radiating cable system will have a larger signal level than the incoming signal from a simulcast tower; and it may cause a reduction in signal level, but not the canceling type.

There is one more communication system overview to analyze. That is the ferry system, which consists of the ferry communications as it travels along the normal route. This is depicted in Figure 6.4. As can be observed in the figure a Yagi antenna is implemented, because in this instance the route is set and a directional antenna is necessary to reduce the gain required at the ferry. The radio is more efficient when the communications between the two end docking points and the ferry as it travels is along a specific route. A GPS track of the travel with check points is made. The radio antennas are located on buildings at both docking points, as shown in the diagram of Figure 6.4.

Since the ferry is a passenger type, a suite of equipment similar to what is provided on the bus is also necessary on the ferry. After the necessary equipment is added other equipment is added for environmental considerations. The equipment necessary to provide safe travel on the ferry is as follows:

1. Navigational aids because the ferry travels on the sea and not on roads.
2. A radar system to check for vessels and obstacles in the local area of the ferry travel route.
3. Engine and other essential equipment such as pumps and hull integrity are necessary; and passenger comfort during rough seas is also necessary.

The time schedule of a ferry can be rather stringent. In times of rough weather ferry operations may be canceled; this must be decided by the authority. Also schedule cancellations must be handled to ensure that people that work and depend

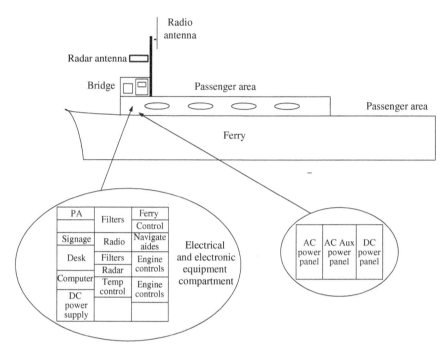

FIGURE 6.5 Ferry onboard communications layout

on the ferry for commuter information must be notified through public television and radio communications. Alternative bus routing is generally done by the authority that may require longer scheduling times than a short trip across the bay. These types of scheduling anomalies are handled by the bus authority bus dispatchers.

A simplified view of the ferry (Figure 6.5) indicates where some of the equipment is located. A ferry definitely has a very good PA system: it must cover the entire ferry as a matter of safety in the event of a collision. Signage is required for indicating times of arrival, which is very important to commuter travelers. A computer that is much more elaborate than the computer used on the bus is located in the equipment compartment with a desk. It must have Internet communications access for a laptop and WiFi for passenger access. A DC power supply is necessary because much of the equipment requires 48 V DC. The radar and radio equipment both require filtering. Ferry controls equipment is rack-mounted along with the navigational equipment. A large portion of the rack may contain engine monitors and other controls and monitors for maintaining generators, engine functions, pumps, ventilation systems, hull integrity, environmental controls and other types of equipment on board related to safety and fire prevention and suppression.

Ferry EMC/EMI issues generally come to light during integration testing; this occurs just before commissioning of the system. Each contractor must be available during integration tests to determine if there are issues with any particular subsystem that has been implemented during construction. The ferry must go through a battery of functional tests during the integration phase testing, generally a test run or several test runs may be made across the bay to determine how well the ferry system functions. This may be several runs depending on environmental factors to determine how well the equipment handles environmental anomalies. Generally a report is required by the federal government when federal money is involved, that is in most cases. The State that controls the bus and ferry system will require a DOT appraisal of the equipment and functioning for that particular State.

6.3   REFLECTIONS (FERRY AND BUS)

The height of the antenna mount at both ends of the docking is 25 m above grade and the height above the sea level and the ferry antenna is 5 m. An analysis is required to determine if reflections will impinge on the ferry antenna that may disrupt or reduced performance of communications. Saltwater has a relative permittivity of 81 and a conductivity of 1 S/m. The assumption is the ferry is in the middle of the bay, 3 km from the dock. The reflection occurs at 0.5 km from the dock. The analysis is to determine if the radiation from the reflection area is within the Fresnel zone. The layout for the calculation of the Fresnel zone is shown in Figure 6.6.

The following are the values used to make the calculations for radiation impinging on the Fresnel zone: $x = 0.5$ km, $d = 3$ km, $d_2 = 0.501$ km, $d_1 = 2.5$ km, $\lambda = 0.143$ m at $f = 700$ MHz. The minimum height above the point of reflection is as calculated

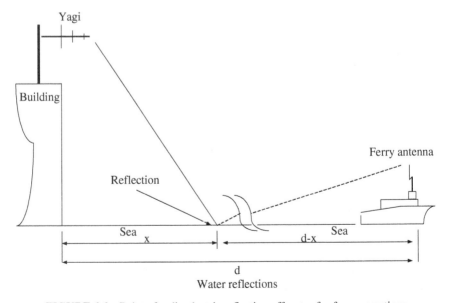

FIGURE 6.6   Point of radio electric reflection off water for ferry operations

using Equation 6.1. The actual height is 5 m above the calculated value or 18.33 m because this was subtracted off when making the calculations.

$$\tan\left(90^{\circ} - \theta_i\right) = \frac{25}{x}, \quad \theta_i = 88.6^{\circ}$$

$$d = 3\,\text{km}, d_2 = 0.501\,\text{km}, d_1 = 2.5\,\text{km}$$

$$h = 17.3\sqrt{\frac{d_1 d_2}{f\left(d_1 + d_2\right)}} = 17.3\sqrt{\frac{0.501 * 2.5}{0.7(.501 + 2.5)}} = 13.33\,\text{m} \tag{6.1}$$

Line of sight height at the reflection point $\approx 2.5\,\text{km} * \sin(90^{\circ} - 88.6^{\circ}) = 61.1\,\text{m}$.

Height at 18.33 m (including the height of the antenna mast) is $(13.33 + 5)$ m = 18.33 m. The distance at the reflection point is $(61.1 - 18.33) = 42.77$ m above the reflection point in the first Fresnel zone. As a rule of thumb Equation 6.2 may be used as a rough estimate for determining loss due to grazing by the Fresnel zone [8]. The 0.5 km reflection is not a problem.

$$d = 4\frac{h_1 h_2}{\lambda} = 4\frac{25 * 5}{0.143} = 1.129\,\text{km} \tag{6.2}$$

This distance calculation implies that a 6 dB loss must be added to the free space loss calculations. If the distance is increased beyond this, the loss will become fourth law. The reflection for this calculation adds 6 dB to the free space loss. The free space loss calculation is as follows:

$$L = 32.4 + 20\log\left(f_{\text{MHz}}\right) + 20\log\left(d_{\text{km}}\right) + \text{fade} + \text{Fresnel loss}$$
$$L = \left(32.4 + 56.9 + 9.54 + 40 + 6\right)\text{dB} = 144.8$$

Yagi gain (12) + ferry dipole antenna gain (1.73) = 13.73 dB
Total radiated power = 20 W (43 dBm)
The power at the ferry dipole antenna = 43 dBm + 13.73 dB – 144.8 dB = –88.1 dBm at 1.129 km.

The next item to consider is the reflections from the two patch antennas to the one-quarter wavelength 700 MHz monopole antenna located on the top of the bus. This is shown in Figure 6.7 (a cutaway view of the rooftop). The distance between the patch antennas and a one-quarter wavelength dipole is 2.5 m and the patch antennas are both approximately 1 inch in height. The task is a three-fold determination: (i) if reflections will result in excessive noise at the monopole antenna during transmission, (ii) if the reflections from the monopole antenna for the unlicensed 900 MHz patch spread spectrum antenna will affect radio transmission by the 700 MHz monopole and (iii) analyze the surface currents and fields in the ground plane, that is the bus rooftop.

Currents flowing in the ground plane cannot be filtered or removed easily; however, the reflection problems can sometimes be solved by changing antenna

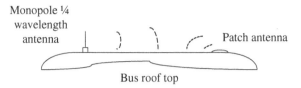

Monopole ¼
wavelength
antenna   Patch antenna

Bus roof top

FIGURE 6.7   Patch antenna on bus roof with an obstruction

TABLE 6.1   Antenna Characteristics

Range of antenna height	Type	Directivity	R_r ohms radiation resistance	A_{eff} (m²)	Electrical length
$L \leq \lambda/50$	Dipole	3/2 (1.76 dB)	$80*\pi^2(L/\lambda)^2$	$3*\lambda^2/8*\pi$	
$\lambda/50 \leq L \leq \lambda/10$	Dipole	3/2 (1.76 dB)	$20*\pi^2(L/\lambda)^2$	$3*\lambda^2/8*\pi$	
$L = \frac{1}{2}\lambda$	Dipole	1.64 (2.16 dB)	73	$(h^2*Z_{air}/4R_y)$ matched (h^2*Z_{air}/R_r) short circuited	L/π
$L = \frac{1}{4}\lambda$	Monopole	3.29 (5.17 dB)	37	$(h^2*Z_{air}/4R_r)$ Matched (h^2*Z_{air}/R_r) short circuited	L/π
		$E = (30*ERP)^{1/2}/distance$			

orientation. Sometimes the gain can be made directional that may eliminate some of these type problems. Also some problems can be solved with conductive or nonconductive paints. Since the GPS antenna is receive only, it will not be considered here because the frequency is higher than that of the unlicensed radio. The RCS of the antennas is calculated to provide the necessary reflection information for free space loss. This is a round trip that has previously done for other radio systems in this book where it is reflected back to the transmitting antenna. The patch antennas generally have their own ground planes included. Often the ground plane is close to the edge of the patch and the E field. For most of these, instead of being close to a perfect hemisphere, the E field is generally distorted and rather jagged due to diffraction at the edge of the ground plane. However, this may not present a problem because the communication is generally to a tower or pedestal along the right of way.

A handy reference (Table 6.1) is provided to assist with reflection calculations. The equation at the bottom of the table has been used extensively for calculating electric field intensity E. It should be noted that, for matched circuits to an antenna, the effective aperture A_{eff} is one-quarter the size of the aperture for the short-circuit version of an antenna or obstacle.

The calculations for the reflection from the patch antennas to the one-quarter wavelength 700 MHz antenna located on the bus top are as follows: the 700 MHz has 5 W of radiated power with a gain of approximately 3.29 electric field intensity E at the 1 inch high patch antenna for the 900 MHz radio is as shown below.

$$\text{Electric field intensity } E = \frac{\sqrt{30 * ERP}}{\text{distance}} = \frac{\sqrt{30 * 5 * 3.29}}{2.5} = 8.89 \text{ V/m}$$

The power impinging on the patch antenna that is reflected is calculated using the equations below.

The ratio of length of obstacle to wavelength 0.05 m divided by 0.443 results in a radiator that falls within the range $\lambda/50 \leq L \leq \lambda/10$. The aperture of the antenna is calculated below using Equation 6.3. To give the reader some perspective, the effective aperture of a bird is 0.01 m² the effective aperture of an adult is 1 m², that is if they are an obstacle radiating reflected power. The Poynting vector P_{Poy} is calculated using Equation 6.4.

$$A_{eff} = \frac{3\lambda^2}{\pi 8} = \frac{3 * 0.443^2}{\pi 8} = 0.0234 \text{ m}^2 \tag{6.3}$$

$$P_{Poy} = \frac{E^2}{4\pi * 377} = 16.6 \text{ mW/m}^2 \tag{6.4}$$

The power radiated by the patch antenna as an obstacle is calculated using Equation 6.5.

$$P_{Poy} * A_{eff} = 16.6 \text{ mW/m}^2 * 0.0234 \text{ m}^2 = 0.3884 \text{ mW} (-4.1 \text{dBm}) \tag{6.5}$$

The free space Friis loss equation is employed to calculate the reflection of the radiated power from the obstacle at the one-quarter wavelength 700 MHz antenna.

$$\text{Loss} = 32.4 + 20\log(f_{MHz}) + 20\log d_{km} = 32.4 + 56.9 - 52.0 = 37.3 \text{dB}$$

The reflected power at the one-quarter wavelength transmitting antenna is the sum of the free space loss and reflected power, that is –41.4 dBm. The carrier to noise ratio at the transmitter is therefore the sum of the ERP – reflected power from the obstacle (49.1 dBm) – (–41.4 dBm) = 90.5 dB. The obstacle should not present a problem for the transmission.

The next calculation is to determine what the reflected power from the one-quarter wavelength 700 MHz will have on a 900 MHz band spread spectrum patch radio. The ERP is fixed because the 900 MHz radio is unlicensed. The calculation is as shown below for the electric field intensity E.

$$E = \frac{\sqrt{30 * ERP}}{\text{distance}} = \frac{\sqrt{30 * 0.1}}{2.5} = 1.2 \text{ V/m}$$

Using the same techniques as previously done with calculating the radiation from the 1 inch patch antenna as an obstacle, the next calculation involves a one-quarter wavelength antenna which is 0.1107 m in length: using the short-circuit version of the one-quarter wavelength antenna to calculate the aperture Equation 6.6 is taken from Table 6.1.

$$A_{eff} = \frac{h^2 Z_{air}}{4R_r} = \frac{0.11075^2 * 377}{4 * 37} = 0.0312 \text{ m}^2 \tag{6.6}$$

The Poynting vector at the one-quarter wavelength 700 MHz antenna is calculated using Equation 6.7.

$$P_{Poy} = \frac{E^2}{4\pi Z_{air}} = \frac{1.2^2}{4 * \pi * 377} = 3.039 * 10^{-4} \text{ W/m}^2 \tag{6.7}$$

The power radiated at the one-quarter wavelength antenna is provided in Equation 6.8.

$$A_{eff} * P_{Poy} = 0.0312 * 3.039 * 10^{-4} = 9.483 \,\mu\text{W} \left(-50.23 \text{ dBm}\right) \tag{6.8}$$

The Friis free space loss equation is calculated between the one-quarter wavelength antenna and the 900 MHz patch antenna.

$$L = 32.4 + 20\text{Log}\left(900\right) + 20\text{Log}\,d_{km} = 32.4 + 59.1 - 52.04 = 39.46 \,\text{dB}$$

Power at the 900 MHz transmitter = −50.23 − 39.46 = −89.69 dBm
The transmitter carrier to noise ratio is 79.69 dB (−10 dBm − (−89.69)]

The analysis is done without regard for cable losses and matching networks. These are generally provided by the radio engineer designing the system. Often the EMC engineer is asked to assist the radio engineer in meeting or exceeding radiation guidelines as presented by the FCC. However, cabling, connectors and connections often require analysis to determine whether coupling occurs between the various system components. Therefore, one should always be vigilant when examining installation and check wiring very closely. As frequencies of equipment such as computer integrated circuits provide higher data rates they also provide larger problems with preventing electromagnetic radiation from impinging on wiring and other equipment. During this study, the authority was only interested in providing an analysis to show work has been done to determine if EMC/EMI violations are possible. Generally most comments of possible EMI problems are taken as constructive criticism and considered during the design phase. However, some issues may be overlooked during the design phase but appear at integration testing before commissioning of a particular system. EMC issues may occur as previously shown from ground plane edge radiation, which often may be overlooked and only discovered during the integration phase of a system. As can be observed from previous endeavors something as innocent-looking as a small post or pole sticking up near radio antennas can reflect radiation not only to the transmitter but also to other antennas in the local area in the antenna farm.

The next topic of interest is the radiation impinging on the roof top of the bus resulting in circulation currents which appearing on the ground plane, that is the bus top.

The equation for determining the E field is best described by Equation 6.9. The variables are as follows: I_0 = antenna current, η_a = 377 ohms, r = 2.5 m and θ_i = angle of incidence from the vertical axis of the antenna. The one-quarter wavelength monopole radiates 5 W with a 37 Ω radiation resistance, then I_0 = 0.368 A and the incident angle θ_i = 88.2°. The magnitude of the electric field ignores the phase term, that is the exponential.

$$E_\theta = \frac{j\eta_a I_0 e^{-j\beta r}}{2\pi r}\left[\frac{\cos\dfrac{\pi}{2}\cos\theta}{\sin\theta}\right] \text{V/m} \qquad (6.9)$$

$$E_\theta = j\frac{377*0.368}{2\pi*2.5}\left[\frac{\left(\cos\dfrac{\pi}{2}\cos 88.2^0\right)}{\sin 88.2^0}\right] = 8.832*0.9992 \approx 8.825 \ \text{V/m} \quad (6.10)$$

The equation is next converted to Cartesian coordinates using Equation 6.11 as shown below.

$$E_x = E_\theta \sin(\theta)\cos(\phi) + E_\phi \cos(\theta)\cos(\phi) - E_r \sin(\phi) \approx E_\theta \sin(\theta)\cos(\phi) \quad (6.11)$$

The approximation to Equation 6.11 can be made assuming the calculation for the E field is made along the x axis where ϕ is equal to 0° and θ is approximately 90°.

$$\text{Then} \quad E_x = \hat{a}_x E_\theta \sin(\theta)\cos(\phi) = 8.825\sin(88.2^\circ)\cos(0^\circ) = \hat{a}_x 8.82\,\text{V/m} \qquad (6.12)$$

$$H_y = -\hat{a}_y\frac{E_\theta}{377}\sin(\theta) = -\hat{a}_y\frac{8.82}{377}\sin 88.2^\circ = -\hat{a}_y 23.4 \ \text{mA/m} \qquad (6.13)$$

$$J_s = \hat{n}x\overline{H}_y = \hat{a}_y x(-\hat{a}_x 23.4)\,\text{A/m} = \hat{a}_z 23.4\,\text{mA/m} \qquad (6.14)$$

Equations 6.9 and 6.10 are spherical coordinates resulting in the calculation of the electric field intensity E at the connectors at the bus aluminum roof top at the two patch radio antennas. Once these calculations are made it is much easier to calculate the surface currents using Cartesian coordinates. The x-coordinate is normal to the yz plane (the bus rooftop). The y-axis runs across the rooftop and the z-axis runs in the direction between the monopole and the patch antennas along the ground plane. The surface current density J_s ignores the skin effect as can be observed. A current of 23 mA/m seems like a very substantial amount, however the current is only present for a fraction of 1 m. The current surrounding the connector is about 0.5 mA. The patch antennas are generally fairly isolated from the rooftop to provide a magnitude or better isolation. There will always be some type of leakage current between the ground plane, the bus and a patch antenna mounting. This leakage is due to capacitance, inductance, resistive path or combinations of all three. Since the radios with antennas on the bus rooftop are less than 1 GHz, the imaginary part of permeability and permittivity need not be considered. However, future radios are going to be designed at higher frequency bands such as spread-spectrum unlicensed radios that operate above the 5 GHz. For some applications these may result in some extraneous radiation from circuit resonance or wiring on the ground plane. The EMC/EMI engineer must be always vigilant when examining radio installations, not only the fundamental frequency but also the harmonics that can be generated can result in EMI problems.

The shipboard radiation of ground planes on the ferry may have a different consequence. The corrosion from sea air can produce a diode action, even in similar metal joints, as alluded to previously as the "rusty bolt" problem. Sea-going vehicles have several problems of this type because of this phenomenon.

Railway or mass transit installations near the sea where corrosion from salt air will have a large class of problems, especially those areas where ground planes are coated using various chemicals to protect surfaces, such as bituminous paint, or using aluminum alloys that are not conducive to good conductivity. Even stainless steel may have a slight corrosion capability depending on the grade. Often some type of hot dipping or flame spraying is used as a protective coating to other metals such as steel.

The new type of welded steel rails are preferred over insulated joints. The types of rail installation are of particular interest, to be discussed in Chapter 9. Steel rails have a cross-section due change to cracks around bolted joints. Cracks of course (in particular near sea air) will corrode and form diodes near the crack in the bolt holes. Insulated joints can have shorts across the joint.

Thus far no discussion has been made about maintenance vehicles use by transportation departments. They can be pickup and dump trucks, small buses, SUVs and four wheel drive vehicles for driving cross-country. All of these vehicles are equipped with mobile radios or the operators may require the use of mobile two-way radio or the operator may just use his/her cell phone. In most communication applications, short-range mobile two-way radios are used. These are particularly important when vehicles are used at tunnels where reception is dependent on leaky radiating cables. There is a limit to how many channels a leaky radiating cable will support without producing excessive harmonics that result in crosstalk. At the present time, most tunnels support fire departments, police departments, state police and emergency medical services (EMS) as a minimum. The fire department will generally require at least two channels, the police department will also require a minimum of two channels, usually the state police require two VHF radio channels, the authority running the railroad or subway system will require at least two channels and perhaps more and EMS may require multiple channels because of outside municipalities traveling through the tunnels. If each of these services requires only two channels each account is approximately 10 channels of radio in tunnels. However, EMS services may require both UHF and VHF radios because some of the older municipalities may have not installed modern UHF radios; and state police may require an additional UHF channel or two to handle the municipality.

The problem with mixing VHF and UHF low- and high-band radios is that, in the combiner network installed in tunnel services, radio harmonics may be generated very easily that produce direct hits on several channels. The maximum number of radios that can be supported in tunnel applications is approximately 17 channels, unless of course multiple cables are used. This also presents problems, so the EMC/EMI engineer involved in a project that has more than 17 radio channels supported in the tunnel by leaky radiating cable may have a substantial number of problems with extraneous radiation impinging on the various radios as direct hits. The radio engineer may have done an extensive study of these hits, so one must always be vigilant in examining noise produced by frequencies that are harmonics of any of these radios in a tunnel application.

6.4 REVIEW PROBLEMS

Problem 6.1
The ferry layout in Figure 6.4 has a radio frequency of 700 MHz. Suppose the radio is replaced with: (a) 460 MHz radio, (b) 800 MHz radio. What are the effects on the free space loss for each radio if everything else remains the same?

Problem 6.2
Using A_{eff} for a ¼λ dipole instead of a ¼λ monopole calculate the aperture in the equation in Section 6.3. How does this affect the power at the 900 MHz transmitter and what is the carrier to noise ratio at the transmitter?

Problem 6.3
What is θ in Equation 6.10 if the power is increased to 10 W?

Problem 6.4
Calculate the values for E_x, H_y and J_s for other conversions from spherical to Cartesian and Cartesian to spherical, cylindrical to Cartesian and Cartesian to spherical, and all other collision version factors. See Tables 3.4 and 3.5 in Chapter 3.

6.5 ANSWERS TO REVIEW PROBLEMS

Problem 6.1

(a) $d = 4\dfrac{25*5}{0.674} = 0.742$ km

(b) $d = 4\dfrac{25*5}{0.387} = 1.29$ km

Since both radios are within the Fresnel zone, the free space loss will only be changed by the frequency term in the loss equation.

$$L = 32.4 + 20\log(f_{MHz}) + 20\log(d_{km}) + \text{fade} + \text{Fresnel loss}$$

(a) $L = (32.4 + 53.25 + 9.54 + 40 + 6)dB = 141.19\,dB$

(b) $L = (32.4 + 58.1 + 9.54 + 40 + 6)dB = 146.04\,dB$

Problem 6.2

$$A_{\text{eff}} * P_{\text{Poy}} = 0.0312 * 3.039 * 10^{-4} = 4.741 \mu W \left(-53.32 \, \text{dBm}\right)$$

Power at 900 MHz transmitter = $-53.32 - 39.46 = -92.78$ dBm
Carrier to noise ratio = $-10 - (-92.78) = 82$ dB

Problem 6.3

$$E_\theta = j \frac{377 * 0.5198}{2\pi * 2.5} \left[\frac{\left(\cos \dfrac{\pi}{2} \cos 88.2^\circ \right)}{\sin 88.2^\circ} \right] = 12.477 * 0.9992 \approx 12.467 \, \text{V/m}$$

Problem 6.4

$$E_x = \hat{a}_x E_\theta \sin(\theta)\cos(\phi) = 12.477 \sin\left(88.2^\circ\right)\cos\left(0^\circ\right) = \hat{a}_x 12.46 \, \text{V/m}$$

$$H_y = -\hat{a}_y \frac{E_\theta}{377} \sin(\theta) = -\hat{a}_y 33.05 \sin 88.2^0 = -\hat{a}_y 33.04 \, \text{mA/m}$$

$$J_s = \hat{n}x\overline{H}_y = \hat{a}_y x\left(-\hat{a}_x 33.04\right) A \, / \, m = \hat{a}_z 33.04 \, \text{mA/m}$$

Another technique for calculating surface current is finding the current density (A/m^2) and using equation $J_s = \delta*J$, where δ = skin effect and J = current density in A/m.

7

HEALTH AND SAFETY ISSUES WITH EXPOSURE LIMITS FOR MAINTENANCE WORKERS AND THE PUBLIC

7.1 ELECTROMAGNETIC EMISSION SAFETY LIMITS

Radiation limits are of primary concern to the public and the transportation mainte-nance workers that may be exposed to non-ionizing radiation. The United States Department of Transportation (DOT) and the Federal Transit Administration (*Guidance on the Prevention and Mitigation of Environmental, Health and Safety Impacts of Electromagnetic Fields and Radiation for Electric and Transit Systems*) have research documents showing various exposure levels to humans. The testing is an ongoing effort to ensure public safety from sources of non-ionizing radiation. Non-ionizing radiation can be produced by several sources; some have been discussed previously but they will be reiterated here because they are analyzed to some extent.

The following are some of the best sources of non-ionizing radiation:

1. Power substations or transit systems. These produce 60 Hz high voltage levels. They operate at 13.6–25 kV, depending on the authority operating the transit system.
2. DC power supplies that operate at 680 V DC and up to 1000 A. depending on the make-up of the transit trains.
3. AC 150 hp traction motors are installed on each truck of transit cars.
4. Large ventilating fans that operate in tunnels near stations.
5. Catenary and hot rail arcing that occurs in the pantograph or hot rail shoe.

Electromagnetic Compatibility: Analysis and Case Studies in Transportation, First Edition.
Donald G. Baker.
© 2016 John Wiley & Sons, Inc. Published 2016 by John Wiley & Sons, Inc.
Companion website: www.wiley.com/go/electromagneticcompatibility

6. Harmonics that occur from DC power supply inverters.
7. Radio frequencies for narrowband FM communications.
8. Spread spectrum radio that operates on vehicles and wayside locations.
9. Diesel electric railroad locomotives that have very large traction motors that allow them to pull large loads; or multiple locomotives that run in tandem.
10. Step down transformers along the right of way for transit systems.

Some of these sources have a very large range of broadband harmonics. But it is not just exposure to the fields, that is magnetic field intensity H or electric field intensity E that must be controlled, it is the energy which has a time element that must be observed also for maximum exposure. Some of these sources are of little consequence because the levels are so low, but a table of values has been prepared indicating suggested exposure limits. Due to the ongoing nature of the research, the table is only meant as a guide. The Department of Transportation should be consulted for accurate figures.

Planners and designers have a flexibility of choice of where to install communication and signal houses, traction power substations and traction power supply houses. This is done using software for placing the sources of radiation. In this section, information is provided as a guide to EMC/EMI design and analysis. Electric charges produce electric E fields due to differential voltages and magnetic H fields are produced by moving charges or electric currents flowing. Electric fields are shielded or weakened by materials that conduct electricity (e.g. trees, buildings, human skin, conductive plastics, conductive rubber, metallic paints and metal enclosures), while magnetic fields pass through most materials and are difficult to shield. They are of more concerned for their potential biological effects. Magnetic fields can penetrate biological tissue (human body) without attenuation. Both E and H fields decrease with distance from the source but the decreases are dependent on the size and shape and whether the sources are linear, compact or multiple power lines, and whether the sources are at ground level (this affects the field shape), underground (where earth attenuation occurs) or overhead catenary.

Focus on public and scientific concerns has been on biological adverse health effects from power line frequency magnetic H fields. No dose metric has been agreed upon that pertains to human health as a proposed guide (see Tables 7.1, 7.2 and 7.3). The power frequency gives a variable electric field intensity, that is the E field attenuation by human body cell walls and skin. There is ongoing research to determine the effect of E fields on the human body. However, magnetic intensity H fields at power line frequencies may or may not have an effect on the human body because of the long wavelength. There are still ongoing studies about the health effects of both E and H fields on the human body, but there are several organizations that have recommended standards or guidelines on exposure limits for human bodies. All of the limits have a time element, that is the duration of exposure, which also relates to energy expended on a particular area of the human body. The FCC regulations for exposure limits for the general public vary with frequency; and the durations in Table 7.1 is for the general population and uncontrolled exposure limits. These are FCC standards.

**TABLE 7.1   Electric and Magnetic Field Exposure Limits for 30 min.
F = Frequency (MHz)**

Frequency range (MHz)	Electric field intensity (V/m)	Magnetic field intensity(A/m)	Power density (mW/m²)	Average time (min)
0.3–1.34	614	1.63	100[a]	30
1.34–30.0	824/f	2.19/f	180/f²	30
30–300	27.5	0.073	0.2	30
300–1500	—	—	f/1500	30
1500–100 000	—		1.0	30

[a] Plane wave power density.

TABLE 7.2   Exposure Limits as Suggested by organizations

	Other exposure limits		
IEEE C 95.6 (2002) Reference levels, external	Static (DC) field	AC power at 60 Hz field	Remarks
B field (flux density)	1.18 kilogauss	9 gauss	
E field	5 kV/m	5–10 kV/m	
International Committee for Non-Ionizing Radiation Protection (1998) (ICNIRP)	Static (DC) field	AC power at 60 Hz field	Remarks
B field (flux density)	400 gauss	0.83 gauss	
E field	25 kV/m	4.2 kV/m	
B field (flux density)	5 gauss	0.2 gauss	Pacemaker
E field	2 kV/m	2 kV/m	Pacemaker

TABLE 7.3   Occupational Control Exposure

Frequency (MHz)	Power density mW/m²	Remarks
0.3–3.0	100	Induction furnaces
3–30	180/f²	AM/FM and CB radio
30–300	1.0	CB and aircraft radio
300–1500	f/1500	
1500–100 000	5	

The exposure limits versus frequency are lower in the 30–300 MHz range. This is due to wavelengths at these frequencies being comparable to human body parts such as arms, legs and body length. These absorb radiation energy at resonance. Body weight and stature in general will determine the amount of energy absorbed. Absorption of energy results in localized heating of tissue. The long-term effects are still under study. This is one of the reasons for the time element to be included so that energy absorption may be observed through the heating. The power absorbed mW/m^2 and time element indicate that the size of the object absorbing the energy, such as skin on a human body, will indicate the extent of local heating.

The exposure limits shown in Tables 7.2 and 7.3 give a summary of the maximum exposure limits provided by two organizations: the Institute of Electrical and Electronic Engineers (IEEE) and the International Committee for Non-Ionizing Radiation Protection (ICNIRP). The magnetic flux density is in Gauss units for large fields. It is sometimes expressed in Teslas (10 000 G = 1 Tesla).

Exposure to RF fields about 10 GHz and power densities of 1 W/m^2 will have adverse health effects due to shock and cause cataracts, burns and body absorption in small body area cross-sections such as wrists, ankles and ears.

Where transit operations elevate urban exposure levels above the norm near schools, hospitals and other sensitive centers, these may be of particular concern and require a community health impact assessment or study. The study may include children in homes, schools or playgrounds near transit right of way facilities; this study will also need to include areas where pregnant women are present, people with electronic implants such as pacemakers and metallic implants that are susceptible to magnetic fields, such as wheelchair controls. The results of the study may require the transit system provider to either reduce the exposure with some form of shielding or provide signage with warnings for that particular anomaly due to radiation exposure.

Some design features for typical electric traction power systems are discussed in the ongoing pages of this book. It is not meant as a designers guide for traction power systems; it only makes the EMC engineer aware of some of the design features that may require radiation measurements after the design is done in a right of way survey. There are several handbooks that have resources that may be used to determine how traction power systems are modeled. They are as follows:

1. The American Railway Engineering and Maintenance of Way Association (AREMA). This manual for engineering is annually updated. The AREMA Committee 12-Rail Transit prepared Chapter 12 with a summary guide and procedures for the construction of track structures, infrastructures (including utilities), modeling for passenger design and systems management. Chapter 11 on Commuter and Intercity Rail also includes information on power and propulsion subsystems.

2. The American Public Transit Association (APTA) Rail Task Force produces the APTA Manual, all Standards and Recommended Practices for Rail Transit Systems.

3. The Transportation Research Boards (TRB) Transit Cooperative Research Program (TCRP) produced in 2000 the TCRP RPT-57, Light Rail Track Design Handbook. In part D of Chapter 11 Transit Traction Power, it describes in detail the complex electrical Traction Power System (TPS) components, such as overhead catenary system or third hot rail, Traction Power System Substation (TPSS), connecting cables, wayside distribution systems and corrosion control system design to redirect stray currents in tracks to a substation.

These handbooks indicate how TPSS locations along the route are chosen using training modeling performance software to allow operation at peak power demand, given the topography of the proposed work route and the local power utility demand to ensure power supply for substations. After the route has been chosen all the geometry features of track installation layout, including support poles, span length and tension, will ensure constant contact with the vehicle pantograph. The substations locations in size, cable access conduit and duct banks systems and manholes must meet the constraints of electrical safety in the urban environment and be shielded to minimize EMI with nearby conductors.

The state in which the transit system is operating or newly installed must comply with local state regulations for electric E field intensity and magnetic flux density B field at the center and edge of the right of way. States may have standards for rail crossings where private vehicles pass. These may vary from state to state; again the EMC engineer must look up state regulations for both E and B fields (magnetic field density). The range of the various fields is: E field for the right of way 1.6–9.0 kV/m and for rail crossings 1.0–11.8 kV/m and B fields as high as 200 mG.

Table 7.2 should not be construed as a regulation when dealing with state transit systems it is only a guide to determine safety issues. It should be looked at as a starting point when dealing with E and B fields. The state has jurisdiction and state statutes should be followed when dealing with transit systems. The whole objective here is safety issues.

Electromagnetic proliferation of the environment and rising occupational exposure is due not only to urban transit systems but personal wireless data and voice devices such as cell phones, base station transmitters, TV, radio, satellite broadcast radio navigational devices such as GPS, weather radar, law enforcement radar, in vehicle wireless devices, military transportation, power lines and wireless devices in the home such as wireless speakers, telephones and other types of devices. Magnetic field devices with long wavelengths include such as power line transformers located near bedroom windows, public housing and large-scale switch yards for high voltage switching, large motors in manufacturing areas that operate such items as shears, sawmills, Meglev vehicles and electric furnaces in steel mills. The list is very extensive and the damage to the human body due to exposure from large magnetic fields is still in question.

A typical power distribution system with data for B magnetic flux densities shown in Table 7.4 can be compared with Table 7.2. The limits are less than the proposed exposure limits in this table for low frequencies, that is the peak values in parentheses. This also holds true at distances of 10 m from the right of way where peak values are no more than 100 mG, rolling off rapidly to 0.01 mG at 1000 m. It must all be kept in mind that these wavelengths are extremely long so that resonance does not occur in the rooms or enclosures where these installations are housed. All of these areas may have other types of shielding which may reduce the magnetic flux density B field even further, that is where reinforced concrete structures block some of the H field radiation, as was alluded to in Chapter 2.

Values for sources of electromagnetic B field flux density is provided in Table 7.5.

The main exposure to magnetic B fields to passengers inside a bus are in the engine compartment and rack mounted communication equipment. Emergence of

TABLE 7.4 Emission Limits for Magnetic Fields for Transportation Facilities. Values are Averages (with Maximum Values in Parentheses)

System and facility	Static mG	5–45 Hz mG	50–60 Hz mG	65–300 mG	305–2560 Hz mG
Control room UPS vault	669 (1176)	1.9 (2.8)	37.4 (48.4)	18.3 (21.3)	37.2 (44.6)
Substation	349 (358)	0.4 (0.9)	10.9 (34.3)	0.7 (2.1)	0.6 (1.5)
Relay room	326 (464)	0.3 (0.5)	1.4 (2.8)	0.8 (1.8)	0.6 (1.1)
Traction power station	841 (2750)	1.3 (18.4)	9.6 (110.7)	3.8 (78.4)	3.4 (55.8)

TABLE 7.5 Values for Sources of Electromagnetic B Field Flux Density

mG	0–10	10–20	20–30	30–40	40–50	50–60
Ferry boat	1					
Escalator	2					
People movers	5					
Electric cars + light trucks	6					
Conv.Cars + light trucks	6					
ATV	6					
Jetliner		15				
Airport electric shuttle		15				
Commuter train (AC electric)					50	
NEC electrified 60 Hz						55
PR-07 Meglev						60

AC power commuter and high-speed trains such as Amtrak Acela passenger rail on a Northeast corridor, New Jersey Transit Coastline, and Metro-North commuter rail have shown that an average 60 Hz magnetic field exposure for passengers and workers may exceed 50 mG. However, the public has accepted any excess electromagnetic field as a trade-off against economic and environmental benefits, commuting convenience and time savings.

Transmission and distribution lines are designed with some of the following features:

1. The power distribution lines can be either overhead or underground. Overhead power lines produce both AC electric fields and magnetic fields at 60 Hz frequency and their higher harmonics. Electric E field intensity is dependent on the line voltage and magnetic H field intensity. When lines are routed underground through duct banks, they produce lower E fields but may produce magnetic H fields

above ground level. The power transmission is at high voltage levels and step down at the substation to lower usable voltages for the transit system.

2. Transmission system lines at distances of 300 feet were averaged electrical demand may produce magnetic fields similar to those in the home.

3. Power distribution lines have typical voltage ranges from 4 to 24 kV. One of the most common is 13.6 kV. E fields directly beneath the power distribution lines can vary from 3 to 200 V/m and B fields directly beneath the power lines or main feeders 10 to 20 mG. These levels are also present for underground installations but peak magnetic B fields are highly dependent on H fields which are dependent on current; 40–50 mG can be present above the underground lines. This is dependent on the current flow that affects the H field in the construction of the duct bank.

4. Electrical power at the substation: the strongest electromagnetic levels outside the substation is due to the power lines, that is cables entering and leaving the substation. The strength of the electromagnetic fields from equipment within the substation, such as transformers, reactors and capacitor banks, decreases rapidly with increasing distance. The B field decreases by the reciprocal of the square of the distance and it is a function of the permeability multiplied by the magnetic H field intensity. This produces the magnetic flux density or B field ($B = \mu*H$). A wall or enclosure surrounds the substation. Outside the wall the noise is indistinguishable from noise in the environment.

The standards provided in Tables 7.1 and 7.3 are voluntary standards used as guidelines. However, the FCC requirements enacted as 47 CFR 1.1307 (b) are legal standards that require compliance. When an EMC engineer is taking measurements, he/she must comply with these standards because of the legal aspects.

The current standards for electromagnetic electric and magnetic fields prevent induce body currents in excess of normal levels and RF heating of the whole body or body parts, such as limbs, ears, eyes, hands, head and other body parts. Existing standards address only safety, not health, since they protect from short-term acute heating due to RF radiation exposures. These standards do not address the long-term potential to produce chronic effects on health due to the level of exposure. There are many other organizations that have standards for exposure limits some without regard to long-term health adverse effects.

Standards are designed which a large safety margin to prevent potentially damaging induced currents that affect body core temperature rises of 1 °C. A 1 °C increase in core temperature is considered harmful to cells, tissue and organs. The standards provide a margin of safety for both occupational and public exposure, including sensitive targets such as pregnant woman, children, handicapped persons with pacemakers, and other health issues that electromagnetic radiation may affect. It should be then noted that uncontrolled exposure limits for the public are much lower than those for controlled occupational limits.

Table 7.6 is provided with other organizations that have standards that pertain to transportation issues these may or may not be obsolete because standards are a living

TABLE 7.6 Organizations that Set Emission Standards

American Public Transportation Association Standard for the development of an electromagnetic compatibility plan (EM CCP), APTA SS-E-010-98
Association of American Railroads, Railway electronics environment requirements, AAR-5702
Association of American Railroads, Remote control locomotive standards, standard tests, AAR-5507
Association of American Railroads, Specification for remote control locomotive communication systems operations at 220 MHz
DOD Standards MIL-STD-461E Requirements for the control of electromagnetic interference and susceptibility
DOD Standard MIL-Standard-462D Department of Defense test methods standard for measurement of electromagnetic interference characteristics
CENELEC Standard EN 50121 Railway applications. Electromagnetic compatibility. Parts 1–5
CENELEC Standard EN 50155-2001, Railway applications. Electronic equipment used on rolling stock.
European Telecommunications Standards Institute, ETSI EN 300 113-1 V1.4.1 Electromagnetic compatibility and radio spectrum matters; land mobile service; radio equipment intended for the transmission of data (and/or speech) using constant or non–constant envelope modulation and having an antenna connector; part 1: technical characteristics and methods of measurements, 2002

TABLE 7.7 Body Contact Standards

Condition	General public (mA rms)	Controlled environment (mA rms)
Both feet	2.5	6.0
Each foot	1.35	3.0
Contact, grasp[a]		3.0
Contact, touch[b]	0.5	1.5

[a] Grasping contact limit pertains to control environments where personnel are trained to affect grasping contact and to avoid touch contacts with conductive objects that present the possibility of painful contact burns.
[b] Limits apply to current line between body and grounded object that may be contacted by the person.

document they are updated from time to time. The EMC engineer analyzing or designing transportation equipment should always be vigilant and check with the latest standard. Often new limits are imposed that may be less than those that appear in the examples provided in this book.

Table 7.7 gives the limits on current that result in burns to the general public or maintenance personnel. The burns are different for various types of contact, for example burns from radio frequency are painful because they burn from inside out: what appears to be a small burn on the outside actually greater inside under the skin and they are very difficult to heal. Often when working with high voltage the person (maintenance personnel) must always ensure that any capacitance in the system has been discharged.

The author's experience with high-voltage discharge has been rather extensive. These are some incidents that occurred during a research project. The equipment built had a capacitive discharge system that when discharged produced a high current high voltage at a gap similar to a spark plug. Only, the energy required in the gap was very large and it was immersed in saltwater. The equipment was built for forming metals in a mold with a rubber bladder with a gap inside with saltwater. The technician working with the equipment made an error that caused the arc that should have occurred inside the rubber bladder to arc on the outside of the enclosure. It was equivalent to a small lightning bolt. He lost hearing in one ear due to the sheer noise of the arc striking close by.

Another incident occurred in the same laboratory with the same capacitors that were used for capacitor discharge. A discharging stick was always used before performing maintenance on the discharge system for the spark gap. The maintenance person forgot to discharge the capacitor bank. These capacitors were huge, four farads; each could store a large amount of energy. The person touched the top of the capacitor with his hand and his elbow grounded against the enclosure; the arc that occurred split his arm from wrist to elbow. After observing this accident the author has always been aware to discharge any equipment that may have capacitors or energy-storing devices.

The author had to give a speech about electrocution safety and one thing that stuck in his mind is that it only takes 7 mA across the heart to cause death. The reason most people do not get electrocuted easily is due to the 10 kΩ of resistance in the skin on the fingers. However, wet hands or other body parts have reduced resistance and present much more of a shock hazard. The shock hazard for touching rails is 55 V RMS. This can be due to induction from magnetic fields or ground return currents. This does not include the hot rail, it is the running rails only.

Electromagnetic interference prevention and control:

1. Passive engineering controls where EMI is observed, such as shielding that can be as simple as copper screening or as stringent as μmetal shielding.
2. Wire loops used for canceling H fields can be implemented in the opposite direction to the current, causing EMI similar to a twisted-pair.
3. Active shielding that requires a power supply and feedback loop can control current, magnetic field and direction.
4. Changing equipment location or separating equipment that results in the EMI problem.
5. Use lower resistance components, larger circuit breakers, higher values of insulation buried wiring in a steel conduit.
6. Implement filters active for small print applications such as PA systems, non-active filters and large component applications such as traction motors.

7.2   EMI PREVENTION AND CONTROL

Before any work is done, a survey must be done along the right of way. This requires a heavy-duty vehicle such as a four wheel drive SUV with sufficient capacity to carry test equipment. The test equipment suite is: antennas sufficient to cover the spectrum of frequencies of 30 Hz to 1 GHz, spectrum analyzer capable of covering the same frequency range, an inverter, laptop computer, attenuator with at least five decades of attenuation and other miscellaneous tools that may be necessary. EMI measurements for electromagnetic interference must be taken and recorded as a baseline.

These measurements are taken and recorded generally at every 100 meters depending on the authority requesting the measurements. Some authorities may require measurements on both sides of the track bed out as far as 30 m. The terrain and weather measurements to be taken may require the removal of brush and ground cover but usually the roadbed will be barren. However, both sides of the roadbed out to 30 m may not.

After the track and all equipment is installed prior to commissioning (this is generally during the integration phase when all contractor equipment has been installed) the same measurements must be taken along the right of way. These measurements are taken while the subway is operational. The measurements include electromagnetic E and H levels versus frequency, time and distance, the full range of speeds of the particular light rail, the number of cars both minimum and maximum (such as a single car, married pair, or ten unit string of cars), transfer trip operations (i.e. the substation feed is disabled and the load must be carried by a single sub-station), during braking (automatic and emergency stop breaking) and acceleration (automatic and hard acceleration due to motorman control).

Exposure to EMI fields, in particular the E and B fields, determines if the safety standards to the public and occupational personnel are compromised. Areas where EMI fields are compromised will require warnings and in some cases fences to prevent personnel from being injured by EMI sources. A case in point is where high-power radio antennas may be located along the right of way. These are surrounded by a fence with warning signs; this may also be necessary near TPS locations. Most TPS houses are adequately shielded outside for the public and inside for maintenance personnel.

Generally prior to any work done during the design phase of the project an EMC plan is required (Chapter 8, Sections 8.1–8.4). It must have the following:

1. Identify potential EMI sources and hazards to transit rail operations and communications equipment.

2. Consider all the options for the best EMI prevention, control and mitigation techniques using the simple flowchart in Chapter 1 that includes shifting locations of equipment, shielding and filtering. Adding warning signs for EMI fields that are hazardous to humans.

3. Consider best case practices for EMI susceptibility control procedures such as shielding, corrosion protection, surge protection, failsafe circuit design, backup communications, fire protection and suppression locations,

surveillance equipment,. rack or enclosure mounted equipment (communications and signals equipment), wayside installations in cabinets with air-conditioning and heating, control room workstation layouts and track display board layouts.

4. A safety and failure analysis of transit systems is required by most authorities and should be a part of EMC analysis. It need not be in the plan but it is usually a separate document.

5. All grounding methods should be studied and presented as part of the document they may be included as a separate section or may be included for each particular item in the section.

6. Earth ground return currents, grounding to metallic sections such as rebars, return rails, communication house ground grids, ground halos used in communication houses, lightning protection grounds, analysis of various types of power supplies that are non-linear that use frequencies that may be found on the neutral. Control rooms that have a false floor that allow wiring and grounding beneath the control room floor, whether rack mounts are insulated from the floor of communication houses are grounded in another fashion and soil conditions for earth grounds.

7. Frequency bands that generate noise by the pantograph and catenary during normal operations and transient conditions when arcing occurs.

8. Along the right of way and stations analyze the frequency spectrum, (E, H, and B) field strengths, wireless communications, intermodulation analysis, power propulsion system, auxiliary HVAC power, emergency lighting, and the signage and security systems.

The author has elected not to use best case and worst case analysis as presented in the DOT document for obvious reasons. Since ongoing testing and research and health issues are continuing it is more prudent for an EMC engineer to look up the DOT documents for the latest research and analysis. To prevent litigation it is always good practice to look up the best radiation exposure that deals with health at the time when EMC plans and tests are performed. Provided here is a short list of organizations that may be of interest to the EMC design engineer. The entire list can be found in the document alluded to at the introduction.

1. National Capital Institute of Environmental Health Sciences, www.niehs.nin. gov/emfrapid/booklet/home.htm

2. FCC questions and answers on RF safety, www.fcc.gov/oet/rfsafety/rf-fags. html

3. World Health Organization. Fact Sheet 183, Electric Fields and Public Health. June 2007 update, www.who.int/mediacentre/factsheet/fs322/en/index.html

4. AREMA Manual for Railway Engineering, www.erema.org

5. IEEE Safety Levels with Respect to Human Exposure to Radio Frequency Electromagnetic Fields, 3 kHz to 300 GHz, 2005

6. IEEE C 95.7 Recommended Practice for Radio Frequency Safety Programs, 3 khz to 300 GHz

7. ICNIRP Guidelines limiting exposure to time-varying electric, magnetic and electromagnetic fields (up to 300 GHz) *Health Physics* **74**:494, 1998

8. OSHA radio frequency and microwave radiation, www.osha.gov/SLTC/ radiofrequency radiation

9. ANSI C 6318-1997, *"Recommended Practice for an On-site Test Method for Estimating Radiated Electromagnetic Immunity of Medical Devices to Specific RF Transmitters"*

10. Electric Power Research Institute, American Association of Railroads, and AREMA, *Power Systems and Railroad Electromagnetic Capability Handbook*, EPRI revised 1st edn, 10102652, Final Report, November 2006.

This list contains some of the more important references. The original document has 45 references.

7.3 ANALYSIS OF RAILS AS A SHOCK HAZARD

Rails can be best characterized as a transmission line. This is discussed in Chapter 9; however, the analysis is conducted on an individual running rail that may be a touch shock hazard under certain conditions. The assumption is made on a 700 ft (218.5 m) section of rail. It is 100 LB/yard rail with a DC resistance of approximately 54.5 μΩ/m and inductance at the seventh harmonic of the power supply (i.e. 420 Hz). This is approximately 0.8 μH/m, including the inductance of the rail plus the leakage inductance due to the second rail separated by air. At the fundamental frequency of the AC drive of the traction motor at 218 Hz, the DC resistance and inductance are 107.5 μΩ/m and 1.507 μH/m respectively.

The worst-case condition is three cars starting at the fundamental frequency of 218 Hz with the peak value for the current at 80 A. The calculations are shown in Equation 7.1. The two evaluations of Equation 7.1 for the fundamental and harmonic; as can be observed the harmonic is insignificant. Do not confuse 218.5 m with 218 Hz in the calculations.

$$V = \sqrt{\left[R_{\text{rail}} \left(\mu\Omega / \text{m} \right) * \text{dist} \left(\text{meters} \right) \right]^2 + \left[2\pi * f \left(\text{Hz} \right) * L \left(\mu\text{H/m} \right) * \text{dist} \left(\text{meters} \right) \right]^2 } * I_{\text{Peak}}$$

(7.1)

$$V = \sqrt{ \left(107.5 * 10^{-6} * 218.5 \right)^2 + \left(2\pi * 218 * 1.507 * 10^{-6} * 218.5 \right)^2 } * 80 = 36.13 \ \text{V} \left(\text{fund} \right)$$

$$V = \sqrt{ \left(642 * 10^{-6} * 218.5 \right)^2 + \left(2\pi * 420 * 0.8 * 10^{-6} * 218.5 \right)^2 } * 0.4 = 0.2 \ \text{V} \left(\text{harmonic} \right)$$

Measuring the touch voltage on the hot rail would be very precarious to say the least. It is not attempted here for obvious reasons. The analysis here is more for a catenary type system.

Each traction motor is 150 hp (11.19 kVA). This makes the peak value that the motor may draw 164 A for a 680 V supply. From the calculations above, it should be noted that acceleration is very slow and not instantaneous or the peak currents could rise much higher because these motors are AC induction type. The running speeds, frequency and voltage magnitude vary because the drives are the pulse width modulated (PWM) type. The type of modulation, such as a multistep drive, provides lower harmonic content. For these multistep types the filtering is smaller than the larger step modulation. The manufacturers of the cars use the drive appropriate for the application; and it should not concern the EMC engineer unless employed in subway car design. When the emission tests are made, the car must pass the emission tests for railway systems. The test results for subway cars should be included in the EMC plan if possible. This will generally be demanded by the authority initiating the project.

7.4 LIGHTNING AND TRANSIENT PROTECTION

Protection against lightning strokes is not only the touch type application but it should also be applied where charge buildup can occur. The charge buildup must be bled off to prevent serious accidents due to static charge buildup, as was alluded to previously for the 4 F capacitor banks. Stored charge can be extremely dangerous. In another case in a 33 000 V transmission was disconnected from the power company but connected to the insulators and open at both ends. A thief decided to steal the copper lines, noting that they were disconnected at both ends. However the insulators had managed to allow the copper to store charge and when the man touched the copper lines he was killed instantly because of the high charge stored on the copper line. If the power company grounds their lines during a storm when lightning is present, the high voltage during a lightning stroke to the lines would possibly damage the tower if the lines are grounded using the tower. This example is to warn anyone deciding to work on metallic objects that have high capacity to ground that care must be taken to discharge the materials before touching them.

Another type of safety that is not electrical but has a serious possibility of death occurring is communication houses and remote areas that use fire suppression such as Halon gas. This type of suppression will also suppress oxygen from getting into the lungs of maintenance personnel. Signs and placards are required as warnings to leave the house immediately if the gas escapes or is used to suppress flames. These types of systems are similar to sprinkler systems; however this type of suppression does not destroy electronics equipment when used as automatic sprinklers would.

All communication and signal houses have automatic alarm systems that connect directly to the fire department and the central control room. These types of alarms are both audible in the communication or signals house itself and also have some alarm method to warn the operations control room that a fire has occurred. The fire alarm system has a battery backup for complete loss of power so that it will function. The battery is continuously monitored and must be replaced periodically, usually every five years; but this is set by the authority and determined to be adequate during the design phase of the communication or signals house.

Signs are necessary at railroad stations to warn passengers that that may be hearing-impaired. These signs operate automatically at the station and are periodically tested to make sure that the software is functioning properly. The same warning system is necessary but with audio instead of signage for people that are vision-impaired. These messages are considered canned messages because they are automatic and prepared offline and do not change with time and are protected from hackers or other types of personnel from altering the content.

Security phones with easy access are located on platforms. They only require the user to pick up the phone and generally press one button for assistance. This assistance may be in the form of security police or may be a call for assistance for a physically impaired person. These are connected directly to the central control room and in some cases the security police department of the particular area being served.

7.5 POWER LINE SAFETY CALCULATIONS

Example 7.1

The following are the parameters for the power line shown in Figure 7.1:

$L_{1-2} = L_{2-3} = 6$ in (0.1525 m), $r_3 = 13$ ft (3.96 m), $r_2 = 4$ m, $r_1 = 4.1$ m, d = 19 ft (5.796 m), power line voltage (V) = 13.6 kV, power line frequency = 60 Hz, three phase.

$$V = 13.6\,\mathrm{kV} = -\int \overline{E} * d\overline{r} = -\frac{\rho_L}{2\pi\varepsilon_0} \int_0^L \frac{dr}{r} = \frac{\rho_L}{2\pi\varepsilon_0} \ln[r]_{r=0}^{r=0.1525} = \frac{\rho_L}{2\pi\varepsilon_0} 1.88$$

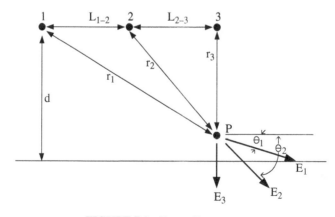

FIGURE 7.1   Power line exposure

Then the linear charge on a conductor ρ_L is as calculated below.

$$\rho_L = \frac{2\pi\varepsilon_0 V}{1.88} = \frac{2\pi * 8.854 * 10^{-12} * 13.6 * 10^3}{1.88} = 0.4024\,\mu C/m$$

Assume the y direction is vertical from point P and x the horizontal direction from point P the calculation for the three electric fields E are as follows:

$$\bar{E}_3 = -\hat{a}_y \frac{\rho_L}{2\pi\varepsilon_0 r_3} = -\hat{a}_y \frac{1.422 * 10^{-6}}{2\pi * 8.854 * 10^{-12} * 3.96} = 1.825\,kV/m$$

$$\bar{E}_2 = \frac{\rho_L}{2\pi\varepsilon_0 r_2}\left(-\hat{a}_y \sin\theta_2 + \hat{a}_x \cos\theta_2\right)*10^3 = 1.825\left[-\hat{a}_y \sin\left(87.9^0\right) + \hat{a}_x \cos\left(87.8^0\right)\right]*10^3$$

$$\bar{E}_2 = \left[-\hat{a}_y 1.825 + \hat{a}_x 0.0669\right]*10^3 \text{ V/m}$$

Three phase lines are spaced fairly close; a correction for phase is rather small. This is a more conservative value as shown for the maximum. For widely spaced lines the phase terms must be used as well such as in 765 kV lines where spacing can be over 15 m between them. The calculations include correction for the three phases $0°$, $120°$ and $-120°$.

$$\bar{E}_1 = \frac{\rho_L}{2\pi\varepsilon_0 r_1}\left(-\hat{a}_y \sin\theta_1 + \hat{a}_x \cos\theta_1\right)*10^3 = 1.8\left[-\hat{a}_y \sin\left(85.6°\right) + \hat{a}_x \cos\left(85.6°\right)\right]*10^3 \text{ V/m}$$

$$E_y = [-\hat{a}_y 1.78 + \hat{a}_x 0.138]*10^3 \text{V/m}$$

$$E_y = -\hat{a}_y 1.78 \text{ kV/m}$$

$$E_x = -\hat{a}_x 0.205 \text{ kV/m}$$

$$E_{Max} = \sqrt{1.78^2 + 0.205^2} \text{ kV/m} = 1.79 \text{ kV/m}$$

This electric field violates many of the exposure limits for maintenance personnel and would warrant some correction to the height of the poles to reduce the electric field intensity. Particularly this would be hazardous to people with handicaps such as a person with a pacemaker. Warning signs would need to be installed if the height of the poles could not be increased for this application, for example if the lines are near stations.

High magnetic H field intensity is another danger that should be considered. It is 17.69 A/m [(1790 V/m)/377 Ω] and the B field is 5.97 Wb/m ($4.748*4\pi*10^{-7}$). This value translates to 59.7 mG which can be considered a fairly low level.

Example 7.2

Figure 7.2 depicts power transmission in duct banks with manhole access. This is a common method for routing cabling between substations and from substations to power supplies. The duct banks are usually arranged in a 3×3 tube fashion; sometimes

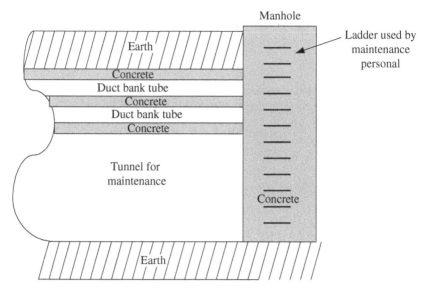

FIGURE 7.2   Manhole layout for power and communications

larger ones are used such as 3 × 6 arrangements. However, this can vary depending on the demands of the system layout. Often two sets of duct banks are installed to separate power from communication lines and signals. The duct bank tubes are usually made of a composite of steel and fiberglass, or the high-voltage lines are installed in steel inter-ducts, that is similar to a conduit with a fiberglass liner. The voltage on the power lines is 13.6 kV.

The calculation for finding the charge per linear foot is done using Equation 7.2:

$$V = -\frac{\rho_L}{2\pi\varepsilon_r\varepsilon_0}\int_0^{0.0256}\frac{dr}{r}\,\overline{a}_r = \frac{\rho_L}{2\pi\varepsilon_r\varepsilon_0}\ln 0.0256 = -\frac{\rho_L}{2\pi\varepsilon_r\varepsilon_0}\left(-3.66\right) \qquad (7.2)$$

A duct bank liner consisting of 1 inch thick fiberglass with a relative permittivity of three lines the inter-duct to prevent corrosion. The charge per linear foot is as calculated with Equation 7.3

$$\rho_L = \frac{2\pi\varepsilon_r\varepsilon_0 V}{3.66} = \frac{2\pi * 3 * 8.854 * 10^{-12} * 13.6 * 10^3}{3.66} = 2.189\,\mu\text{C/m} \qquad (7.3)$$

The calculation for the voltage at the steel in the duct is done using Equation 7.4

$$V = \frac{\rho_L}{2\pi\varepsilon_r\varepsilon_0} = \frac{2.189 * 10^{-6}}{2 * \pi * 3 * 8.854 * 10^{-12}} = 13.1\,\text{kV} \qquad (7.4)$$

The next equation for attenuation through from the steel of the inter-duct has the following parameters [2]: μ_r = relative permeability of the metal shield, t = thickness, r = shield radius, r = 2 inches, t = 1/4 inch, μ_r = 1000.

$$A_{Total} = 20 Log\left(1 + \mu_r t / 2r_{radius\ of\ shield}\right) \tag{7.5}$$

The result is shown below.

$$A_{Total} = 20 Log\left(1 + 1000 * 0.25/2\right) = 42\,dB$$

Applying attenuation to the steel conduit results in a voltage at the surface of 103.96 V.

V = 103.96 V; the resistance of the concrete is calculated at 217.5 Ω with Equation 7.6.

$$R_{concrete} = \sqrt{\frac{\mu}{\varepsilon}} = \sqrt{\frac{4\pi * 10^{-7}}{3 * 8.854 * 10^{-12}}} = 217.5\ \Omega \tag{7.6}$$

The voltage at the outside of the duct bank is 38.0 V using Equation 7.7.

$$\frac{R_{concrete} * V}{R_{concrete} + R_{air}} = \frac{217.5 * 103.5}{377 + 217.5} = 38.0\ V \tag{7.7}$$

The voltage at the surface of the duct bank is not a shock hazard to the touch. However, manholes are generally damp and can be full of water. It is advisable to warn personnel to wear rubber gloves and rubber boots when going down into manholes. Generally when repairs occur, the circuits are shut down and locked out using lockout tag out safety to prevent electrocution; however, personnel must always be vigilant not to make assumptions about the power having been shut off.

7.6   FCC REGULATIONS

This section of the book contains excerpts from the FCC regulations of occupational exposure and exposure to the public of various types of radiation. Tables 7.8 and 7.9 provide the limits on human exposure for occupational and public personnel. The time element for exposure is in terms of controlled and uncontrolled exposure. The tables provide the range of frequencies, the electric field intensity E, the magnetic field intensity H, the power density S and the average exposure time. Given the exposure time and the various fields the energy can be calculated easily in joules per square meter by multiplying S by the average time in seconds.

7.6.1   Occupational/Controlled

Limits apply in situation in which persons are exposed as a consequence of their employment, provided the persons are fully aware of the potential for exposure and can exercise control over their exposure. Limits for occupational/control exposure also apply in situations when an individual is transient through a location where occupational/controlled limits apply, provided he/she is made aware of the potential for exposure. These limits apply to amateur station licenses and members of their immediate household, as discussed in Reference [8].

TABLE 7.8   Limits for Occupational/controlled Exposure. F = Frequency (MHz)

Frequency range (MHz)	Electric field intensity E (V/m)	Magnetic field intensity H (A/m)	Power density S (mW/cm²)	Average exposure E² or H² (min)
0.3–3.0	614	1.63	100[a]	6
3.0–30	1842/f	4.89/f	900/f[a]	6
30–300	61.4	0.163	1.0	6
300–1500	—	—	5	6
1500–100 000	—	—		6

[a] Plane wave equivalent power density.

TABLE 7.9   Limits for General Population/uncontrolled Exposure. F = Frequency (MHz)

Frequency range (MHz)	Electric field intensity E (V/m)	Magnetic field intensity H (A/m)	Power density S (mW/cm²)	Average exposure E² or H² (min)
0.3–1.34	614	1.63	100[a]	30
1.34–30	824/f	2.19/f	180/f[a]	30
30–300	27.5	0.073	0.2	30
300–1500	—	—	f/1500	30
1500–100 000	—	—	1.0	30

[a] Plane wave equivalent power density.
Table 7.4.2

7.6.2   General Population/uncontrolled Exposure

Exposures apply in situations in which the general public may be exposed, or in which persons that are exposed as a consequence of their employment may not be fully aware of the potential for exposure or cannot exercise control over their exposure. As discussed in Reference [8], these limits apply to neighbors living near amateur radio stations.

7.6.3   Specific Absorption Rate in the Human Body Due to Wireless Communication Devices

Tables 7.10 and 7.11 are tables taken from Reference [8] for radiation from cordless telephones and wireless devices. These tables are meant as a guide to the engineer taking measurements and not as a design guide because data could have possibly changed since the tables were generated. The prudent engineer doing design work will always look up the latest radiation limits. The document alluded to in Reference [8] provides data and a description of the testing done to accumulate this data on tissue, with techniques for doing specific absorption rate (SAR) testing. The document also provides radiation data for the ears, feet, eyes, ankles, hands and entire body length. It also provides testing techniques for all wireless equipment with the limitations given.

The transmitter output power must be clearly defined when comparing the radiation levels in Tables 7.10 and 7.11 for wireless devices. The tables are generated for mobile and fixed transmitters. The exposure limits are dependent on the antenna configuration and the power output of the device itself.

TABLE 7.10 Cordless Phone Standards and Repeaters

Transmitter or device type	Output	Application method to ensure compliance
Cordless phone handsets and most other transmitters using monopole or dipole type of antennas as an integral part of the device	≤0.3 W at 915 MHz or ≤0.2 W at 2450 MHz	If the device or its antenna operates at less than 2.5 cm from a person's body (excluding hands, wrists, feet and ankles), the potential for exceeding SAR limit is dependent on the operating configuration and exposure conditions of the device. Operating and warning instructions in the operator's manual may be used to ensure compliance. If such instructions are ineffective for ensuring compliance, especially when output is greater than 50–100 mW, it may be necessary to demonstrate compliance with respect to SAR limit.
Cordless phone handsets and other transmitters that are carried next to the body of the user or operate at a distance closer than approximately 5 cm to the body of the user or nearby persons.	≤0.3 W at 915 MHz or ·≤0.2 W at 2450 MHz	Generally at above 300 mW (200 mW at 2450 MHz) the potential for exceeding SAR limits is dependent on the antenna design and device operating conditions. Warning instructions and warning labels may be used to limit the exposure durations and/or conditions to ensure compliance. However if manufacturers believe that such warning instructions and labels will not be effective in keeping persons at specific distances necessary to ensure compliance, especially when output is greater than 400–500 mW, it may be necessary to demonstrate compliance with respect to SAR limit.
Transmitters using monopole or dipole type antennas as an integral part of the device normally operate at closer than 20 cm to users or nearby persons but more than approximately 5 cm away from such persons.	≤0.3 W at 915 MHz or ≤ 0.2 W at 2450 MHz	Operating and warning instructions in the operator's manual indicating the minimum separation distance between the antenna and nearby persons in order to avoid extended periods of exposure at closer than this distance to ensure SAR compliance. When operating and warning instructions are ineffective, the use of warning labels on a transfer element may also be necessary to caution nearby persons to limit their exposure duration and/or conditions to ensure compliance. If warning labels are not desired, SAR evaluations (even though they may not be required) may be used to demonstrate compliance to obviate the need for any other warning label that might otherwise be necessary.

TABLE 7.11   Wireless Emission Standards

Transmitter or device type	Output	Application method to ensure compliance
Transmitters using external antennas, including Omni, patch, logarithmic, parabolic reflector and dish type antennas. For outdoor operating, antennas generally mounted at remote locations such as the top or side of most buildings where antennas are at least 20 cm away from nearby persons.	≤2.5 W at 915 MHz ≤2.5 W at 915 MHz or ≤ 4 W at 2450 MHz	Professional installation provides installers with instructions indicating the separation distance between the transmitter/antenna and nearby persons to ensure RF exposure compliance and to inform installers to ensure compliance to proper installation. Professional installation is preferred for these types of operations. However end-user installations may require certain additional information to allow persons who do not have professional skills to properly install antennas to ensure compliance. Transmitters operating at 2.5 W EIRP (1.5 W EIRP) or less at 915 MHz, or at 4 W EIRP (2.4 W ERP) are less at 2450 MHz and generally are not expected to exceed MPE limits when nearby persons are 20 cm or more from most antennas. Therefore, special instructions and warnings are normally not necessary to ensure compliance.
Transmitters using indoor antennas that operate at 20 cm or more from nearby persons.	≤2.5 W at 915 MHz ≤2.5 W at 915 MHz or ≤ 4 W at 2450 MHz	If the MPE distances are greater than those required for normal operation of the device, instructions, warning instructions and or warning labels may be used to ensure compliance by indicating the minimum separation distance to comply with MPE limits. If the antennas are professionally installed to ensure compliance, warning instructions and warning labels are not necessary. Transmitters operating at 2.5 W E IRP (1.5 W ERP) or less at 915 MHz or at 4 W E IRP (2.4 W ERP) or less at 2450 MHz, generally are not expected to exceed MPE limits when nearby persons are 20 cm or more from most antennas. Therefore, special instructions and warnings are normally not necessary to ensure compliance.
Transmitters using high gain and antennas for indoor or outdoor operations	≤4 W at 2450 MHz	If the MPE limits may be exceeded in the main beam of the antenna, installation procedures, warning instructions and or warning labels as described above may be used to ensure compliance by providing professional installers and end users with instructions to point the main beam of the antenna at locations not occupied by persons and to warn others to maintain a specified distance from the antenna.

The EIRP levels are for fixed and mobile operations as defined in 47 CFR 1.1307 and 2.1091 FCC regulations. Portable operating configurations as defined in 47 CFR 2.1093 for both conducted and radiated EIRP output power could be considered for near field exposure conditions. EIRP is the product of the maximum output power available at the antenna terminals of the transmitter and antenna gain. Also, when applicable, a source based average time duty factor must be considered.

When simultaneous transmission occurs at multiple frequencies, all within the same physical location such as an antenna farm, the total radiated emission is considered for the RF exposure limit. As an example, for two fixed radios transmitting, each operating at 1 W, the radiation exposure is 2 W. More information on these exposure limits is provided in 47 CFR 1.1.307 (b, 3). See Reference [8] for more information on the text presented above. The information provided is to make the EMC engineer aware of the effect of single and multiple radiating devices that pertains to health issues. If for any reason the engineer suspects an antenna and an antenna farm with wireless devices are health hazards, the FCC title 47 CFR regulations should be consulted.

Limits alluded to in Table 7.10 are for lower level transmitters and are applicable to radios or devices operating more than 2.5–3.0 cm from a person's body, excluding wrists, hands and ankles. The information provided in these tables is a starting point. The complete set of regulations must be examined by the EMC engineer to determine if exposure limits have been violated. However from the radiation levels an engineer can at least decide whether or not it is necessary to pursue health issues.

Reference [8] has a host of references or organizations that are testing for radiation exposure of various types and is a good source to provide the engineer with various testing techniques. The federal government produces a great deal of information on the radiation level of exposure. Another source of information is the FDA, which also has resources for such an investigation and has test results.

There is a great deal of information in the European Union on wireless devices and cell phones because of the mass proliferation of these devices. Many are in the throes of testing for health hazards of these devices. Therefore, the information presented in these sections of the book are not cast in stone but are a snapshot of the regulations at present. As most engineers working on problems that require regulations will know, they are living documents and susceptible to change on a yearly and sometimes on a monthly basis.

7.7   REVIEW PROBLEMS

Problem 7.1
The frequencies that have the most effect on the human body are the following?

Problem 7.2
If a person is exposed to an electric field intensity of 25 V/m, is this in violation of the exposure limit at 300 MHz?

Problem 7.3
The electric field intensity exposure limits for DC and AC fields are?

Problem 7.4

If a maintenance person is touching a power line and holding onto the frame of a grounded object with the other hand, the current when shocked may travel across the heart. What is the current that would cause death to the maintenance person? Most electricians will not have this situation because they are trained not to ground one hand while touching live or suspected live wiring.

Problem 7.5

A touch voltage hazard is?

Problem 7.6

This review problem has parameters for illustration. The result of the final calculation determines whether exposure limits have been violated for the maintenance worker.

A high-voltage three phase power line 746 kV at 60 MHz with line spacing of 16 m has conductor diameters at 0.6 m. The distance of the power line grade is 15 m. The assumption is that the maintenance person is approximately 2 m tall.

7.6.1   Find the line charge ρ_L in µC/m

7.6.2   Find the electric field intensity contribution from each phase.

7.6.3   Find the electric field point P where a maintenance worker would be underneath the power lines.

7.8   ANSWERS TO REVIEW PROBLEMS

Problem 7.1

Frequencies between 30 and 300 MHz have the most effect on human body because the wavelengths approach the length of the limbs and act as antennas. At these wavelengths the body will absorb more of the radiated energy that can produce heating in body tissue. At frequencies above 300 MHz other parts of the body such as ears, eyes and head will absorb radiation at smaller wavelengths, that is in the centimeter range 1–10 GHz.

Problem 7.2

No, exposure limits are 27.5 V/m and 300 MHz.

Problem 7.3

Electric field intensity limit for DC fields is 5 kV/m and AC limit is between 5 and 10 kV/m.

Problem 7.4

Death will occur if 7 mA flows across the heart.

Problem 7.5

The touch hazard for voltages is 55 V.

Problem 7.6.1

$$V = 746\,\text{kV} = -\int \bar{E} * d\bar{r} = -\frac{\rho_L}{2\pi\varepsilon_0} \int_0^L \frac{dr}{r} = \frac{\rho_L}{2\pi\varepsilon_0} \ln\left[r\right]_{r=0.3}^{r=15.5} = \frac{\rho_L}{2\pi\varepsilon_0} 3.944$$

$$\rho_L = 2\pi\varepsilon_0 V/3.944 = 2*\pi*8.854*10^{-12}*746*10^3/3.944 = 10.52\,\mu C/m$$

Problem 7.6.2

$$\bar{E}_3 = -\hat{a}_y \frac{\rho_L}{2\pi\varepsilon_0 r_3} = -\hat{a}_y \frac{10.52*10^{-6}}{2\pi*8.854*10^{-12}*3.944*13} = 3.68\,kV/m$$

$$\bar{E}_2 = \frac{\rho_L}{2\pi\varepsilon_0 r_2}\left(-\hat{a}_y \sin\theta_2 + \hat{a}_x \cos\theta_2\right)*10^3 = 2.32\left[-\hat{a}_y \sin\left(39°\right)+\hat{a}_x \cos\left(39°\right)\right]*10^3\,V/m$$

$$\bar{E}_2 = \left[-\hat{a}_y 1.46 + \hat{a}_x 1.8\right]*10^3 \quad V/m\,at\,r_2 = 2.6\,kV/m$$

$$\bar{E}_1 = \frac{\rho_L}{2\pi\varepsilon_0 r_1}\left(-\hat{a}_y \sin\theta_1 + \hat{a}_x \cos\theta_1\right)*10^3 = 0.17\left[-\hat{a}_y \sin\left(22.1°\right)+\hat{a}_x \cos\left(22.1°\right)\right]*10^3$$

$$\bar{E}_1 = \left(-\hat{a}_y 0.063 + \hat{a}_x 0.157\right)*10^3 \quad V/m$$

$$E_y = -\hat{a}_y[3.68 - 1.46\underline{|120°} + 0.065\underline{|-120°} = 2.2\,kV/m$$

$$E_x = \hat{a}_x\left[1.8\underline{|120°} + 0.157\underline{|-120°}\right]*10^3 = \hat{a}_x\left[-0.9 - 0.0785\right]*10^3 \quad kV/m$$

Problem 7.6.3

$E_{Max} = \sqrt{2.2^2 + 0.97^2}\,kV/m = 2.4\,kV/m$. This field intensity does not violate the standards for exposure limits.

8

MISCELLANEOUS INFORMATION TEST PLANS AND OTHER INFORMATION USEFUL FOR ANALYSIS

8.1 INTRODUCTION

In this chapter, Sections 8.2–8.5 are elements of an EMC test plan. These sections are meant as a guide so that the engineer working on the design will not overlook particular elements. The test plan must always include an introduction, which is the first segment discussing the extent of the plan, that is the scope of work to be performed. EMC design is across many disciplines, for example how wiring may be installed in duct banks to prevent EMI sources from coupling into the signal lines and power supplies and motor drives. Occasionally mistakes are made or power lines are too close to sensitive signal lines in the duct bank and it is easier to inspect these during installation rather than wait until commissioning or integration testing is performed, at which time removing culprit lines becomes very expensive.

The EMC test plan covers most topics and not all are necessary for a particular subsystem. The test plan generally covers a part of the project that is assigned to the contractor and the engineer may only need to analyze the part that pertains to the contractor assigned. However, in many cases there is an overlap between the various subcontracts were EMC issues may be outside the realm of the particular contract being executed. For example, the communications contractor may be checking the emissions of all the communications equipment and subsystems but may be impacted by signals equipment or the power supply contractor in the form of harmonics from

Electromagnetic Compatibility: Analysis and Case Studies in Transportation, First Edition.
Donald G. Baker.
© 2016 John Wiley & Sons, Inc. Published 2016 by John Wiley & Sons, Inc.
Companion website: www.wiley.com/go/electromagneticcompatibility

the DC power supply. Generally, during commissioning all the lines of the various contracts are crossed and EMC issues will cross the various contracts. When writing a test plan, it should include all aspects of the testing and for each of those sections not covered by the particular contract to make the plan complete, "does not apply" should be inserted. This is necessary so that other contract personnel will know it is not being covered. This is also a checklist of all items that may be an EMC issue. As an example, the contractor installing the PA and security camera subsystem may ask to be included in the communications EMC test plan due to interaction between PA and security subsystems with communication systems. The EMC for fire alarm and communication systems (including PA and security subsystems) must be negotiated with a fire alarm subsystem contractor to ensure that all EMC issues are covered between the two.

In most cases, the fire alarm subsystem is designed by a fire alarm PE, that is the placement of smoke detectors and heat sensors and the method for communications covered in several drawings. The installation of the fire alarm is generally handled by the communications contractor; this is not to say this is always the rule. The author has been involved in several instances where the entire fire alarm system is both designed and installed by a fire alarm contractor. These are some of the rules that will be covered in a requirements document that is produced by the authority in charge of the project. The review problems added at the end of each section are to assist the engineer or technician reading this to more fully understand the implications of each EMC design and analysis.

Section 8.6 in this chapter provides a treatment of the Fresnel zone – that is useful for calculating the clearance necessary that will prevent line of sight loss. If an obstruction is discovered, such as a stand of trees or a hill, the person doing the survey may have available a GPS with altitude included. A rough estimate can be made using a balloon filled with helium and a string with graduation marks as a tether. Observations made by eye can be made to determine the clearance available; these of course are rough estimates. In the past before GPS many crude tools had to be used to determine clearances; this was among them. When estimating distance using a GPS, planar trigonometry may be used to calculate distances due to the short range. Longitude is 60 nautical miles per degree and latitude distance changes are somewhat less, depending on the latitude angle, that is at $40°$ latitude distance is less than at $50°$.

8.2 EMC PLAN

The document entitled the Electromagnetic Compatibility Control and Test (EMC) Plan examines the equipment and methods for compliance. EMC is the methods for Electromagnetic Interference (EMI) control to obtain FCC Part 15 Regulation compliance. The major divisions are the actual methods for EMI Control and Management Control, that is organizational capabilities to implement EMI control. The source of the EMI is legally liable for any damages incurred due to conducted or radiated emission induced interference.

This document is a presentation of the various methods employed for analysis and design of Electromagnetic Compatibility EMC. The information herein describes tests for susceptibility of equipment to conductive EMI and radiated RFI/EMI. Calculations to perform conductive and radiated EMI analyses are described herein. Mechanical, electrical and electronic design features are presented that provide EMI integrity to systems and equipment.

8.2.1 EMI General Characteristics

1. Conducted
2. Narrowband
3. Continuous
4. Intersystem
5. Common mode
6. Broadband frequency range DC to microwave
7. Radiating (non-ionizing)
8. Broadband (outside the measurement range)
9. Susceptibility
10. Intermittent
11. Intra-system
12. Differential mode
13. Immunity

As may be observed, the general characteristics of EMI are formidable. Each piece of equipment will not be susceptible to all of the interference. The scope of investigation is thus narrowed to something more manageable. Manufacturers have no formal standard for immunity to RFI. In practice however, equipment is tested for field strength of 3 V/m (129.5 dB μV/m). European Union EC standard EN50082-1(92) is similar for light industry.

EMI has an impact on performance of equipment therefore; any necessary testing will be a part of Factory and Field Acceptance Tests. The full impact of EMI is monitored when the traction power substations (TPSS) and unit substations are applied to the system. These both have near fields that can couple into various equipment and cable.

8.2.2 Definition of Terms

EMC – The ability of a device to function satisfactorily in its electromagnetic environment without introducing intolerable disturbance to that environment (or other devices).

EMI – Impairment of electromagnetic signal by an electromagnetic disturbance.

Conducted and radiation emitting sources (non-ionizing) of EMI – These are disturbances due to power lines, motors, fluorescent lights, dielectric heaters, arc welders, lightning, galactic noise, electrostatic discharge, engine ignition, radio transmitters, etc.

TABLE 8.1   Units of EMI Measurement

Power units	Voltage	Current	Magnetic flux density (B field)
Watts (W) dBW = 10logP/1 W dBm = 10logP/1 mW dBm/kHz	Voltage (V) dBV = 20logV/1 volt dBμ = 20logV/(μV) dBμV/MHz	Amperes I dBA = 20logI/1A dBμA = 20 log I/μA dBμA/MHz	
Power density	Field strength (E)	Magnetic field (H)	Magnetic flux density (B Field)
W/m² dBW/m² dBm/m² dBm/m²/kHz dBm/m²/MHz	V/m dBV/m dBμV/m	A/m dBA/m dB|μA/m dB|mA/m/MHz	Tesla (weber/m²) Picotesla (pT) dBpT dBpT/MHz

Transfer or coupling EMI – Radiated antenna to antenna, case radiation, case penetration, field to wire, wire to field and wire to wire is some of the radiated coupling methods of EMI. Conducted coupling occurs through common ground impedance, power line and interconnecting cable.

EMI receptors – Receptors are radio receivers, analog sensors, high gain amplifiers, computers and humans (biological hazards).

Inter-system EMI – This type of EMI is computer to UPS, power line to computer and other types of coupling previously addressed.

Intra-system EMI – This type of EMI occurs between the system components such as board to board, power supply to board, system ground and so on.

Narrowband EMI – This type of EMI is generated within a narrow spectral bandwidth and is common in radio receivers, as an example, intermodulation products that fall within the pass band of receivers.

Broadband EMI – This type of EMI produces noise components across the spectrum, such as avalanche effects in semiconductors during breakdown.

8.2.3   Units of EMI Measurement

Units of EMI measurement are given in Table 8.1.

8.3   EMC/EMI PERFORMANCE EVALUATION OF COMMUNICATIONS EQUIPMENT

All the communications equipment is required to meet or exceed the emission regulations in FCC CFR 47 Part 15 before it can be installed. After installation, equipment may radiate or be susceptible to radiation from both intentional and unintentional radiation. A survey will be conducted to determine the extent of radiation surrounding the equipment.

Induction and capacitive coupling of extraneous signals into sensitive equipment based on performance and studies made of the potential culprit sources. Grounding will be examined to determine if common impedance paths result in conducted interference signals. MIL-STD-461E can be used as a guide.

8.3.1 Public Address System

The following is a deduction that can be made from abnormal sound emanating from public address (PA) systems. If clicks are heard on a PA system, this represents power spikes one would expect from relays switching or the power mains. If they are very loud, they are most likely from power mains. The PA system has a peak voltage near 70 V, which is the peak voltage swing of the audio amplifiers. If a loud frequency is heard (often called singing) it is most likely attributed to feedback from amplifier output to input through the microphone lines. The interference to noise ratio is defined as shown in Equation 8.1.

$$PEMI = 10 \log P_{nt}/P_n \tag{8.1}$$

$P_n = -134$ dBm thermal noise floor for the audio at the microphone.

PMJC $= -68.24$ dBm audio signal power level at microphone.

PAD $= 53$ dBm audio amplifier output power full load.

PAD $= 7.8$ dBm audio amplifier output noise power (non-objectionable).

Audio amplifier output signal to noise ratio (SNR) $= 45.13$ dB.

The noise level in the audio amplifier is 100 dB down from the audio amplifier output or at -47 dBm. The recommended INR $= 7.8 - (-47) = 52.5$ dB.

Recommended INR $= 52.5$ dB

$P_{mput} = -40$ dBm audio amplifier input power at full load

$P_{NI} = -90$ dBm audio amplifier input noise power at full load

If the interference is not objectionable, the resultant interference to noise ratio (INR) is 52.5 dB electrical. The noise levels alluded to in the equations also include hum, phase noise, shot noise and so on. The EMC measurements will be made as part of the acceptance tests. Corrective measures as well as measurements will be made on a case by case basis.

The Clicking may be the result of magnetic or capacitive coupled signals on the signal lines to the audio power amplifiers. This type of abnormal operation can also be the result of poorly installed signal line shielding and grounding.

In band oscillations in audio amplifiers can result from positive feedback from the output to the amplifier input signal lines. The remedy for this type of behavior is to break the feedback loop. This may not always be possible. Amplifier compensation can be added to shift the poles until the output phase shift is sufficient to prevent oscillation. All previously discussed abnormal operations would appear during factory and field acceptance testing. The factory tests are considered the first test for EMI. If performance is deficient and the equipment does not meet specifications, then EMC testing is the next logical step.

RF coupling of AM radio to the signal or ground lines due to poor shielding can result in the radio broadcast being heard at the loud speakers. AM detectors are composed of simple diodes with an RC circuit at the output. Diodes can be formed from poor connections, dissimilar metals, dirty connectors and so on. The capacitances and resistances are from wiring. This type of EMI is particularly annoying if it occurs on the input to audio power amplifiers. The audio will be amplified along with the PA announcement and it will be heard as faint background music or conversation.

8.3.2   LCD Monitors

These devices will radiate RF or will couple to stray electromagnetic EMI fields, if not installed correctly. The monitors are compliant with FCC Part 15 subpart J class B computing equipment rules. Noise in the video field can occur from magnetic or electrostatic fields coupled to the power and signal lines. The characteristics are a point where the video scene is distorted. The magnetic field can be measured with a current loop and the electrostatic field measured with antennas.

Coupling of EMI into the signal or power lines will result in: vertical roll of the video scene; interference patterns in the display; ghosts in the background (crosstalk); ghosts moving through the video display. EMI problems will be revealed during system factory and field acceptance tests. Any abnormal functioning of equipment can be corrected during the testing.

8.3.3   Variable Message Board

The variable message board (VMB) equipment has controllers at each sign with a modem link to the central control unit. EMI may occur in the transmission cables. These anomalies will be investigated and the condition corrected. Radiated EMI can occur from fast turn-on of the LEDs in the sign. This is considered a design problem. It will be investigated and corrected by the manufacturering company. Diodes approximate a square law device, that is they perform modulation with sidebands. If the rise time is sufficiently fast, the LEDs will produce sidebands that can affect radio transmission.

8.3.4   Video Display Unit

The display unit and local controller should not require EMC testing if the equipment is installed per manufacturer installation instructions. However, the communications lines and power mains may radiate or produce EMI spikes on the power mains. If the performance of the equipment is adequate to meet the necessary test criteria during field-testing, the EMI survey will be sufficient to satisfy EMI requirements. Spikes on the control lines will produce a blackout of pixels and distortion of the scene of each frame that looks like large or small dark spots depending on the duration of the spike.

8.3.5   Controller

The central control unit is populated with PC workstation. These devices must pass FCC Part 15 subpart J class B computing equipment rules. But the connections must be made with strict adherence to the manufacturer's installation instructions to prevent EMI emanations. The power mains and signal leads can be a source of RFI/EMI or conducted EMI to the controller.

8.3.6   Fire and Intrusion Alarms

The alarm circuits are generally differentially wired circuits with twisted pair leads. Often the sensors are part of a bridge circuit. Common mode signals across the bridge will be summed out at the bridge output. Differential mode signals can produce errors at the bridge output with sufficient signal levels to produce false alarms. During factory and field tests these abnormalities will be apparent and corrective action will be taken.

8.3.7   Sensors

Should field-testing of the various sensors (i.e. smoke detectors and heat detectors) reveal any deficiency in performance due to EMI, corrective action is necessary on a case by case basis.

8.3.8   Control Panels

The same precautions taken with microprocessor-controlled equipment must be taken with control panels. Both fire alarm control panels and intrusion alarm control panels that are susceptible to EMI/RFI. These panels will not emanate sufficient EMI to result in equipment failure, if the manufacturer's installation instructions are strictly adhered to. EMI susceptibility and immunity is done on a case by case basis during performance testing (factory and field acceptance tests).

8.3.9   Communications Transmission System

These system main components include SONET OC-12 terminals, PCM channel banks, fiber optic modems and digital access cross connect (DACS). The system will be tested during factory and field acceptance tests for performance, if the equipment meets or exceeds FCC part 15 B; this guarantees a limit on emissions. However, the testing will be conducted during acceptance testing for susceptibility and immunity from extraneous emissions of non-licensed sources such as power lines, traction power, lighting and so on. This testing will be conducted during the site survey and field test phase. Licensed emitters such as two-way portable radios and cell phones will also be used as a test for susceptibility. Any deficiencies in performance will be corrected using appropriate means as described in the design section. The radiation of the stratum clock and its harmonics from the SONET equipment can affect the radios due to the radio receivers sensitivity (0.5 μV/m). If performance of the SONET equipment, due to susceptibility and lack of immunity to extraneous signals is apparent during radio or cell phone transmission, the bit error rate (BER) and other performance will suffer. If any deficiency in performance due to EMI occurs, the culprit circuitry will be identified and corrected on a case by case basis as part of the factory and field acceptance tests.

8.3.10 Central Control System

This system has microprocessors and controllers with system clocks. All of the equipment will be FCC part 15 B compliant and will have as a minimum a 3 V/m immunity. Non-licensed RFI sources and power line conducted interference are tested during factory and acceptance testing of the system. The need for additional shielding and filtering will be identified on a case by case basis.

8.3.11 Workstation LCD Monitors

These devices will radiate RF or couple to stray electromagnetic EMI fields, if not installed correctly. The monitors are compliant with FCC Part 15 subpart J class B computing equipment rules. Distortion in the video field can occur from magnetic fields coupled to the monitor. The characteristics are the same as described for monitors in Section 8.3.2.

Coupling of EMI into the signal or power lines will result in: vertical roll of the video scene; interference patterns in the display; ghosts in the background (crosstalk); ghosts moving through the video display. EMI problems will be revealed during system factory and field acceptance tests. Any abnormal functioning of equipment will be corrected during the testing.

8.3.12 Remote Terminal Unit

These devices must pass FCC Part 15 subpart J class B computing equipment rules. But the connections must be made with strict adherence to the manufacturer's installation instructions to prevent EMI emanations. The power mains and signal leads can be a source of RFI/EMI or conductive EMI to the remote terminal unit (RTU). The RTUs are tested during the factory and field acceptance tests for performance. Deficiencies, due to EMI/RFI, are identified and corrected on a case by case basis.

8.3.13 Radio Subsystem

Transmitter   Radios are licensed radiators. A two Watt radio or 600 mW cell phone transmitting 18 inches from a piece of equipment produces a field of 20 or 11 V/m respectively. The 20 V/m requirement is a stringent requirement; it can result in a greater than 50% malfunction in equipment without shielding.

Receivers   This equipment has a narrow band of tuning; however, mixing in the receiver front-end or the transmitters, or signal levels that are very large can result in intermodulation and desensitization problems. The intermodulation study conducted in the radio section will provide data sufficient to design filtering to remove the culprit frequencies. Large unlicensed radiators that result in receiver desensitization will be identified and reported to appropriate to authority for action by the FCC.

8.3.14   Telephone Equipment

The telephone equipment is tested during field acceptance testing for quality of speech. It must be kept in mind this type of test is very subjective. The lines will be monitored for noise at the analog inputs of the channel banks. Signal levels and noise levels are measured. Field tests are the most stringent due to the proximity of strong near fields. Magnetic field coupling from substation and traction power is the area that EMI has the most effect. EMI can enter due to open shields, poor grounds or installation deficiencies, which can result in poor performance.

8.3.15   EMC/EMI Performance Evaluation of Signals Equipment

All signals equipment is required to pass FCC regulations CFR 47 Part 15 before it can be installed. After installation, equipment may radiate or be susceptible to radiation from both intentional and unintentional radiation. A survey will be conducted to determine the extent of radiation surrounding the equipment; it also has an immunity regulation due to safety of 10 V/m.

Inductive and capacitive coupling of extraneous signals into sensitive equipment is based on performance and studies are made of the potential culprit sources. Grounding will be examined to determine if common impedance paths result in conducted interference signals. MIL-STD-461F (which is more stringent) is generally used as a guide.

8.3.16   Vehicle Emissions

The harshest vehicle emissions are the result of inductive and capacitive coupling from the propulsion system. The 750 V DC rectifiers to provide power to the vehicle AC traction motors, inverters, air conditioning and motors are of the most concern. The power lines from the substations have harmonics of 60 Hz; the PWM inverters have harmonics of the chopper frequency that can affect the TWC loops, track circuits and AF track circuit coding; and the DC rectifier circuits have harmonics of the 60 Hz substation frequency. These will be addressed in the following subsections.

Vehicle propulsion to track circuit   The propulsion system has several frequencies that can mix due to galvanic action that produces a weak diode. The frequencies are provided in Section 8.5 with all calculations. The calculations are made using MathCad 11.0. Calculation are made using the data provided or assumptions that are labeled. The calculations are data from the substation, rectifiers, AC inverter, motor and 100 Hz track circuit power supply. A method of testing are proposed for near field radiated emissions and conducted emissions. The emissions are measured where necessary to provide a safe environment.

Vehicle to audio frequency coded track circuits   The audio coded baseband signaling commonly implemented have frequencies that are mixed due to galvanic action that will be analyzed to determine the spectral content and magnitude. These circuits present a varied spectrum of frequency components. See Section 8.5 for the analysis.

Vehicle to TWO   Loop detectors will be analyzed to determine EMI affects from the vehicles. Measurements of the noise due to mixing for propulsion and power supply harmonics are analyzed to determine if filtering is adequate. See Section 8.5 for the analysis.

8.3.17   Signals and Propulsion Equipment Effect on Communications

This is of particular concern. The communications system is similar to a nervous system in humans. It provides the path to control the movements of light rail equipment. Particular care must be taken not have common impedances with signal or propulsion equipment. This is a difficult task at times due to methods of grounding.

Signal houses   Signal houses have an abundance of magnetic devices that can affect the operation of sensitive equipment. The manufacturer is responsible for susceptibility of field strengths of up to 3 V/m (130 dB μV/m). The equipment manufacturers will be required to provide the necessary test documentation. The equipment is measured during the site survey to determine if radiation is present. Both far and near field will be measured. The survey will be conducted on the interior and exterior of the signal house.

VHLC processor   The VHLC processor must meet or exceed FCC part 15 emission regulations. However, after all connections are made to the equipment, radiation, inductive/capacitive coupling or common impedance paths may be present. The installations will be examined on a case by case basis. Fiber optic modems are recommended for long cable runs. One of the weak areas of fiber optics is the optical to electrical conversion of the transimpedance amplifier. This is a very high gain element with high input impedance (mega-ohm region). Stray signals are easily coupled into receivers. Therefore, placement of these devices in areas with low emissions and shielding is imperative. However, present day modules are available that are self-contained and present a low probability of problems. But, circuit boards that have the conversion components soldered to a mother board can have some of the problems alluded to.

Track circuit relays   Relays have an inductive kick this is controlled by a snubber network or varistor. However, harmonics may be present that couple into circuits from radiation, inductance or capacitance. During the survey, these devices are included in the measurements using loop antennas and spectral analysis techniques.

TWC circuits   The survey will include measurements for compliance to FCC part 15 for all installed wayside controllers.

TWO loops   These loops will be measured for FCC part 15 and for near field radiation of magnetic fields. Modulated waveforms will have harmonic content depending on the rise time of the modulating waveform.

8.3.18   LRVs

All measurements will be taken while LRVs are passing and without the LRV present. These are the worst-case and best-case measurements. Near and far field measurements are taken of the LRVs from the wayside to use as a reference. The readout is then examined for potential EMI related problems.

8.3.19   Signals Equipment

Switch machines   All switch machines are measured for FCC part 15 compliance and the near field will also be measured. This will reveal if any magnetic H fields present that can couple to copper wiring such as power and signal lines. This is part of the site survey.

8.3.20   AF Track Circuits Interference Control Program

The objective of the control program is to protect the victim or receptor from disturbances in the electromagnetic environment. This is done through the control of predicted EMI from components within and external to the system. Intra-system EMI, due to manufacturing deficiencies, will not be examined in detail, but any discrepancies will be noted and reported to the Authority. Only those components that are part of the integration effort, when warranted, will be examined, such as wiring or items fabricated by a company.

Control of interference is a three- or four-step process, as described in the flow diagram in Figure 8.1 after the source of the EMI is identified; the coupling method to the victim must be determined. The objective is to determine the source of EMI by means of electrical measurements. In certain instances, the source and coupling of EMI can be determined very easily. For example, radio intermodulation products can be identified from the local transmitter sources. However, cross-modulation may often is mistaken for intermodulation products, which of course is not the case. These latter signals are generated by the victim radio internally. Potential sources of EMI are investigated and analyzed, if necessary, such as the traction power supply, AC power distribution system, vehicle chopper system, train control, grounding system, nearby facilities, lightning protection, illumination control systems and so on.

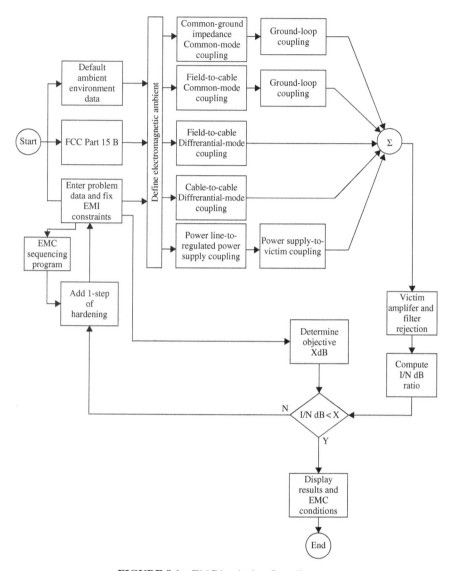

FIGURE 8.1   EMC hardening flow diagram

The method of EMI coupling must be determined to perform any type of analysis. Only a few of the coupling paths are described in the document. The specific coupling method will be analyzed particular to the EMI present. This may or may not also produce the method of entry. Occasionally the victim circuitry will produce partial EMI, which can be difficult to separate from external sources.

The emission specifications that the equipment must operate under are examined, when warranted, to determine the extent of EMI disturbance. FCC CFR 47 part 15 B regulations will be followed and testing methodology will use MIL-STD-461 as a guide.

8.4 EMC/EMI DESIGN PROCEDURES

The easiest method for describing the design procedure is provided in a flow diagram, Figure 8.1. The starting point in the diagram is the data used to define the electromagnetic environment. FCC Part 15 B regulations/requirements provide for allowable emissions; and default data is EMI from intentional and unintentional emissions such as radios and arcing from catenaries. If a problem exists, it must be defined. The value of X, an acceptable value for the interference to noise ratio (INR) = I/N dB, is determined. In PA or other analog system the noise floor N given by the manufacturer and the level that is slightly objectionable to the listener is the interference (I). For digital systems, the performance is calculated based on the maximum allowable bit error rate (BER). The noise level is either provided by the manufacturer or an equivalent can be calculated after the environment is defined. The method/methods of coupling are determined as indicated (see the blocks in Figure 8.1 up to the summing junction). After a step of hardening and I/N ratio computation, a comparison is made with the objective I/N shown as X. The amount of rejection for the device being analyzed can be calculated. A calculation is made of the I/N dB. If the I/N dB is inadequate then a step of hardening is added. The hardening is added until the regulations are met. EMC sequencing is the type of hardening done.

The three blocks in Figure 8.1 after *Start* are as follows:

- *Default ambient environment data*: this is maximum emission allowable that is different from FCC part 15B that is, for both licensed and unlicensed emission that the equipment has sufficient immunity against EMI. Equipment manufacturers' test for immunity is 3 V/m depending on the application certain transportation equipment where life and safety are concerned the immunity may be much higher such as signals equipment that requires 10 V/m.
- *FCC part 15B* is provided for all unlicensed emission for equipment in the System.
- *Problem data*: this is derived from equipment FCC part 15 B and default ambient that is greater than 3 V/m in field strength.

The electromagnetic ambient is now defined. The blocks into the summing junction determine how the EMI is being coupled into the system (the most important step). This is determined through performance tests or subjective performances that depend on human senses, such as PA systems. After a mode of EMI rejection is determined and I/N dB is calculated, it is compared with the desired I/N dB. If the I/N dB is sufficient, the EMC analysis is completed. If not, a step of hardening is added and the I loop begins again. The EMC sequencing block determines the types of hardening to be used, as described below.

The hardening alluded to in the diagram is the addition of re-orientation of equipment, shielding, filters, shielding with μmetal or other techniques for reducing the I/N ratios. Some of the hardening techniques can be inexpensive, such as reconfiguring the placement of equipment to take advantage of the normal attenuation of air.

Other types of hardening such as shielding, grounding, balancing, filtering, isolation, circuit impedance level control and cable design shall be considered, based on a cost and benefit analysis. Radio system hardening, for example, such as the addition of tuned cavities and filters, can be quite expensive. The first line of hardening will always be the most inexpensive, that is re-orientation and separation of critical signal lines and components. Low-cost solutions to EMI problems will not sacrifice the performance of the equipment. The display of results and EMC conditions will be done in a program or added as a conclusion to the analysis and design.

8.4.1   Mechanical Design

The first type of EMI control is generally mechanical, that is reorientation of signal or power leads, adding shielding material and other means discussed in this section. Mechanical design may range from the simple low cost approach, alluded to previously, to the complex and costly, such as machined components with special gaskets or other special EMC enclosures. Magnetic and electric field shielding are not identical. Material suitable for one field may not produce effective shielding for the other. The two fields are related by Maxwell's equations, but they have completely different characteristics.

8.4.2   Shielding Materials

The two types of fields are defined and discussed before providing information about shielding. Figure 8.2a, b gives a pictorial description of magnetic and electric fields respectively. The most difficult field to shield is the low frequency magnetic fields due to the low impedance.

The near and far field inequality is provided below in Equations 8.2 and 8.3.

$$d \leq 2 * L^2 / \lambda \quad \text{Near field} \tag{8.2}$$

$$d \geq 2 * L^2 / \lambda \quad \text{Far field} \tag{8.3}$$

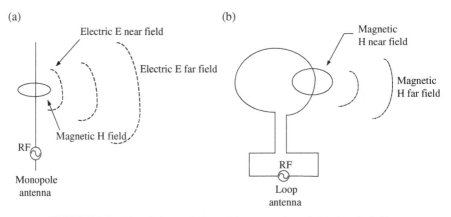

FIGURE 8.2   Pictorial description of (a) magnetic and (b) electric fields

Where L = max length of source antenna, λ = frequency wavelength.

The far field impedance is constant at 377 ohms, the approximate impedance in a vacuum. Air of course is slightly higher however, for most shielding calculations this impedance can be used. The H field in Figure 8.2a is only shown as one ring. Each E field has an H field ring orthogonal path that is not shown in the illustration. The near field acts as a storage medium and it is inductive for low impedance fields and capacitive for high impedance fields. As progression radiation out from the near field occurs, a transition slowly from the reactance occurs to the far field where the impedance is resistive. Mechanical shields must suppress these fields through absorption, reflection (external/ internal) or both. Copper, aluminum, μmetal and hypernick are some of the common shielding materials. The latter two have high magnetic permeability, and are therefore generally used in magnetic shielding or for near field applications. Copper and aluminum are more frequently used in electric field applications. An equation that describes the relationship of absorption and reflection is Equation 8.4.

$$\text{Shield effectiveness,} \quad S_{eff} = \frac{40\log(K+1)}{4K} + 131.4 * t_{mm}\left(f_{MHz} * \mu_r * \sigma_r\right)^{1/2} \text{ dB}$$

$$(8.4)*$$

Where t_{mm} = thickness in millimeters, f_{MHz} = culprit frequency to be shielded in megahertz, μ_r = permeability relative to copper, σ_r = conductivity relative to copper and K = voltage standing wave ratio. All of these variables are easily measured; this allows a quick analysis of any shielding materials that may be used for a particular application. The measurements are as follows: the thickness of the material in millimeters many times gauge, a standing wave ratio and frequency in megahertz. The other variables are available in common lookup tables or from manufacturers' specifications.

At high frequencies 1 GHz, 0.003 mm of copper has the same effective shielding as 0.01 mm of iron for electric E fields. Note, the far field for all shielding should be calculated and entered in a table to calculate the distance between the shield and source (Equation 8.3). The thin shielding can be a coating of copper or aluminum paint/flame spray for high frequencies, while iron or cold rolled steel would be a sheet of material of choice for magnetic fields. Steel is a less effective shield than iron because the permeability is 10% lower than iron, μmetal has a permeability 100 times higher than iron and 18 dB more effective shielding properties than iron for near field magnetic EMI. However, it is much more expensive. When producing an EMC/EMI test plan, it should include suggested shielding methods and materials and specification sheets for materials that provide good shielding with a range of cost for each material.

8.4.3　Electrostatic Field Shielding

Far field and low current electrostatic fields are the sources considered in this section. Electrostatic fields are generally easy to shield. Thin sheets of copper, aluminum or conductive plastic can be used. In some applications, EMC shielding tapes can be

---

*μ_r and ε_r are functions of frequency and they become imaginary above 1 GHz.

used for wire or other areas where RF leakage occurs. The actual design will be on a case by case basis.

Conductive paints are relatively inexpensive and easy to apply. Conductive paint surface scratches or cracks are not difficult to repair. Considerations that must be made for conductive paints are the following:

- *Electrical properties:* surface conductivity, shielding effectiveness, and pitting due to high current density.
- *Mechanical properties:* resistance to scratching, temperature range of paint, resistance to corrosive environments, and required surface preparation.
- *Application method:* spraying, dipping, brushing, roll coating, silk screening, and equipment requirements
- *Miscellaneous items:* number of components to mix and apply; the number of coats to obtain the required shielding thickness; curing technique either oven or air drying; and storage conditions.

An assortment of resin bonded conductive paints are available with a silver, nickel, graphite and copper, just to name a few. A sample of the conductor manufacturers is provided in Table 8.2.

8.4.4   Magnetic H Field Shielding

This shielding will be a magnetic material in either foil or sheet form with the form and materials based on observations by the EMC engineer. Once decided on a case by case basis (depending on the severity of the EMI) some calculations should be made as backup during factory and field acceptance tests. As emphasized throughout this book the earlier the EMI can be discovered and corrected the less costly will be corrected action. The worst-case scenario is when the problems are found during the commissioning phase of any project.

Mechanical integrity enclosures   EMI/RFI cabinets are readily available from various vendors such as Equipto, Hoffman and AMCO. These enclosures are fitted

TABLE 8.2   Conductive Coatings

Manufacturer	Trade name	Surface resistivity (ohm/square)
Acheson Colliods Co.	Electrodag 424	0.050
	415 440 503 504	0.015
		2.000
		0.050
		0.035
Cal-Metex	Xecot	0.050
Emerson Cummings	Ecoshield	0.100
Glidden Co.	RGL-3897	1000 ohms
ACME Chemical and Insulation Co.	E-Kote 3063	1 ohm

with gaskets and various seals to prevent emissions from the equipment inside. The equipment inside will also be immune to emissions from outside the enclosure. Standard cold-rolled steel or aluminum cabinets can be fitted with gaskets and seals to shield against RFI/EMI should the need arise. The integrity of enclosures is generally fairly good; however often electricians and other installation technicians will compromise the shield effectiveness by cutting holes when installing wiring without a conduit or any means of gasketing for the wiring installed. Some enclosures with air-conditioning will require screening as a means for shielding around the ventilators. These type of anomalies are often overlooked. Air-conditioning for example may be installed inside or outside of a large cabinet. Outside the cabinet is the preferred method because air-conditioners have motors and motors generate very low-frequency noise. Some installations use open racks for installing equipment that operate at frequencies below 10 MHz. These may be installed in close proximity to enclosures that are meant as shields. These types of installations may have disastrous effects on unshielded cabinets because of their long wavelength. EMC/EMI engineer must be vigilant in conducting an analysis of this type of installation.

Gasket and spring-loaded contact shielding   The methods employed will be best if the manufacturer of the shielding device is consulted. As shielding methods change over time the mechanical means for installation will also change. Conductive rubber seals are sometime employed around doors and openings. Spring-loaded contacts begin to loose tension over time which will result in leaks. Figure 8.3 provides an insight into how gaskets perform when the shielding seal has uneven joints. The soft materials are rubber or other organic substances impregnated with metal to provide a pliable gasket. These types of gaskets may require only moderate pressure with screw or door locks to provide the pressure for sealing. The hard material can be pure aluminum or copper that is pressed into the uneven surface under high pressure, such as closely spaced bolts under high torque. A typical spring-loaded contact application has a finger stock; it is readily available at fairly low cost. However, labor to install this shielding component can be expensive as compared to gaskets. This application is useful where the enclosure requires that the doors or covers be removed frequently. Gaskets will wear and generally need to be replaced more often. When air purifiers or other devices that generate ozone (including electric motors) operate near rubber gasket materials it has a destructive effect on the rubber.

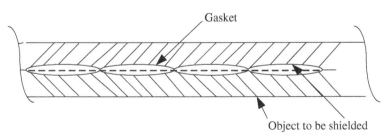

FIGURE 8.3   Uneven compression of gasket

TABLE 8.3 **Compression for Types of Gaskets**

Uncompressed (%)	Pressure (psi)	Type	Uncompressed (%)	Pressure (psi)	Type
48	75	Soft gasket	80	100	Hard gasket
50	60	Soft gasket	84	80	Hard gasket
62	40	Soft gasket	85	70	Hard gasket
70	30	Soft gasket	90	40	Hard gasket
82	15	Soft gasket	92	30	Hard gasket
94	5	Soft gasket	96	10	Hard gasket
100	0	Soft gasket	100	0	Hard gasket

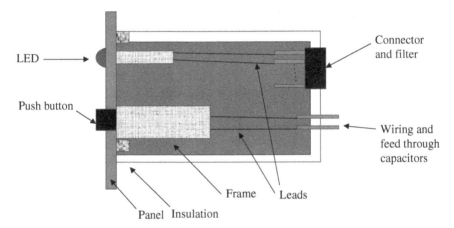

FIGURE 8.4 Typical shield applications for a pushbutton and lamp with filtering and shielding to maintain the integrity of the enclosure

Table 8.3 gives typical test data for soft gasket materials as can be observed in the table require a wide range of pressure of compression as opposed to hard gaskets. These are only typical values. For example, soft aluminum can actually be compressed as compared to a hard aluminum alloy. Some gaskets are a combination between hard and soft, such as gaskets made with honeycomb that when tightened are crushed to form the seal, such as may be found in automotive gaskets. Anyone who has worked with automotive head gaskets for example will have found that the gaskets are made of composite laminates. The hard and soft gaskets are at both ends of the spectrum for gasket materials.

Figure 8.4 is a depiction of typical shield applications for a pushbutton and lamp with filtering and shielding to maintain the integrity of the enclosure. This would not include equipment manufactured that exhibits RFI leakage in or out of the equipment, that is provided with a cabinet or enclosure. Typical installations of screen over vents and leaded glass in viewing areas are techniques employed as a means of shielding against RFI/EMI by industry. Workstations must meet or exceed CFR Regulations part 15 for RFI/EMI emissions. The responsibility is with the vendor to insure compliance. However, workstations can be the victim of RFI/EMI rather than the culprit. It would be the contractor's responsibility to identify the equipment emitting the RFI/EMI. Correcting an emission problem from equipment supplied by others or existing equipment is not part of this contract. When designing workstations for example, panels may be installed that are manufactured by the installation firm. Many times the panels are installed and found to have RFI/EMI leakage. These panels should always be tested to FCC standards. Underwriters Laboratory (UL) has a special section for testing control panels, or they may be brought to some of the private laboratories that do this testing. However, whatever method is decided upon, certification that emissions are under control may be required as part of an EMC report. This same methodology is done on a larger scale by vehicle manufacturers of locomotives and subway cars. The panels may be tested separately but the whole workstation in most applications is not tested so the engineer must be vigilant about any emissions that have been uncovered. Sometimes simple hand wands for sniffing out electric and magnetic fields may produce sufficient results. The problem with workstations is generally internal coupling due to wiring because all the components installed must meet or exceed FCC standards.

Mechanical connections Typical EMI/RFI shielding methods for coaxial cable joints are shown in Figure 8.5. This provides some of the techniques for both magnetic and electrostatic shielding that can be employed, if warranted.

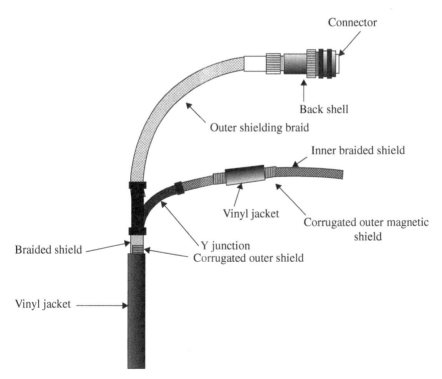

FIGURE 8.5   Double magnetically shielded cable with Y junction

8.4.5   Electrical Design

This section of the report addresses electrical RFI/EMI control. A limited description of common electrical design practices is presented. A description all of the design practices would require a very large publication of approximately 1000 pages. The actual design will be on a case by case basis. The electrical analysis techniques, site surveys and similar approaches to EMC on other equipment applications will be addressed in this section. An examination of various RFI/EMI signal coupling modes to power mains and signal lines are presented in this section. The section does not deal with the equipment only, but also its connections to the outside world.

Capacitive coupling   The first method of coupling is cable to cable capacitive coupling as shown in Figure 8.6, which depicts the circuit diagram between two parallel wires.

The capacitance relationships are provided in Equation 8.5 below.

$$C = e_0 e_r A/S \tag{8.5}$$

Where e_0 = approximate permittivity of air, e_r = relative permittivity dielectric of air, A = area of capacitive plates, S = spacing between plates. If the parallel wires are well grounded, then capacitance $C_2 = 0$. The surfaces of the parallel wires are similar to the plates of a capacitor. Therefore, changing the orientation of the parallel wires to reduce the effective area A, such as placing them orthogonal to each other, will

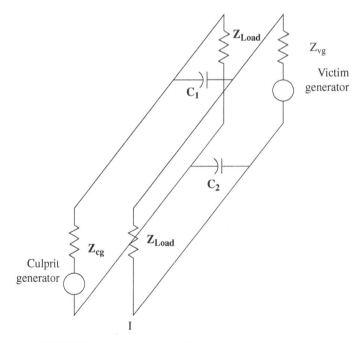

FIGURE 8.6   Capacitive coupling between culprit and victim

reduce the capacitance and the coupling of the culprit signal. For air, the relative permittivity is approximately $e_r = 1$. Dielectric materials will increase the value of C is $e_r > 1$ for other insulators. Note, if reorientation of the wiring increases the spacing S, C will decrease.

Variation of these three parameters will control the capacitance between wires once the wires are installed. This may be a difficult task in that the spacing and insulation are most likely the only options the EMC designer has to correct EMI problems. Another more costly method to reduce or eliminate the coupling can be accomplished by adding a wire between the two parallel wires and grounding it. This technique is often used in ribbon cable to decouple signal lines. The (culprit) signals can be reduced by operating at the lowest frequency possible, that is do not use 1 GHz transmission if 100 MHz is adequate for the application. Do not use a high frequency for power (400 Hz) if 60 Hz can be used. Keep wiring as short as possible; long lengths of wire result in larger equivalent capacitor plates. Graphs illustrate this, with several tables available in the *EMI Control, Methodology and Proceedings* section of Reference [5], by D White. Calculations can also be made using P-Spice modeling.

When designing an EMC test plan, this analysis need not be done in detail or with equations. Written simple statements that describe the capacitive coupling that may occur and the methodology for correcting the anomalies are all that is necessary. Analysis equations are presented that may be an attachment to the EMC test with a notation in the body of the plan. This is to satisfy the more astute engineer looking at the plan. Anything that can be presented in tables or graphic form without the use of equations is the suggested method to present EMC test data. Most engineers and technical managers do not like to page through large numbers of equations or tables of test data. But graphics give an immediate visual representation of the data that make it easy to digest since humans get 90% of their information through visual means.

Inductive coupling   This type of coupling is the result of mutual inductance between two parallel wires that are similar to the single turn transformer. The equation for mutual inductance is given in Equations 8.6, 8.7 and 8.8 and provides the calculations for self-inductance. Figure 8.7 is an illustration of the model employed for the calculations. The capacitance to ground is not included in this model as it describes inductance only.

$$L_{cv} = \frac{\mu_0}{2\pi} L_n \frac{\left[\left(h_1 + h_2 \right)^2 + s^2 \right]}{\left[\left(r_1 + r_2 \right)^2 + s^2 \right]} \tag{8.6}$$

$$L_v = \frac{\mu_0}{2\pi} L_n \left[\left(\frac{2h_2}{r_2} \right)^2 - 1 \right] \tag{8.7}$$

$$L_c = \frac{\mu_0}{2\pi} L_n \left[\left(\frac{2h_1 \cdot}{r_1} \right)^2 - 1 \right] \tag{8.8}$$

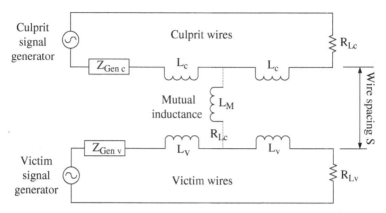

FIGURE 8.7   Mutual inductance coupling between culprit and victim

The inductive effects of EMI are very similar to the capacitive effects. The various parameters are as follows:

L_v, L_c, L_m are the victim self-inductance, culprit self-inductance and mutual inductance respectively. The variables are: L_n is the length of the wires that are in parallel, μ_0 is the magnetic permeability of free space = $4n \times 10^7$ henrys/meter (H/m), h_1 is the height of the culprit wire above the ground, r_i is the radius of the culprit wire, h_2 is the height of the victim wire above the ground and r_2 is the radius of the victim wire.

For larger cable, the estimate of crosstalk will be very conservative and lead to a more sound design.

Area correction = $40\log(d/0.6)$, d = wire diameter in mm.

Twisted pairs   Twisted pairs are one of the most important topics in the analysis. Shielding is the next order of protection against RFI/EMI. Mechanical means of shielding equipment have been discussed previously but a cable shield may require removal of the culprit and victim wiring and replacing with a twisted pair, which can be costly. Some of the methods have already been discussed indirectly. As an example, as the spacing of wire between a ground plane will decrease the crosstalk and isolation, the same principle is employed in shielded wire.

When two wires are routed without twisting the pair, a loop is formed by the two wires, which will produce differential mode noise currents. Twisting the pair results in a series of smaller loops that allow differential mode current to cancel at the loop nodes, as shown in Figure 8.8. The amount of canceling that occurs depends on how symmetrical the twists in the wire are. Each loop will node is identical and currents will not cancel completely. The electromagnetic wave does not have uniform density therefore; the noise currents do not completely cancel. Figure 8.4.7 illustrates the effects of twisting a pair of wires and the ability to reject the noise generated by and electromagnetic wave. Even for a small number of twists in the pair of wires, rejection can be 10 to 20 dB. As the rejection increases, signal-to-noise (SNR) will also increase. Twisted pair wiring is relatively inexpensive and easy to do, if the wire has been purchased without twisting.

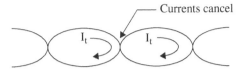

FIGURE 8.8   Twisted pair wiring loops

Cable shielding   The tighter the shield braid the more effective the shield, with the exception that VHF and UHF RFI/EMI shields must always have very tight braid (appearing as a sheet for VHF and UHF RFI). Other methods of shielding are as follows: (i) add multiple shielding and grounding, for example two braided shields, one grounded at each end; (ii) provide thin wall tubing shields; (iii) use heavy walled conduit as a shield. Methods for shielding coaxial cable are as follows: (i) the use of triax where the outer braid and inner braid are insulated from each other; (ii) the use of semi-rigid coaxial cable; (iii) the use of solid tubular shields. These methods of crosstalk rejection are not the only means of solving RFI/EMI problems. The methods described are some of the most commonly implemented methods for shielding. Isolation transformers and opto-isolators are also other means of protecting equipment from RFI/EMI signals. Occasionally the solution to these problems is more a physics problem then electrical or electronic, and the solutions may be unique to the application. Therefore, all possible solutions cannot be covered in this document, only the highlights.

8.4.6   Grounding and Bonding

The two types of grounds are system and equipment grounds. In all cases, the grounding of systems and equipment is a safety issue. All conductive materials must have sufficient low impedance to ground, thus preventing shock hazards to personnel. Only an electric current of 7 mA through the heart for approximately 5 s will result in death.

Low impedance grounding   Type 1 (Figure 8.9a) is one of the most common types of grounding used on many of the signal houses and power distribution centers. It is for moist clay soil which is common in many parts of the country where transit systems are installed. The alternative to using Type 1 is Type 2(Figure 8.9b) for sandy soil. Other types of grounding can be found in the National Electric Code (NEC) or local electrical codes. They should always be consulted before making an assumption of the grounding technique used. Type 3 (Figure 8.9c) is commonly used on communication houses or the grounds must be separated to prevent noise on the communication returns. The ground symbol as shown is a series of ground rods. That may be a single ground rod or a cluster of ground rods connected in a star configuration to provide a good ground return for power. These ground returns may be situated at all four corners of the communication house. The safety ground as shown encircles the communication house and is connected not only to the power and green wire safety grounds of equipment but also to the rebar matrix that composes a Faraday cage around the communication house. Internally

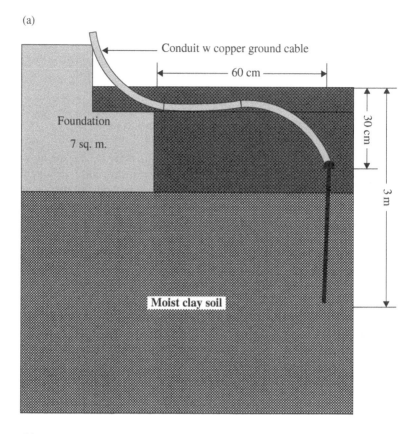

(a)

Conduit w copper ground cable

60 cm

Foundation
7 sq. m.

30 cm

3 m

Moist clay soil

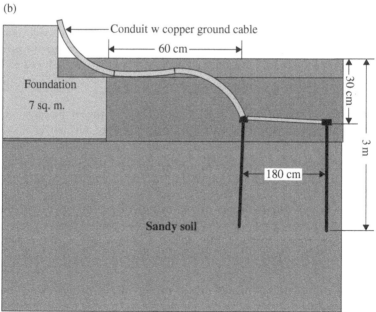

(b)

Conduit w copper ground cable

60 cm

Foundation
7 sq. m.

30 cm

3 m

180 cm

Sandy soil

FIGURE 8.9 Grounding layouts, not to scale. The ground cable and ground rod size are dependent on the service. (a) Type 1. (b) Type 2

(c)

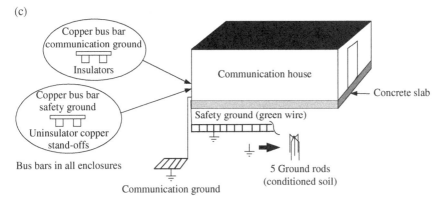

FIGURE 8.9 (Continued)   (c) Type 3

this is connected also to the ground halo within the communication house. The telecommunication ground is connected to a separate grid. As shown in the diagram this grid is the central point of all communication returns. The Type 4 grounding scheme (Figure 8.9d) is found in most operation control centers (OCC). The ground grid is the safety ground or green wire ground that most people are familiar with. It is used to ground all cabinet workstations, computers and other equipment. The floor has a space between the subfloor where the ground grid is located and the actual floor of the control room. All air-conditioning ducting, water pipes, heating ducts, power and auxiliary equipment connections are made beneath the flooring. All of the metallic ducts, water pipes, conduit and other metallic objects are grounded to the subfloor grid.

The typical grounding methods illustrated in Figure 8.9 are meant as a guide. The National Electrical Code (NEC) or local electrical code should be consulted for the actual application. The electrical code is continuously changing and being improved by the industry.

The results of poor bonding are shown in Figure 8.10a, b. Figure 8.10a depicts the results of a poorly bonded filter across the power mains to the equipment. Impedance from the filter to ground is high, as shown with dashed lines. The filter is essentially ineffective as a line filter. This condition will result in EMC problems. One of the problems with poor bonding is certain frequencies may be filtered, leading one to believe that it is operating correctly. This situation could result in the poor bond being completely overlooked during testing.

The bonding strap will appear as a tuned circuit at certain frequencies. This will result in a high impedance to ground on or near the resonant frequency, f_0. Figure 8.9c is an illustration of the effect; the inductance and resistance are due to the bonding strap. The capacitance is the result of equipment case/enclosure, input connector and output connector parasitic capacitance to ground. The objective is to make the resonance frequency as high as possible so that it occurs far beyond the operational frequency range. Low-inductance bonding straps, high-quality connectors with low

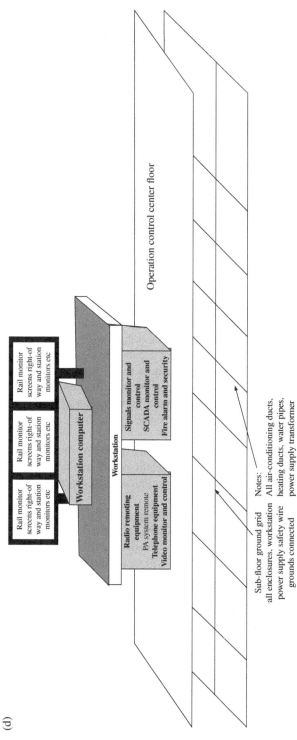

Rail monitor screens right-of way and station monitors etc

Rail monitor screens right-of way and station monitors etc

Rail monitor screens right-of way and station monitors etc

Workstation computer

Workstation

Signals monitor and control
SCADA monitor and control
Fire alarm and security

Radio remoting equipment
PA system remote
Telephone equipment
Video monitor and control

Operation control center floor

Sub-floor ground grid all enclosures, workstation power supply safety wire grounds connected

Notes:
All air-conditioning ducts, heating ducts, water pipes, power supply transformer safety grounds, etc are connected to this ground grid

FIGURE 8.9 (Continued) (d) Type 4

(d)

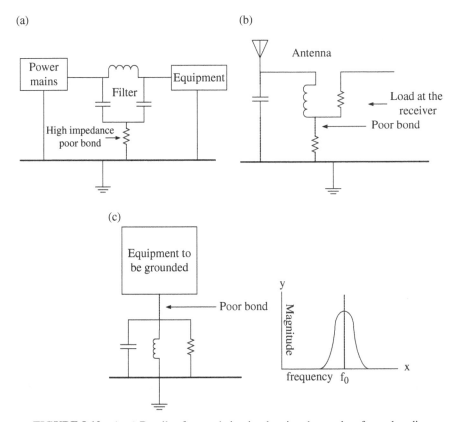

FIGURE 8.10 (a–c) Details of ground circuits showing the results of poor bonding

parasitic capacitance and tight connections with highly conductive plating will result in the desirable effects.

8.4.7 Non-Ionizing Radiation Coupling and Potential Antennas

The emphasis of this section is RFI/EMI effects due to outside disturbances and emissions. Most of the equipment inputs and outputs allow some RFI/EMI to affect performance. The quipment installed also produces emissions but the manufacturer is responsible for these emissions. However, if installation procedures are not followed to the letter, emissions may be produced that compromise the equipment EMC. Equipment is designed to function when EMI emissions are present and all equipment will emit small amounts of EMI. EMC will examine these EMI emissions to determine if equipment is appreciable effected or if the equipment produces appreciable amounts of EMI.

The emission from various equipment is due to poor installation or power sources that are not very clean, that is they have transients or noise on them that produce radiation. RFI/EMI is the culprit and the victim may be equipment located in the system or external to the system. RF radiation can be from very insignificant-looking antennas, as an example, or a slot cut in a metal case to install a non-metallic component such as a rubber foot. The slot can be an aperture antenna under certain conditions. The manufacturer of the equipment may have passed emission tests but the slot would compromise the EMC. Even enclosures with screws missing can allow the emission of internal signals to the outside world, which will compromise the equipment EMC.

Magnetic H fields are almost always present to some degree because of the long wavelength in UPS inverters that use frequencies from 4 to 10 kHz. For single phase power the Wye connected transformer may have one leg where these frequencies appear at neutral. Since the transformers are not balanced at this particular frequency it may appear on grounds of the other two legs. Additional filtering is sometimes required to remove these anomalies. Motor-driven circuits such as air conditioning and vent fans produce transients that may appear on grounds. The EMC engineer must be vigilant to examine all possibilities for low-frequency noise on power grounds.

8.4.8   Propulsion and Signal Equipment Analysis

Subway cars and freight engines along with any wireless equipment installed on cars are tested by the manufacturer prior to delivery. They must meet or exceed DOT and/or FCC standards for admissions. This type of testing can be found in DOT documents that will not be covered in any detail in this book. One of the reasons for this is that DOT testing changes periodically and should be consulted prior to any EMC analysis of equipment. Signals equipment is continuously changing because of the technology used along the right of way. In particular, as technology changes the EMC engineer working with wireless devices in transportation must be aware of all these changes at the time the test plans are written. The test plan produced in most cases need not be elaborate, that is it does not contain tutorial-like information. However, it is handy to have this information at one's fingertips because often questions in meetings may become more technical in detail than described in the test plan. Subway car and locomotive field test results should always be included if possible in EMC test plans. They can usually be obtained from the manufacturer. For signals equipment, immunity testing is also necessary, as it is for medical equipment for safety reasons. This should be included in the test plan or at least with a reference to it.

8.4.9   Professional Engineer's Sealed Documents

A professional engineer (PE) is required to review or write the EMC test plans. It generally must be signed, sealed and dated; this is required because it is a public safety document. The final test results generally must be observed, sealed and signed off by a PE. This is a requirement not only by the Authority initiating the project but also by State regulations for all PEs. Municipalities will require PE seals on almost every document. Special documents such as fire alarms and fire suppression systems require a seal by a PE with a fire alarm specialty. The fire alarm will also be signed off by the fire marshal designated to check the fire alarm installation. Once the inspection by the fire marshal is made and signed off, modifications cannot be made to the fire alarms and fire suppression systems without consulting the fire marshal and getting the installation re-inspected.

8.5   FRESNEL ZONE CLEARANCE

The first Fresnel zone is shown in Figure 8.11. Several equations have been provided in the event that calculations must be in feet or meters. As can be observed in the diagram, the dimensions are the distance from the radio antenna to the obstruction. The minimum clearance h is from the line of sight to point P. A general equation for clearance is shown as Equation 8.9.

$$h = 2\sqrt{\frac{\lambda d_1 d_2}{d_1 + d_2}} \qquad (8.9)$$

This is a general form. The more useful form of Equation 8.9 is given in Equation 8.10, for the clearances are given in meters, f is in GHz, d_1 and d_2 are in km and h is in m.

$$h = 17.3\sqrt{\frac{d_1 d_2}{f(d_1 + d_2)}} \qquad (8.10)$$

Another version is also provided here as Equation 8.11, where f is in GHz, d_1 and d_2 are in statute miles and h is in feet.

$$h = 72.1\sqrt{\frac{d_1 d_2}{f(d_1 + d_2)}} \qquad (8.11)$$

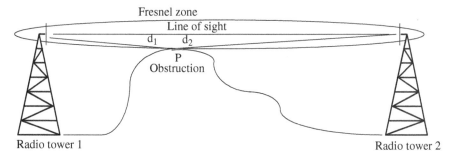

FIGURE 8.11   Fresnel zone with obstruction

Example 8.1

The following distances have been measured from radio tower 1 to the obstruction using a rangefinder or GPS in order to do the calculations: d_1 is 2 km and d_2 is 8 km. The frequency of the radio is 812 MHz. Putting all the values into Equation 8.10, the result is shown below as 24.3 m. This is the minimum clearance. Anything less than this results in attenuation of the signal level. Since this is an approximation, the equations break down when the endpoints are too close to the obstruction, that is about less than 1 km.

$$h = 17.3\sqrt{\frac{2*8}{0.812(2+8)}} = 24.3\,\text{m}$$

Figure 8.12 shows how the far affects reflections, as can be observed point of impact where a reflection occurs, inducing a field as shown in the diagram.

The distance where the planar wave impinges on a surface from the transmitter where the reflection occurs is provided using Equation 8.12.

$$d = \frac{4h_1h_2}{\lambda}\,\text{m} \tag{8.12}$$

Where h_1 and h_2 are the heights of the antennas in meters at the endpoints and λ is the wavelength, also in meters.

Example 8.2

If the source antenna is 10 m high and the point of observation (i.e. the receiver) is 20 m high with a radio frequency of operation at 812 MHz, the calculation for the distance where the reflection occurs is shown below using Equation 8.12.

$$d = \frac{4h_1h_2}{\lambda} = \frac{4*10*20}{0.3817} = 2.095\,\text{km}$$

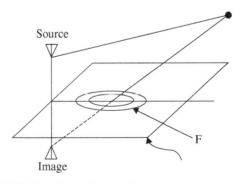

FIGURE 8.12   Wave reflection from antenna on ground plane

Example 8.3

If the source antenna is 10 m high and the point of observation (i.e. the receiver) is 20 m high with a radio frequency of operation at 460 MHz, the calculation for the distance where the reflection occurs is shown below using Equation 8.12.

$$d = \frac{4h_1 h_2}{\lambda} = \frac{4 * 10 * 20}{0.6739} = 1.187 \, \text{km}$$

From these two examples, it should be noted that, as the frequency increases, the point of reflection also increases. This is a means for correcting problems that occur due to reflections off water in particular. If possible use the highest frequencies that can be obtained when operating across open areas such as bays or lakes.

8.6   DIFFRACTION LOSSES

Diffraction losses have been discussed previously. That was for a knife edge; in most cases dealt with previously this analysis sufficed. For the situation with not a knife edge but a rounded top of a hill the loss is different that it is for the knife edge. An extra term is added to the knife edge calculation to allow for the rounded surface. If the surface is covered with vegetation and not smooth, the calculation requires an additional term to compensate for the roughness; this of course will increase the loss. To do the calculations the first part requires calculating the knife edge and then, in addition to the knife edge, an access loss is added (L_{ex}) and an additional term to that is required for the surface roughness.

There are some observations that can be made. As the frequency goes up the wavelength would be shorter; and the loss would be lower with longer wavelengths from observations made from Equations 8.12 and 8.13; and D_s get smaller. As the roundness of the diffraction edge gets closer to a knife edge the loss gets smaller because the excess loss approaches zero. Generally the diffraction loss over a top of the ridge that is covered in vegetation and is not smooth will be a bit less than a smooth surface. In areas where the signal path is over several rooftops or an area where the obstacle can be estimated as a cylindrical area and an estimate can be made of the loss as a smooth surface, a reduction can be made in the loss calculation of about 50%. This gives a conservative estimate for a rounded rather than a knife edge path. Therefore, modeling the surface as smooth has the advantage of giving the model a 50% margin for excess loss.

$$v = 2\sqrt{\frac{\Delta d}{\lambda}} \tag{8.12}$$

$$L(v) = 6.9 + 20 \log \left[\sqrt{v^2 + 1} + 1 \right] + L_{ex} \, \text{dB} \tag{8.13}$$

$$r = \frac{2D_s d_1 d_2}{\theta \left(d_1^2 + d_2^2 \right)} \text{ m} \tag{8.14}$$

$$L_{ex} = 11.7\theta \sqrt{\frac{\pi r}{\lambda}} dB \tag{8.15}$$

Example 8.4

The calculation for a knife edge following an examination of Figure 8.13 assumes a height of 300 m and d_1 is equal to 1 km. The hypotenuse from Tx through the triangle peak is 1040 m and the hypotenuse from Rx to the peak of the triangle for d_2 is 761 m. The frequency for the knife edge calculation is 700 MHz.

$\Delta d = (1040 + 761) - (1000 + 700) = 101$ m, λ at 700 MHz is 0.4428 m. Insert these values into Equation 8.12. Then

$$v = 2\sqrt{\frac{101}{0.4428}} = 30.2 \text{ A unit-less ratio}$$

Insert v into Equation 8.16. The result is as shown below.

$$L(v) = 6.9 + 20 * \log\left[\sqrt{30.2^2 + 1} + 30.2 \right] + L_{ex} = 37.97 \ dB + L_{ex} \tag{8.16}$$

To calculate the value of θ, the angle where the radius r is shown is calculated using trigonometry; and arctangent θ is 0.737 radians (diagram in Figure 8.13 is not to scale). The angle θ must always be in radians rather than degrees. If D_s is 15 m, it is calculated using Equation 8.14. The result is shown in the equation below.

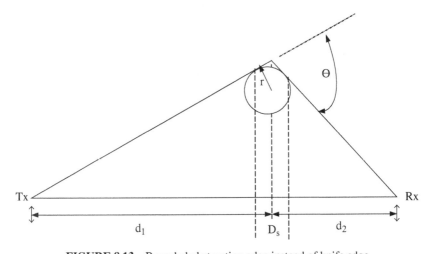

FIGURE 8.13   Rounded obstruction edge instead of knife edge

$$r = \frac{15*1000*700}{0.737\left(1000^2 + 700^2\right)} = 9.55\,\text{m}$$

The access loss can now be calculated using Equation 8.15 the result is shown below:

$$L_{ex} = 11.7*0.737\sqrt{\frac{\pi*9.55}{0.4428}} = -70.978\,\text{dB}$$

This value is now inserted into Equation 8.16 for a total loss of L(ν) = 108.9 dB. This must be added to the free space loss to get the total loss and transmission over the path. The free space loss is calculated using the Friis equation [L_{fs} = 32.4 + 20Log(f_{MHz}) + 20Log(d_{km}) + L(ν) + fade].

The total loss = 32.4 + 56.9 + 64.6 + 108.9 +20 (fade) = 282.8 – 70.978 = 211.82 dB

To conclude this chapter several review problems have been added to allow the audience to test their knowledge of the material. These tools may assist the EMC engineer in determining many aspects of radiated RFI and EMI. In many cases these will not show up until the integration testing begins before commissioning of the systems.

In one particular instance a radio link set up for communications in the site design was completed with all the loss budget calculated. However, the project lasted approximately one year and in the meantime a building was constructed that proved to be an obstruction directly in the path of the communication link. Since the communication link was dormant during the year there was no knowledge of this until testing began during the integration phase. The engineer who did the original design had tried to use a building in the construction path as a reflector. However it did not work too well and caused interference for radios on other buildings. The FCC required that the culprit (i.e. the communication link) be shut down until the problems were corrected. Since the tower had been constructed antennas mounted on the entire structure had to be redesigned, which became a very costly endeavor.

From the example stated in the above paragraph some construction during the installation of subway systems can take one to two years. In the meantime the topography may change somewhat, which can cause costly delays in commissioning. Obstructions to communications can become very challenging to circumvent and produce a viable highly efficient communications system. As one might surmise, something as serious as high-powered switching or a high-tension power line can be very disruptive to communications if installed after the communications system has been designed and tested prior to subway system commissioning.

8.7 REVIEW PROBLEMS

Problem 8.1
A transmission line has a ground with the following characteristics: the capacitance is equal to 50 pF, resistance 0.1 Ω and inductance 1 μH. Find the resonant frequency and power radiated at the ground if the current is 1 mA.

Problem 8.2

A 10 m tower is fitted with a dipole antenna located 10 miles from an obstruction, a hill. It is communicating with a tower that is 8 miles on the other side of the obstruction. The radio frequency is operating at 702 MHz. Find the minimum height to the first Fresnel zone.

Problem 8.3

A 10 m tower is fitted with a dipole antenna located 2 km from a tower 5 m high. Find the distance to the reflection point.

Problem 8.4

To increase the reflection point in Problem 8.3, does it require increasing the tower heights to 15 m each, increasing the frequency of operation to 812 MHz or does it require both to increase the point of reflection.

Problem 8.5

Find the knife edge loss for Example 8.2 for a frequency of 460 MHz.

Problem 8.6

Find the excess loss for Example 8.2 with the change in frequency to 460 MHz.

Problem 8.7

Find the total free space loss in Example 8.2 for a frequency change to 460 MHz. Include the following: knife edge loss, access loss, free space loss and fade.

8.8 ANSWERS TO REVIEW PROBLEMS

Problem 8.1

$$f = \frac{1}{\sqrt{LC}} = \frac{1}{\sqrt{50 * 10^{-12} * 1 * 10^{-6}}} = 141.4 \, \text{MHz}$$

$$\text{Power} = 0.1 * 1 * 10^{-6} = 0.1 \, \mu\text{W} \left(-40 \, \text{dBm} \right)$$

Problem 8.2
Using Equation 8.11, the calculations are shown below:

$$h = 72.1 \sqrt{\frac{d_1 d_2}{f * (d_1 + d_2)}} = 72.1 \sqrt{\frac{10 * 8}{0.702 * (10 + 8)}} = 181.4 \, \text{ft}$$

Problem 8.3

$$d = \frac{4 * 10 * 5}{0.4428} = 451.6 \, \text{m}$$

Problem 8.4

$$d = \frac{4*15*15}{0.3817} = 2.357\,\text{km}$$

Problem 8.5

$$v = 2\sqrt{\frac{\Delta d}{\lambda}} = 2\sqrt{\frac{101}{0.674}} = 24.48, \quad L(v) = 6.9 + 20*\log\left[\sqrt{24.48^2 + 1} + 24.48\right] = 40.7\,\text{dB}$$

Problem 8.6

$$L_{ex} = 11.7*0.737\sqrt{\frac{\pi*9.55}{0.674}} = 57.53\,\text{dB}$$

Problem 8.7
The total loss = 32.4 + 53.25 + 64.6 + 40.7 + 57.53 + 20 (fade) = 268.48 dB

9

TRACK CIRCUITS AND SIGNALS

9.1 INTRODUCTION

The two types of track circuits most commonly used are the type at interlockings where crossovers in track usage occur and audio frequency (AF) for mainline running rails before interlockings. These are commonly used for automatic train control (ATC). These advances in track circuits are all in use with AC propulsion traction motors. Some of these AC propulsion drives use a combination of frequency sweep and pulse width modulation, while others may use switched thyristors. All of these systems produce harmonics of 6× and 12× the power frequency. The harmonics produced depend on how the transformer and inverter are constructed. Some of the DC power supply chopper and inverters are wired to eliminate even harmonics and odd harmonics to theseventh. They then filter the 7th, 9th and 11th that are the remaining harmonics to control noise on the running rails.

ATC systems can be installed on insulated joint rails or continuous welded rails. A combination of train detection, train cab information, maintenance information and other ATC related functions are part of train control. There are other stray magnetic fields from any auxiliary equipment on subway or freight trains, such as air conditioning, heating, lighting and wayside wireless devices. These are inductively coupled to running rails. For both hot rail and catenary systems arcing will result in stray transient magnetic fields that couple into the running rails. The third rail (i.e. the hot rail) inductively couples to the loop formed by the running rails for catenary system. This depends on placement in relation to the running rails. If the catenary is implemented it is placed directly down the center of the track between the running rails and the magnetic coupling will cancel any stray magnetic H fields from the catenary within the running rail.

Electromagnetic Compatibility: Analysis and Case Studies in Transportation, First Edition.
Donald G. Baker.
© 2016 John Wiley & Sons, Inc. Published 2016 by John Wiley & Sons, Inc.
Companion website: www.wiley.com/go/electromagneticcompatibility

The best way to visualize a subway track system is as a transmission line with LRC (inductance, resistance, capacitance) components. The rails provide both inductance L and resistance R, while the ballast (the gravel beneath the rails) provides a conductive and capacitive path (C) between the rails. The 100 Hz transmitter, a vane type balanced relay, has a set of points that are held open with the power on. When a train pulls into the interlocking area it shorts out the relay through the axles of the car, it breaks the power to the relay, thus allows closing the relay contacts. This failsafe system will determine when the interlocking is occupied by a train. The signal is sent to the ATC center control, or operation control center (OCC) as it is commonly called.

The AF track circuits are similar but, instead of a single frequency, multiple frequencies are used to determine which segment of the track a train is on. This consists of an AF transmitter and receiver so that, when the train's axles short out the transmission of the particular AF frequency, the receiver at the end of the track will sense the loss of signal and produce an indication that is sent to the OCC. Often, to communicate with the engineer in the cab of the train a loop detection system is installed at the front of the engine (whether subway or freight). This can also be used as a speed control method for the operator. Often on freight systems a wayside wireless radio located on the right of way interrogates the cars as they pass the sensor and sends a message back to the OCC indicating a car ID. These types of systems are shown in Figure 9.1.

This section is meant as an introduction and it gives the reader some feeling of subway system signaling and other communications using the rails as transmission lines. Included in this treatment are the necessary differential equations for solving electromagnetic compatibility (EMC) issues should they arise from track circuits, substations, DC power supplies or AC propulsion units. Many of the equations are from Section 9.3. This is a general section for vector analysis of electrical engineering for Maxwell's and the wave equation. As one may surmise, most of the analysis is due to the inductance and conductivity of the track circuit loops. The analysis provided in this section will not contain measurement apparatus. This is done to prevent the book from becoming obsolete because instrumentation is improving so rapidly with today's technology.

Figure 9.1a gives an example of an insulated joint; and Figure 9.1b shows AF continuous audio frequency track circuits. In some cases the track circuit frequency is 110 V AC instead of the 100 V AC shown in Figure 9.1a. The 100 V AC is preferred because it is generated locally just for these track circuits. It requires extra expense; however, 110 V AC is not as clean or reliable as the latter. A step-down transformer is shown in the diagram supplying the track circuit voltage at much lower than 55 V. This is considered a shock hazard on rails and it is a safety issue for maintenance personnel and the public. The voltage should overcome the track impedance with sufficient power to the vane balanced relay. A step down in the voltage to the other side of the vane relay will allow it to maintain balance and hold the points open. The resistor shown is a combination of track resistance an a physical resistor added to the track circuit. R_2 is composed of the physical resistor and potentiometer to balance the vane relay. The signal rails have insulated joints as shown in the diagram. This track circuit controls a block so then when it is empty, that is no train or subway car

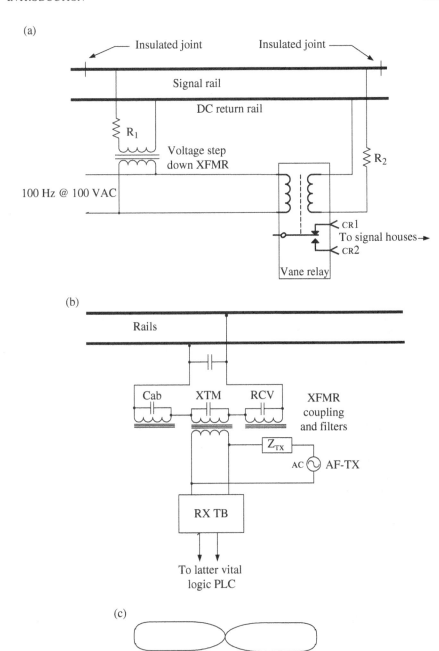

(a)

Insulated joint   Insulated joint

Signal rail

DC return rail

R_1

Voltage step
down XFMR

100 Hz @ 100 VAC

R_2

CR1
To signal houses →
CR2

Vane relay

(b)

Rails

Cab   XTM   RCV   XFMR
coupling
and filters

Z_{TX}

AC ⟳ AF-TX

RX TB

To latter vital
logic PLC

(c)

Loop detector for engine
in interlocking

FIGURE 9.1   Track circuit details

occupying the block, the indication is green or yellow depending on whether the previous block is occupied. At high speeds several blocks must be vacant before the signal will go from yellow to green.

One of the major problems with insulated joints is that they may scab over and short to the next block due to wear on the rails. They must be inspected periodically to determine the condition of the joints in the rails. As a subway car or train enters the block, the axle will short out the signal to the receiver, thus causing the relay dropout closing the points. The condition of the blocks is monitored in the local signal house or bungalow that has a board similar to the one shown in Figure 1.1 in Chapter 1. This controls the local interlocking and the controls are key operated for security purposes. When the block information at the signal house is to be locally controlled, permission must be granted from the OCC that has a large display which monitors the entire system. The operator maintenance person must have a key; however, if the signal house control is not exercised, information is displayed on the local board of the signal house or bungalow.

The crossover tracks at the interlocking have a loop detector, as shown in Figure 9.1c. This will detect a subway car or train entering the crossover (interlocking) and the local vital logic unit will indicate that vehicle is in the crossover running against normal traffic flow or awaiting permission to continue the crossover against traffic. The codes generated by the block information on both sides of the interlocking is sent to the OCC for control by the train control dispatcher. The vital latter logic presents information to the dispatcher and blocking traffic before the interlocking and passing traffic through the interlocking safely. The codes are sent when the vital logic is interrogated and the code is generated and sent to the communication house for transmission to the OCC. The entire system is designed to be failsafe, that is if no one does anything all trains will be blocked from the interlocking. To enter a crossover the operator transit car or train must ask permission and it is the dispatcher at the OCC who will make this decision. When an interlocking is entered and the train is proceeding against traffic, all of the traffic is under strict control to prevent head-on collisions at the interlocking or along the right of way. The information presented here is only an overview and should not be construed as design information. Signals engineering is much more complex than presented here.

9.2   AF TRACK CIRCUITS

The details of an audio track circuit are shown in Figure 9.1b. The transmitter AF-TX generates an audio burst signal that is injected into the running rails via a step-down transformer. The transformer not only steps down audio frequency but it isolates the transmitter and receiver from the rails. The transmitter transformer, receiver and cab filters form a network that are allows burst audio to the rails and accepts audio at a different frequency from another track or cab communication circuit (coded message). A transmission line is connected between the track and receiver, and the transmitter and filter network. A shunt capacitor is installed across a transmission line to control the ballast capacitance (crushed rock beneath the rail bed).

TABLE 9.1 F_0 = 1590 Hz Operating Frequency (all Values are in Millivolts)

	Frequency (Hz)												
	1490	1510	1530	1550	1570	1590	1610	1630	1650	1670	1690	1790	1710
Code rate (Hz)						F_0							
Minimum (5.0)		411	249	102	74.5	70.2	78.1	162	335	461			
Intermediate (10.09)	474	329	201	98.1	69.1	65.4	74.2	139	253	345	439		
Maximum (20.4)	498	329	208	103	71.7	63.0	78.0	147	255	328	450	490	

The sensitivity of the track circuit relay, that is the point closure when no vehicle axle has shorted out the block uses [magnitude $\leq 0.5\ V_{RMS}$ ($1.4\ V_{peak\ to\ peak}$)]. The test frequency range of a particular receiver should be 300 Hz bandwidth and 1–10 kHz carrier. These are not all operational frequencies; this is to ensure the relay does not pick up due to extraneous frequencies coupled into the rails. A chart must be prepared by the signals engineer doing the testing showing the range of pick ups of the signals relay. An example of such a chart is shown in Table 9.1. This table is only provided as a sample to assist the EMC engineer's analysis when determining if, for external signals coupling into the rail, there will be an EMC problem. The geometric mean of the maximum and minimum code rate provides the intermediate code rate, that is 10.9 Hz = $(5 \times 20.4)^{\frac{1}{2}}$.

The maximum sensitivity at 1590 Hz from Table 9.1 is between 63.0 and 70.2 mV and, looking at the roll off with frequency (i.e 1490 and 1690 Hz) it is at the minimum code rate; greater than 15.3 dB (lower frequency end) and less than 16.3 dB (upper frequency end) respectively. The same argument can be made for the intermediate (17.2 and 16.5 dB) and maximum (17.9 and 17.1 dB) code rates for the lower and upper end frequencies, respectively.

Example 9.1

If the frequency of 1590 Hz is used in the signal block, the seventh harmonic ($218.5 \times 7 = 1529.5$) traction motor within the block is 235 mV would be sufficient to indicate the block is occupied. This condition of course presents a hazard that cannot be tolerated; therefore a frequency greater than the seventh harmonic coupled into the rail by the hot rail must be used in the signal block. This calculation is made later in this section using field theory.

There are some other anomalies that are not covered that may or may not impact EMC on track circuits. During a subway car's travel along the system rails, occasionally the subway will stop or slow down and arcing may be observed. The rise time of a normal arc is 10 μs and the fall time is about 20 μs. The peak current that occurs at the peak of the rise can be as high as 80 A. This pulse is a transient and generally results when crossing over interlockings where the hot rail contact shoes bounce on the hot rail and a catenary system response on the wire occurs. These particular events are analyzed later on a case by case basis under case studies. There is also one transient that produces burnout and equipment failure if it occurs, that is a lightning stroke near or directly on the rail. A lightning stroke has sufficient energy to burn out electronics equipment. It has a rise time of approximately 10 µs with a peak current of 100 kA. A lightning stroke need not be direct but only close to a hot rail or catenary. The structure or wires carrying the substation currents are at most risk of a lightning stroke. Other types of transients that can occur are due mainly to man-made activities, such as arc welding, the use of electrical vehicles near tracks such as new automobiles that are hybrid with electric/gasoline engines and all vehicles operating with electric propulsion motors. Some of these have not been experienced yet but will be in the future, such as buses that operate on fuel cells and are powered solely by electric motors. When such a vehicle is operating on or near a railroad crossing, it may have frequencies in multiples of 60 Hz generally and start currents of 80 or 90 A. The EMC engineer must be vigilant for the new technology emerging for electrical drive systems or motors that may impact older equipment on rail systems.

There is another type of AF transceiver, where a step-down transformer is connected via transmission lines to a filter located near the track. The transmitter and receiver connections are made at the input of the transformer. All of these transceiver connections are considered when designing signal systems. The transceivers are arranged on PC boards for a number of transmitters or receivers and are in single or multiple racks. Signals equipment in general must be highly reliable and designed for failsafe operation. Signals equipment generally takes a great deal of time to get approval and it does not change as rapidly as other electronics equipment. Often some types of relay logic, although minor, still exist but will be gradually replaced until everything is solid state.

Signals testing is beyond the scope of this book. However, the report alluded to in Reference [10] as an *"Introduction to Conducted Interference Mechanisms in Rail Transit Systems using Solid-State Propulsion Control"* discusses the following topics: track circuits, solid-state propulsion control and conductive interference. Section 2 has report formats, recording and documentation procedures. Section 3 consists of conductive suggested test procedures such as: (i) conducted susceptibility of audio-frequency rate-coded signaling systems from 300 Hz to 10 kHz; (ii) conductive emission test of substations; conductive emissions test of vehicles; (iii) audio frequency rate-coded track circuit receiver operating characteristics from 300 to 10 kHz. Appendix A has definitions and systems. Appendix B has sample test outputs using inductive suggested test procedures. Also, references are included at the end of this report.

It is expected, however, for vehicle manufacturers of freight engines, cars or subway cars to provide all emissions from the equipment, including air-conditioning, heating equipment, engine and motor controls. The demarcation between areas of responsibility is not always completely clear; however, during commissioning testing it may be evident that some EMC issues exist. These must be solved before commissioning is complete and it may be up to the EMC engineer to solve the problem. The manufacturers of equipment are often very helpful in determining design features, but the engineer must always be on guard against specsmanship (a colloquial term coined by the engineering community meaning do not allow a manufacturer's representative overextend a feature of equipment performance). This has been done in the past with disastrous consequences that resulted in a very expensive resolution to a simple problem. A particular case was when a heat pump was designed into a system instead of an air conditioner: the heat pump failed because the temperature difference between outside and inside the cabinet must not exceed more than 20–25 °C to operate efficiently whereas an air conditioner has no such restriction. The cost of replacement was quite high to retrofit the cabinets.

A system is shown in Figure 9.2 of how AF transceivers can be connected to form a network for a bungalow or signal house. The connections to the track are shown, with the filtering not shown. The connections to the track go to a resonant bond at the AF track frequency for the block.

The block is a segment of track between adjacent AF frequencies at each end of the frequency used for the particular block. An example of the frequencies used is as follows:

1. $f_1 = 3.1$ kHz (block D)
2. $f_2 = 3.5$ kHz (block B)
3. $f_3 = 3.9$ kHz (block C)
4. $f_4 = 3.1$ kHz (block A)

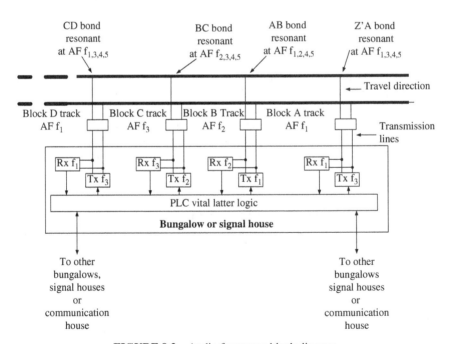

FIGURE 9.2 Audio frequency block diagram.

The resonant bond is at C D (3.1 and 3.5 kHz), B C (3.5 and 3.9 kHz) and A B (3.1 and 3.5 kHz). Blocks A, B, C and D are all loops with audio frequencies. When a subway car or train enters a block the axle will short out the audio frequency to the receiver relay. This results in a failsafe relay dropout, indicating the block is occupied. This indication is built into the signal's vital logic. The indication has several effects: (i) the block indication is calculated by the latter logic that determines the block is occupied, (ii) the speed of the car being monitored determines whether the block will be yellow or red, based on the speed and position of any subway car or train behind it and whether a track obstruction occurs ahead of the advancing vehicle. The obstructions may be another train or subway car in an interlocking with vehicles in the crossover or approach to the station. The dispatcher from the OCC may have maintenance crews busy along the right of way; in turn the yellow caution signals are for trains approaching the work area. Caution means that a train may move into a block at a very restricted speed so as not to produce a safety hazard. Also, crossing gates occasionally require caution with a speed restriction because of traffic congestion in the area. It can always be assumed the public may not always be alert to the hazard of the train approaching.

The diagram in Figure 9.3 shows how power in the third rail is coupled into this signal rail and its effect on signal frequencies. The diagram describes a third rail (hot rail) installation; as can be observed the magnetic H field of the hot rail couples to the running rails. The hot rail is installed on both left and right sides of the running rails. This has a tendency to lower the value I_3 that would normally keep increasing along the running rails. The connection shown where the crossover from one side of the track to the other is done with a heavy cable that is buried. In this configuration, the running rails have continuous welded joints with AF track circuits.

For DC power on a catenary down the center of the running rails, instead of using a hot rail configuration, the inductance coupling of the H field would be balanced in both running rails. The mutual inductance coupling would be zero, if the catenary is exactly positioned down the center of the track. Some subway installations have a combination of catenary and hot rail; this may be prevalent in some tunnels or where overhead space is limited. The track circuit receivers are shown in the blocks at the end of the loop in the diagram. If the hot rails are in a position too close to one of the receivers, the sum of the coupled H fields at the receiver is unbalanced that may affect their performance.

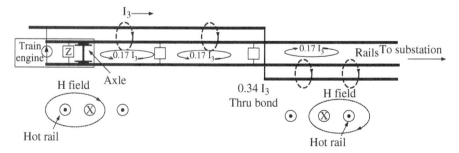

FIGURE 9.3 Magnetic coupling of hot rail to running rails.

The substation harmonics are multiples of the fundamental AC powers frequency. These are from 6× to 12× the fundamental frequency; this depends on how rectifiers are arranged. In some substations the even harmonics are canceled with only the odd harmonics prevalent from seventh and above. The magnitude of these harmonics will vary depending on the DC load presented to the substation.

The traction power of some motors has harmonic frequencies of 1100 Hz. This is a frequency that can interfere with some of the signals equipment located along the right of way if the magnitudes are large. Added filtering may be necessary to the substation or along the area to suppress culprit frequencies that can degrade the performance of the particular sensor. These anomalies are caused by pulse skipping or frequency sweeping during vehicle startup from a complete stop. Traction motor test results should always be examined with the possibility that the harmonics can interfere with signals equipment performance. This analysis should be undertaken before installing signaling equipment at an early stage in the project to prevent costly modifications during commissioning tests.

The diagram in Figure 9.3 shows the engine or train at one end of the running rail configurations has the AC current source and the other end has the signal end. The mutual inductance per loop is 0.9 µH; in the self-inductance of the running rail it is 5.4 µH, as alluded to in Reference [9]. A rule of thumb from this reference is that the running rail ratio to mutual inductance between the hot rail running rail loop is 6 : 1 or $M_{loop}/L_{running\ rail}$ = 0.17 for mutual inductance of 0.9 µH for a loop at the hot rail. These values are not presented as inductance per yard.

Measured values are presented in Table 9.2. These are only for one particular rail size: 100 lb/yd. This is one of the more common rails used in mass transit systems.

This table is taken from Reference [9]. Data used in the calculations are as follows:

1. Distance of the hot rail from the nearest running rail is 67 cm (r_1) center to center.
2. Distance between the running rails is 1500 cm (r_2) center to center.
3. Distance between the hot rail and outer running rails is 212 cm (r_3) center to center.
4. 100 lb/yd running rail at various frequencies.
5. 150 lb/yd hot rail at various frequencies.
6. Carbon steel relative permeability is 1500.
7. Length of the rails between resonant bonds is 2000 feet (610 m) for 70 mph travel.
8. Headway standard is three signal blocks at 19.6 s/block.
9. Stopping distance is 50 s at 70 mph; three block headway is 69.9 s.
10. Industry standard for headway is 94 s. This is for a fixed block system.

An observation that can be made about Table 9.2 is that, as frequency increases, inductance decreases due to rail hysteresis and the resistance component increases due to skin effect. Based on the data taken from Reference [9], this is as follows:

1. $\sigma_{rail} = 4.79 \times 10^6\,\Omega/m$

2. $\mu_{rail} = 2.51 \times 10^{-5}\,H/m$

TABLE 9.2 Rail Impedance Data for 100 lb/yd Running Braille

f (Hz)	DC (A)	Z (μΩ)	Z angle (degrees)	R (μΩ)	X (μΩ)	ΔL (μH)	R_{Rail} (μΩ/m)	R_{tk} (μΩ/m)	L_{tk} (μH/m)
25	0	534	21°	499	191	1.22	54.5	109	1.88
25	410	534	20°	502	183	1.16	55.0	110	1.87
25	680	514	20°	483	176	1.12	53.0	106	1.86
25	730	516	20°	485	176	1.12	53.0	106	1.86
55	0	734	11°	721	140	0.41	79.0	158	1.71
55	560	734	11°	721	140	0.41	79.0	158	1.71
65	0	796	9°	786	125	0.30	86.0	172	1.68
100	0	986	−4°	984	−69	−0.11	107.5	215	1.59
316	0	2520	−40°	1930	−1620	−0.82	211.0	422	1.44
1000	0	8860	−66°	3604	−8094	−1.29	395.0	789	1.33
3160	0	29800	−74°	8214	−28 646	−1.44	899.0	1798	1.30

3. $r_{eff} = 5.23$ cm (effective radius of the rail, i.e. equivalent to a copper tube).

$$\delta = \sqrt{\frac{2}{\omega\mu\sigma}} = \frac{1}{\sqrt{\pi f \mu\sigma}}$$

Using the equation in Table 3.1 from Chapter 3, the skin depth is calculated for the two frequencies being studied 218 and 1526 Hz. Depth in meters is calculated for the hot rail only. The resistance per meter is as shown below, (5) and (6) respectively:

1. $\delta_{218} = \left(\pi * 218 * 2.51 \times 10^{-5} * 4.79 \times 10^{6}\right)^{-1/2} = 3.485 \times 10^{-3}\,\mathrm{m}\,(0.3485\,\mathrm{cm})$

2. $\delta_{1526} = \left(\pi * 1526 * 2.51 \times 10^{-5} * 4.79 \times 10^{6}\right)^{-1/2} = 1.317 \times 10^{-3}\,\mathrm{m}\,(0.1317\,\mathrm{cm})$

3. $A_{218} = \pi * \left[(5.23)^{2} - (5.23 - 0.3485)^{2}\right] = 11.03\,\mathrm{cm}^{2}$

4. $A_{1523} = \pi * \left[(5.23)^{2} - (5.23 - 0.1317)^{2}\right] = 4.27\,\mathrm{cm}^{2}$

5. $R_{218} = 1/\left(A_{218} * \sigma_{rail}\right) = 189.3\,\mu\Omega/\mathrm{m}$

6. $R_{1526} = 1/\left(A_{218} * \sigma_{rail}\right) = 488.9\,\mu\Omega/\mathrm{m}$

The inductance of the rails depends on two entities: the inductance of the material and the geometry of the track. The calculation is made using the following equation as taken from Reference [9]. The data is provided below to do the calculations:

1. $L_{Steel} = 1.617\,\mu\mathrm{H}/\mathrm{m}$

2. $L_{geo} = \Delta L/\mathrm{m}$

3. $L_{Tk} = (1.617 + \Delta L)\,\mu\mathrm{H}/\mathrm{m}$

4. $M_{rr}/L_{Tk} = 0.17$ (rule of thumb 1/6)

Table 9.2 is based on a combination of measurement data and calculations. This data is used later for analysis of the diagram depicted in Figure 9.3. The diagram is also used for calculations for supply harmonics that appear as loop currents and are coupled by the magnetic field of the hot rail. The power supply can be shown to couple into the rails using field equations; however, for this exercise the use of mutual inductance from the loop and the use of circuit theory are beneficial to illustrate the difference between magnetic field coupling through field theory and using circuit theory.

9.3   LOOP CALCULATIONS

For the magnetic field coupling into the loops, as shown in Figure 9.3, the H field from the center of the hot rail the center of the loop is 1.42 m (67 + 75 cm). The peak current for a married pair, that is two subway cars, is approximately 60 A within the first 10 s at 218 Hz, after a travel of 244 m it is 10 A. The peak seventh harmonic of the chopper at 1526 Hz is 0.38 A and at 244 m it is 0.24 A. The seventh harmonic of the power supply, 420 Hz, is 18 A. These are some of the harmonics the signal systems will contend with to function efficiently.

The first loop to analyze in Figure 9.3 is where the engine axle has shorted the resonant bond before it enters the loop. The third rail (hot rail) will couple the peak at 218 Hz and the seventh harmonic (1526 Hz) into the loop. The analysis will be conducted using Faraday's law, as shown below in Equation 9.1.

$$\oint_c \bar{E}_L * \mathrm{dL} = -\frac{\partial}{\partial t} \iint_S \bar{B}_L * d\bar{S} \tag{9.1}$$

Where $\oint_c$ is the linear integration around the loop with the electric field intensity $\bar{E}_L$ and $\iint_S$ is a surface area of integration where the magnetic field density $\bar{B}_0$ is the magnetic field density (webers/m^2). On the right side of the equation the derivative is taken with respect of time because these are peak values in the calculations.

$$B_0 \hat{a}_n \approx \mu_0 H_0 \hat{a}_n = 4\pi * 10^{-7} H/m * \frac{60\,\mathrm{A}}{2 * \pi * 1.42\,\mathrm{m}} \hat{a}_n = 8.45\mu\,\text{webers}/\mathrm{m}^2 \tag{9.2}$$

Equations 9.2 and 9.3 provide a calculation for finding the magnetic field density. It should be noted, a result of the combination of the two vectors is a scalar quantity. The approximation is used because it is not exact but it is an average magnetic field density across the loop and normal to the loop.

$$B_0 \hat{a}_n \approx \mu_0 H_0 \hat{a}_n = 4 * \pi 10^{-7} * \frac{0.38}{2 * \pi * 1.42} \hat{a}_n = 26.76\,\text{nwebers}/\mathrm{m}^2 \tag{9.3}$$

The calculation for the area, Equation 9.4, assumes it is the planar area of the loop and that the magnetic field density is normal to it.

$$\text{Loop area} = 1.5\,\mathrm{m}\,(\text{track width}) \times 610\,\mathrm{m}\,(\text{block length}) = 915\,\mathrm{m}^2 \tag{9.4}$$

The magnetic flux density ψ is found using Equation 9.5. It is in units of webers; and multiplying this value (which is a scalar) by the frequency in radians provides the peak voltage around the loop. The two peak voltages for 218 and 1526 Hz are provided as shown in Equations 9.6 and 9.7.

$$\text{Peak voltage around the loop} = \omega * \psi \tag{9.5}$$

Where $\psi = \bar{B} * \bar{A}$.

$$V_{P218} = 2 * \pi * 218 * 8.45 * 10^{-6} * 915 = 10.59 \text{ V} \qquad (9.6)$$

$$V_{P1526} = 2 * \pi * 1526 * 26.76 * 10^{-9} * 915 = 0.235 \text{ V} \qquad (9.7)$$

Values from Table 9.2 are used in the calculations for the two frequencies 2.18 and 1526 Hz and interpolation is made between frequencies 100 and 316 Hz for the resistance per meter and inductance per meter. The total rail resistance is as shown for two rails. Note the impedance bonds are treated as a short-circuit because of their low inductance value at the frequencies 218 and 1526 Hz. At 1526 Hz the impedance bond is not quite a short-circuit because the resonance is at frequencies of ≥ 3 kHz, that is the AF signal frequencies.

$$\text{Total rail resistance} = (610 \text{ m} \times 2 \text{ per rail} * 353 \mu\Omega/\text{m}) = 0.431 \Omega \qquad (9.8)$$

$$\text{Total inductive reactance} = 2 * 610 * 2\pi * 244 * 1.4 * 10^{-6} = 2.62 \Omega \text{ at } 244 \text{ Hz} \qquad (9.9)$$

$$\text{Total inductive reactance} = 2 * 610 * 2\pi * 1526 * 1.31 * 10^{-6} = 15.3 \Omega \text{ at } 1526 \text{ Hz} \qquad (9.10)$$

$$\text{Total impedance at } 244 \text{ Hz} \, Z_T = 2.65 \underline{|80^\circ} \ \Omega \qquad (9.11)$$

$$\text{Total impedance at } 1526 \text{ Hz} \, Z_T = 15.3 \underline{|88^\circ} \ \Omega \qquad (9.12)$$

The total impedance as calculated using Equations 9.8 through 9.12 the loop current can be calculated as $0.17 \, I_3$. This includes the mutual inductance from the hot rail to the loop. This is using field theory to make the calculations. The difference shown here between circuit and field theory respectively is that one only needs to know the loop area, the current flow through the hot rail and the distance of the hot rail to the center of the running rail. These are all parameters of the geometry with only current necessary to make the calculations.

$$0.17 \, I_3 = 4 \text{ A} (10.59 \text{ V}/2.65 \Omega \text{ with a phase angle} -80^\circ) \text{ at } 244 \text{ Hz} \qquad (9.13)$$

$$0.17 \, I_3 = 15.3 \text{ mA} (0.235 \text{ V}/15.3 \Omega \text{ with a phase angle} -88^\circ) \text{ at } 1526 \text{ Hz} \qquad (9.14)$$

Since the seventh harmonic is not considered a short-circuit across the resonant bond, the 15.3 mA may produce a noise component at the receiver. Alluded to previously, there are more stray currents due to other equipment on board the subway cars or freight engines, for example air conditioning, heating, lighting, traction motors that have heavy start currents, DC supplies for subway systems, power supplies for computers, radios, engine monitoring functions for freight trains and other types of equipment that have currents and voltages coupled eventually onto the rails. However, subway system cars and freight engines are tested by the manufacturer of these vehicles within the emission limits. This information is supplied to the authority implementing the design of the subway or railway system.

9.4   CIRCUIT THEORY IN LOOP CALCULATIONS

The current circulating in the loop is 0.17 times the hot rail current of 18 A at the frequency of the seventh harmonic of 60 Hz, which is the worst case. For a particular project the current flow around the loop is 3.06 A; however, the final loop current is due to measurements and rule of thumb calculations. If the use of field theory is used, only a few geometric measurements need to be made, including the current in the hot rail (this must be done regardless of whether field theory or circuit theory is used). The voltage drops around the loop can be calculated easily because the measured data is available for resistance and inductance. These measurements are rather difficult to make using circuit theory. Once that sum of the voltages around the loop is known only the voltage across the resonant bonds needs be measured. The difference between these voltages and the voltage calculated by field theory is the voltage across the running rails. The mutual inductance is included in the calculation for the field around the hot rail coupling into the running rail loop.

9.4.1   Ballast

The definition of ballast is the crushed stone beneath the ties of the rail bed. The resistance was estimated in Reference [9]. The ballast resistance is quite high, about 15 000 Ω/m or 49 Ω in 1000 ft of running rails. There was no mention in the article about the capacitance per linear foot or linear meter between the rails and between the rail and ground; this may be quite insignificant. However it should be investigated to insure that it is negligible for performing transmission line calculations.

The crossover between the right side hot rail to the left side of the running rail as shown in diagram of Figure 9.3 is very close to the track circuit receiver. This is to be avoided but this is matter of power and signals design features. EMC engineers when analyzing problems with signals are made aware of some of the anomalies that occur. This is not meant as a power and signals section of the book. When the crossover is too close to the receiver of a track circuit it can result in doubling the current and, with phase synchronization between the two hot rail current feeds, 0.17 I_3 may increase to 0.34 I_3 or greater because of the phase component not shown here. As a subway train enters the running rails, as shown in Figure 9.3, it acts as a current source to the running rails; the substation at the other end resembles a perfect voltage source because of its low impedance. The short-circuit across the running rails due to the axle and wheels results in a relay dropout that will acknowledge the block is occupied. As the train progresses along the running rails and leaves a block the AF frequency is sensed by the receiver and its relay. Points are closed indicating the train has left the block. This progression is noted at the OCC, thus the dispatcher will know where the train is at any given time. It should be noted that, even if the dispatcher console fails, information in the cab allows the engineer to know the conditions of the track ahead; this makes the system failsafe.

Headway between subway trains are three signal blocks. This is the distance that must be maintained between the head of one subway train to the end of the car ahead. The calculation for the time span of the head of a car in the signal block at 70 mph (102.67 ft/s) for a 2000 ft block is 19.6 s. The time required for breaking is 50 s minimum. Adding a block to this figure for reaction time is 69.9 s (the industry standard is 94 s). However, subway trains traveling at lower speeds generally require only three blocks because they do not travel at this speed; however, commuter trains may travel at this speed. Most subways only require a three-block headway. On the other hand a freight train pulling 100 cars may require a headway in miles instead of feet, depending on the speed. This is only meant as an example for the audience to understand some of the subtleties involved in rail transportation.

For a three-block headway the signal indication will be green when entering a block if the track is clear two signal blocks ahead. The indication is also provided in the cab for the subway or train engineer. When the axle of the head car enters a block that is two signal blocks from the train ahead, a yellow indication will appear in the cab and at the wayside signal to warn the engineer that a train is two blocks ahead and to reduce speed. When the headway is reduced to one signal block, the engineer is warned with a red indicator. This means stop; if the red indicator is flashing the speed is excessive and the engineer must initiate an emergency stop. If the rules are obeyed and the train is traveling at a normal slow speed a red indicator appears to the engineer. When the signal block at the wayside is red this means stop. Automatic train control (ATC) will perform the breaking as necessary with the engineer observing as a backup if the system has been installed on the equipment.

A great deal of the new freight and subway equipment is being automated to relieve the engineer of the long tedious observations that must be made along the right of way. There are many distractions that can occur, so it is imperative to have automatic braking implemented to prevent accidents. The higher-speed rail systems are structured especially vulnerable to accidents due to breaking. Also speed control at curves is now installed on some systems to prevent the engineer from speeding around curves; this is another feature of automatic train control.

9.4.2 Automatic Braking System

The automatic braking and speed control is depicted in Figure 9.4. The two sensors ahead of the axle in the subway car and locomotive cab as shown pick up code data from a transponder located in the track between the running rails. The sensor, filter and decoding equipment necessary to clean up the waveforms before they enter the decoding relay section consists of ladder logic. The decoded information is displayed in the cab of the train or subway car at the engineer's station. A sensor and speed driver from the axle enters data into the onboard computer that compares code rate and speed information. The computer controls break and throttle settings and determines the position on the running rails to maintain a three-block headway setting. This system has an override if the engineer deems it necessary. As can be observed this is

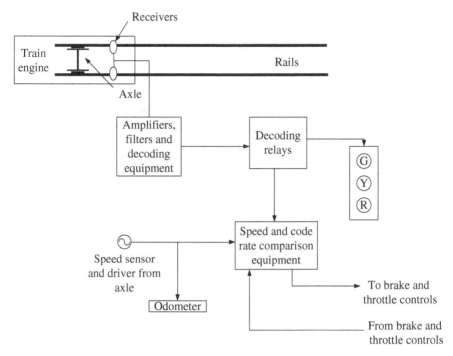

FIGURE 9.4   Cab signal system with automatic speed control.

similar to a dead man's switch, which allows a train or subway to break if the engineer is incapacitated for some reason. This system also has the ability to remove slow reaction time from the equation when breaking automatically. Thus, headway may be reduced due to the removal of human reaction from the automatic breaking system.

The OCC of any subway or railway system is most likely where EMC issues are most common. Most equipment in these areas are in locked enclosures, with only supervisors having access. The security of these areas is very strict. Personnel must wear badges and often swipe cards are necessary to enter these areas. As alluded to previously in Chapter 1, the equipment for workstations is generally a high-performance computer with as many as four screens so that multiple tasks can be observed by the operator. A master workstation allows programming to be modified and must have a locked console. The workstations contain two-way radios and other equipment that may have emissions that can compromise operations. For example, some two-way radios are allowed that are of the stand-alone variety which of course can always produce an EMC problem because the radiation is local in the area when transmitting. Some of the more up to date systems have remote antenna systems so radiation does not impinge on the local equipment at the operator console. The operator may be equipped with a headset for hands-free radio use.

The main system computer stores all the data gathered during daily operations and archives it in large files, usually in a database enclosed in multiple locked cabinets. It is not the computer that has large files but the peripheral equipment such as the storage medium. The networking equipment in these cabinets has its own computer and microprocessor-driven equipment. All of the equipment used in the OCC is off the shelf and can be purchased from either an industrial supplier or the manufacturer directly; however one aspect of these cabinets of equipment is the internal wiring that can have EMC problems if not wired correctly.

EMC issues are generally caused by wiring of the wrong size and type, poor routing in the cabinets, lack of shielding, poor connections, inadequate grounding and other issues that are heat- and air-conditioning related. The wrong type and size of wiring was experienced by the author, with a camera power supply on a pole at about 500 m from the supply source which met the electrical code. However, the resistance of the wire was so high that the voltage drop to the camera was so excessive that the camera would not operate properly. One suggested solution by an electrician was to use an auto transformer to step up the voltage to the camera; of course this was out of the question. This solution would result in a shock hazard to maintenance personnel working on the camera, due to the power supply.

The internal wiring of the enclosures is usually twisted pair shielded or some variation of this type of wiring; in some cases where radio is involved it is coaxial cable. The weak spot in twisted pair wiring is in the connections to circuit boards where the shield is open. At times, adjustments are made to peripheral equipment using components and standoffs; and these may radiate depending on the wavelengths of clocks. Care must be taken when installing microprocessors, radio equipment or telecommunications equipment because these may have fiber-optic receivers and transmitters that are susceptible to radiated fields. Grounding issues are one of the most common EMC problems. Installation electricians have a tendency to daisy-chain grounds. This is not always a problem but it must be considered with new computer installations. Since wiring is a combination of RLC the inductance that may be present in the grounding scheme may be sufficient due to high frequencies being grounded to lift the intended device off ground; and in some cases a sufficient capacitance in parallel with the inductance can result in resonance that will produce a radiating waveform. These of course can couple into other devices such as telecommunications equipment that will result in bits being corrupted during transmission. Often fiber optics cannot be used for short-range telecommunications connections but are generally implemented with high-quality CAT 5 or 6 cable. This is cable commonly used to wire the backplanes of SONET equipment. It is shielded and is also used in network switching equipment at high frequencies.

Some heat-related issues may appear to be EMC-related. When air conditioning fails, for example, and the temperature rises in the enclosure, it may be replaced. However, when circuit boards with sensitive components are heated to an extreme, they may appear to be completely functional but some of the more sensitive components may have an increased leakage current in the transistor collector junction circuits, resulting in a malfunction that is intermittent. This type of anomaly is occasionally very difficult to analyze because of this intermittent malfunction. Another type of problem that occurs is memory leakage. Over a period of time the memory may become saturated, as was the case in one instance; and eventually it becomes overloaded with superfluous data that results in poor performance and eventually a shutdown. This was a particular case for a communications computer. The computer had to be rebooted at regular intervals to prevent the communication system from failing. The reboot had to be done to allowing the backup computer to takeover, with a switchover to the main computer after the reboot. This had to be done at first on a weekly basis and then on a monthly basis as the errors in the software were corrected. At first glance this may seem to be an EMC failure; however, after some observation a software leakage was discovered. The errors seem to occur at the most inopportune time such as during rush hour when communication between subway cars and dispatcher are extremely important due to the increase traffic volume.

9.4.3 Conclusion

The analysis described in this section of the book is only an introduction to the subtleties of signaling and rail configurations on transit systems. There is a host of other configurations too numerous to include in this book. But Reference [9] has a wealth of information on track design and analysis. A list of the major sections in the pamphlet are as follows: (i) comparison of calculated and measured track impedance; (ii) measured rail impedance versus skin effect theory; (iii) circuit analysis track with third rail, track circuits and ballast; (iv) interference source characteristics of multicar train circuit effects; (v) interference source characteristics of multicar trains, statistics and measurements of conductive interference, techniques and data. The appendix has a Fortran 77 program that can most likely be converted to C or C++ for calculating conductive interference transfer functions. Transmission line software for the calculations is available; however, one may choose to modify it for the purpose of analyzing the various rail configurations.

Dr. Kent Chamberlain has designed an algorithm and program for calculating transmission line applications to rail systems [3]. As a suggestion to find further information on these subjects such as testing and transmission line theory applications to rails a search online would be in order, in particular DOT (Department of Transportation). Often construction companies subscribe to DOT libraries. They have a wealth of information on past construction of various transit systems, including the finished construction stamped by a professional engineer at commissioning.

The communications equipment for transit systems has changed rapidly in the last six years, for example fiber-optic OC-3 SONET was considered a very fast network and today's technology is probably $100 \times$ faster (gigabits of data) being the transmission frequencies are very high. This also leads to more EMC issues due to circuit radiation. The fiber was single channel and single wavelength usage; today's technology allows multiple wavelength usage, thus improving the throughput. Therefore, designing and publishing with that particular technology may result in an article or book becoming obsolete quickly.

There are many test documents that can be found in the DOT library. They were not included in this book because testing techniques change rapidly due to new and innovative test equipment. The older techniques for testing used strip recorders and various types of devices that are essentially obsolete because data may be displayed quite easily on a computer. The data can also be put on a flash drive that may be viewed as a very large-scale document.

9.5 REVIEW PROBLEMS

Problem 9.1
A harmonic of a subway train is present on the hot rail of 1630 Hz at 0.25 A and a code rate of 20.2 Hz. Will this harmonic present a problem? Use Table 9.1 to determine the answer. Use circuit theory to make the calculation. Use Table 9.2 at 1000 Hz to calculate the answer for this question.

Problem 9.2
Is the calculated value for 0.25 A enough to pull in the relay and close the points?

Problem 9.3
Calculate the skin effect of the rails in meters and centimeters for a frequency equal to 1630 Hz. $\mu_{steel} = 2.51 \times 10^{-5}$ and $\sigma_{steel} = 4.79 \times 10^{6} = 1.317 \times 10^{-3}$ m (0.1317 cm).

Problem 9.4
Calculate the resistance per meter using the value for skin effect in Problem 3. Radius for the area calculation of 5.23 cm and $\sigma_{rail} = 4.79 \times 10^{6}$ Ω/m.

Problem 9.5
Suppose a larger cross-sectional rail is used that an equivalent radius of 5.5 cm, $\sigma_{rail} = 4.79 \times 10^{6}$ Ω/m. Calculate the skin effect.

Problem 9.6
Calculate the resistance per meter using the skin effect from Problem 5.

Problem 9.7
Determine the inductive reactance per meter at 1630 Hz for $L_{trk} = 1.3$ μH/m.

Problem 9.8
Calculate the total impedance for 2000 ft (610 m) of track.

Problem 9.9
If the hot rail current of 4 A for loop mutual inductance results in 0.17 × the hot rail current, what is the voltage across the loop presented to the receiver?

Problem 9.10
Is the voltage at the receiver sufficient to hold in the relay contact points?

Problem 9.11
If so, reducing the hot rail current to: (a) decrease the hot rail current to 2, (b) increase the running rails size or (c) increase μ_r has what effect on the receiver?

9.6 ANSWERS TO REVIEW PROBLEMS

Problem 9.1
The impedance calculation is:

$$Z_{Rrails} = (789*2 + 1.33*2*2*\pi*1630)*610m$$
$$= (0.1578 \times 10^{-3} + j1.362 \times 10^{-2})*610m$$
$$Z_{Rrails} = (0.96 + j16.6)\Omega = 16.64\,\Omega \text{ at phase angle } 87°$$

The current from the hot rail $0.17*0.25$ A $= 0.0425$ A

The signal level $= Z_{Rrails}*I_{Loop} = 707$ mV

Problem 9.2

This is much higher than required to pull in the relay and can most likely saturate the receiver.

Problem 9.3

The skin effect is 1.274×10^{-3} m $(0.1274$ cm$)$

Problem 9.4

$$A_{1523} = \pi *\left[(5.23)^2 - (5.23-0.1274)^2\right] = 4.123 \text{cm}^2 \left(4.123 \times 10^{-4} \text{m}^2\right), R_{1630}$$
$$= 1/(A_{218} * \sigma_{rail}) = 506.36 \mu\Omega/\text{m}$$

Problem 9.5

$$\delta_{1526} = \left(\pi *1630*2.51\times 10^{-5} *4.79\times 10^6\right)^{-1/2} = 1.407\times 10^{-3}\text{m}\left(0.1407\text{cm}\right)$$

Problem 9.6

$$A_{1523} = \pi *\left[(5.5)^2 - (5.5-0.140)^2\right] = 4.78 \text{cm}^2 \left(4.78 \times 10^{-4} \text{m}^2\right), R_{1630}$$
$$= 1/(A_{218} * \sigma_{rail}) = 437 \mu\Omega/\text{m}$$

Problem 9.7

$$X_{trk} = 2*\pi *L_{trk} = 2*\pi *1.3\mu\text{H}/\text{m} = 8.17 \ \mu\Omega/\text{m}$$

Problem 9.8

$$Z_{total} \approx 437.1 \mu\Omega/\text{m}$$

Problem 9.9

$$V_{Rx} = 437.1*10^{-6} *610*0.17*4 \text{A} = 181.3\text{mV}$$

Problem 9.10

Yes, because the cross-section will increase, therefore reducing the resistance which will load the voltage.

Problem 9.11

a) Reduction of current by a factor of 2 will also reduce the voltage by a factor of 2.

b) Increasing the running rails size will increase the cross-section, thus reducing the resistance per meter. This will also decrease the voltage across the receiver.

c) Increasing the skin thus decreases the cross-section that decreases the resistance and increases voltage at the receiver.

10

USEFUL EXAMPLES

10.1 INTRODUCTION

The examples shown in this chapter can be used for the several purposes. Each one
will be addressed and will have a situation where the example may be of use to the
EMC engineer. The examples are shown with no intermediate steps missing so that
the EMC engineer may use the analysis as a guide when necessary.

10.2 EXAMPLES

Example 10.1

A strip line carrying current has a current density J(t), as shown in Figure 10.1 with
the variables x and y as indicated. The upper and lower sides of this strip line have
the same electric current density but, due to the lossy nature of the strip, the current
is non-uniformly distributed across the strip. The objective is to find the current carried by the cross-section x–y. It should be noted that this is a time-varying field.

$$J(t) = \hat{a}_z 100 e^{-10^{-4} y} \cos(\omega t)\, \text{A/m}, y = 0.25\,\text{mm}, x = 5\,\text{mm} \tag{10.1}$$

The current is calculated using Equation 10.2 by integrating across the cross-section of
the current density, as shown in Equation 10.1. The value 2 is inserted in front of the

Electromagnetic Compatibility: Analysis and Case Studies in Transportation, First Edition.
Donald G. Baker.
© 2016 John Wiley & Sons, Inc. Published 2016 by John Wiley & Sons, Inc.
Companion website: www.wiley.com/go/electromagneticcompatibility

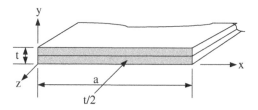

FIGURE 10.1   Strip line current density

integral because the upper and lower halves are integrated in the same way, therefore the cross-section is twice the value of a single one-half thickness of conducting strip.

$$I = \iint_s \bar{J}(t) * d\bar{S} = 2 \int_0^{2.5*10^{-4}} \int_0^{5*10^{-3}} \left[\hat{a}_z 10^2 * e^{-10^{-4} y} \cos(\omega t)\right] * \hat{a}_z dx dy \qquad (10.2)$$

The result of first integration with respect to x is shown below.

$$I = \iint_s \bar{J}(t) * d\bar{S} = 2(5*10^{-3})(10^2) \cos(\omega t) \int_0^{2.5*10^{-4}} e^{-10^{-4} y} dy = \cos(\omega t) \int_0^{2.5*10^{-4}} e^{-10^{-4} y} dy$$

Then collecting all the terms and integrating in the y direction, Equation 10.3 is the result.

$$I = \cos(\omega t) \int_0^{2.5*10^{-4}} e^{-10^{-4} y} dy = 10^{-4} e^{-10^{-4} y} \Big|_0^{2.5*10^{-4}} = 10^{-4}(0-1)\text{A}$$

$$I = 10^{-4} \cos(2\pi * 10^9 t)\text{A} \qquad (10.3)$$

The methodology used to calculate the cross-section current may be useful when evaluating circuit board current flow in the ground plane and the exponential term; and y is the attenuation factor that can be used to do the calculations. For the total thickness of the board the two in front of the integral must be eliminated.

Example 10.2

When an electric field impinges on the surface of seawater, a reflection occurs along with transmission into the seawater. Figure 10.2 depicts this situation. When the surface is flat without a ripple, the relative permittivity ε_r and permeability μ_1 are 81 and 1 respectively. The area above the seawater is air where relative permittivity and permeability are both equal to 1.

The field equations shown are Equations 10.4 through 10.9 and Equations 10.10 and 10.11; and these are taken from Reference [2]. The details of the derivation can be found in this reference.

The first part of the analysis requires the calculation of angles θ_i and θ_r.

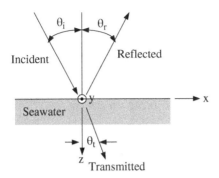

FIGURE 10.2 Transmission and reflection at a seawater boundary

A state trooper using a 160 MHz VHF radio has an antenna that is 2 m high. His point of reception is across seawater, 1 km from the tower. Using Equation 8.12 from Chapter 8, calculate the reflection distance from the tower-mounted 160 MHz VHF radio. The seawater has a relative permittivity of 81. Using this data plus the incident and reflected wave angle, the reflection coefficient and transmission coefficient can be calculated using Equations 10.10 and 10.11. Only one calculation need be made for the transmission coefficient or reflection coefficient because $1 + \Gamma = T$. The ERP equals 100 W and the distance to the state trooper is 1.02 km.

$$d = \frac{4h_1 h_2}{\lambda} = \frac{4*2*200}{1.937} = 826\,\text{m}$$

$$\theta_i = \theta_r = 90° - \tan^{-1}\frac{200}{826} = 76.4°$$

$$\Gamma = \frac{\cos\theta_i - \sqrt{(\varepsilon/\varepsilon_o)}\sqrt{1-(\varepsilon_o/\varepsilon)\sin^2\theta_i}}{\cos\theta_i + \sqrt{(\varepsilon/\varepsilon_o)}\sqrt{1-(\varepsilon_o/\varepsilon)\sin^2\theta_i}} = \frac{0.242 - 9\sqrt{1-\left[1/9(0.97)^2\right]}}{0.242 - 9\sqrt{1-\left[1/9(0.97)^2\right]}} = -0.378$$

$$T = \frac{2\cos\theta_i}{\cos\theta_i + \sqrt{(\varepsilon/\varepsilon_o)}\sqrt{1-(\varepsilon_o/\varepsilon)\sin^2\theta_i}} = \frac{0.484}{0.774} = 0.625$$

$$\beta_o = \omega\sqrt{\mu_o\varepsilon_o} = 2\pi*160*10^6\sqrt{4\pi*10^{-7}*8.854*10^{-12}} = 3.352\,\text{rad/m}$$

$$\bar{E}_o = \sqrt{\frac{30*ERP}{D_s}} \approx \frac{\sqrt{30*100}}{1020} = 53\,\text{mV/m}$$

Inserting the values for θ_i, θ_r, Γ, β_0 and T can be calculated using Equations 10.4 through 10.9.

$$\bar{E}_i = \hat{a}_y E_0 e^{-j\beta_o(x\sin\theta_i + z\cos\theta_i)} \qquad (10.4)$$

$$\bar{E}_r = \hat{a}_y \Gamma_b E_0 e^{-j\beta_o (x\sin\theta_i + z\cos\theta_i)} \tag{10.5}$$

$$\bar{H}_i = \left(-\hat{a}_x \cos\theta_i + \hat{a}_z \sin\theta_i\right) \sqrt{\frac{\varepsilon_o}{\mu_0}} E_0 e^{-j\beta_o (x\sin\theta_i)} \Big|_{z=0}$$

$$\bar{H}_r = \frac{53}{377} (-0.378)(-\hat{a}_x 0.242 + \hat{a}_z 0.97) e^{-j3.352(826*0.97)} \tag{10.6}$$
$$= \left(\hat{a}_x 12.86 - \hat{a}_z 51.4\right) e^{-j3.352(826*0.97)} \, \mu\text{A/m}$$

$$\bar{H}_r = \left(-\hat{a}_x \cos\theta_i + \hat{a}_z \sin\theta_i\right) \sqrt{\frac{\varepsilon_o}{\mu_0}} \Gamma_b E_0 e^{-j\beta_o (x\sin\theta_i + z\cos\theta_i)} \tag{10.7}$$

$$\bar{E}_t = \hat{a}_y T E_0 e^{-j\beta_o^g \left(x\sin\theta_i + z\sqrt{\frac{\varepsilon}{\varepsilon_o} - \sin^2\theta_i}\right)} \tag{10.8}$$

$$\bar{H}_t = \left(-\hat{a}_x \sqrt{\frac{\varepsilon_o}{\varepsilon} - \sin^2\theta_i} + \hat{a}_z \sqrt{\frac{\varepsilon_o}{\varepsilon}} \sin\theta\right) \sqrt{\frac{\varepsilon}{\mu_o}} T E_0 e^{-j\beta_o^g \left(x\sin\theta_i + z\sqrt{\frac{\varepsilon}{\varepsilon_o} - \sin^2\theta_i}\right)} \tag{10.9}$$

Where $\beta_o = \omega\sqrt{\mu_0 \varepsilon_o}$ and E_0 = Constant.

$$\Gamma = \frac{\cos\theta_i - \sqrt{(\varepsilon/\varepsilon_o)}\sqrt{1 - (\varepsilon_o/\varepsilon)\sin^2\theta_i}}{\cos\theta_i + \sqrt{(\varepsilon/\varepsilon_o)}\sqrt{1 - (\varepsilon_o/\varepsilon)\sin^2\theta_i}} \tag{10.10}$$

$$T = \frac{2\cos\theta_i}{\cos\theta_i + \sqrt{(\varepsilon/\varepsilon_o)}\sqrt{1 - (\varepsilon_o/\varepsilon)\sin^2\theta_i}} \tag{10.11}$$

The results of the calculations for the field equations are as follows:

$$\bar{E}_i = \hat{a}_y E_0 e^{-j\beta_o (x\sin\theta_i + z\cos\theta_i)} \Big|_{z=0} = \hat{a}_y 53 e^{-j3.352(826*0.97)} \text{ mV/m}$$

$$\bar{E}_r = \hat{a}_y (-0.378) E_0 e^{-j\beta_o (x\sin\theta_i)} \Big|_{z=0} = -20 e^{-j3.353(826*0.97)} \text{ mV/m}$$

The reflected field is the one of most importance, because it will impinge on state trooper's radio antenna and may or may not affect the reception. This holds true for the H field as well, since the power at the receive antenna may be subject to reflections that may disrupt communications.

$$\bar{H}_i = \left(-\hat{a}_x \cos\theta_i + \hat{a}_z \sin\theta_i\right) \sqrt{\frac{\varepsilon_o}{\mu_0}} E_0 e^{-j\beta_o (x\sin\theta_i)} \Big|_{z=0}$$

$$\bar{H}_i = \frac{53}{377} (-\hat{a}_x 0.242 + \hat{a}_z 0.97) e^{-j3.352(826*0.97)}$$
$$= \left(-\hat{a}_x 34.02 + \hat{a}_z 136\right) e^{-j3.352(826*0.97)} \, \mu\text{A/m}$$

$$\bar{H}_r = \frac{53}{377}(-0.378)(-\hat{a}_x 0.242 + \hat{a}_z 0.97)e^{-j3.352(826*0.97)}$$
$$= (\hat{a}_x 12.86 - \hat{a}_z 51.4)(e^{-j3.352(826*0.97)}) \; \mu A/m$$

The equations $\bar{E}_t$ and $\bar{H}_t$ need not be solved for this particular problem, but they were added just to show the reader the uses they may have for other problems. The exponentials in all cases for e are phase terms; the units are expressed in radians.

$$\bar{E}_t = \hat{a}_y TE_0 e^{-j\beta_o(x\sin\theta_i + z\cos\theta_i)}\Big|_{z=0} = \hat{a}_y(0.826)53e^{-j3.352(826*0.97)} = 43.8 \text{ mV/m}$$

$$\bar{H}_t = \left(-\hat{a}_x\sqrt{\frac{\varepsilon_o}{\varepsilon}-\sin^2\theta_i} + \hat{a}_z\sqrt{\frac{\varepsilon_o}{\varepsilon}}\sin\theta\right)\sqrt{\frac{\varepsilon}{\mu_o}}TE_0 e^{-j\beta\left(x\sin\theta_i + z\sqrt{\frac{\varepsilon}{\varepsilon_o}-\sin^2\theta_i}\right)}\Bigg|_{z=0}$$

$$\bar{H}_t = (-\hat{a}_x j0.72 + \hat{a}_z 0.0863)e^{-j3.353(826*0.97)}$$

The imaginary term for $\bar{H}_t$ indicates the x direction; it is a surface wave. The equation below is a calculation to show the height above the surface of the seawater where the first Fresnel zone occurs. For all practical purposes it occurs at the surface where the reflection is present.

$$h_{\text{reflect}} = 17.3\sqrt{\frac{d_1 d_2}{f(d_1 + d_2)}} = 17.3\sqrt{\frac{0.174*0.826}{160(0.174+0.826)}} = 0.0163 \text{ m}$$

The calculation below is the distance below the line of sight from the radio tower to the state trooper. This calculation is also unnecessary for the problem at hand; it is provided in the event the reader may need complete analysis for later problems.

$$y_{\text{reflect}} \approx 174\sin 13.6° = 40\text{ m}$$

The last item to consider and the most important is the power received at the state trooper's radio antenna through the line of sight losses, using the Friis equation for free space loss. This equation has been used many times. The results are shown below with the distance approximated at 1 km; and fade is not included. The radiated power at the state trooper's receiver is best calculated in Equation 10.12.

$$\text{Loss} = 76.5 \text{ dB}$$

$$\text{ERP}(40\,\text{dBW} - 76\,\text{dB} = -36\,\text{dBW}(-6\text{ dBm}) \tag{10.12}$$

The reflected Poynting vector power density at the state trooper antenna is $P_{\text{Poynting}} = \bar{H}_r \times \bar{E}_r$.

Just taking the magnitudes of the E and H reflected fields produces a value of 1 pico W/m² or −60 dBm/m². This will have a very little effect on the radio.

At first glance this example may seem to be extraordinarily complicated with several calculations unnecessary. However, it is only meant as an illustration of how a problem of this type may be solved. In actuality, the only equations necessary are those for field reflections, the Fresnel zone, reflection coefficient and the Poynting vector for power density.

Example 10.3

See Figure 10.3 for a cavity resonance example.

$$f_{mnp} = \frac{1}{2\pi\sqrt{\mu\varepsilon}}\sqrt{\left(\frac{m\pi}{a}\right)^2 + \left(\frac{n\pi}{b}\right)^2 + \left(\frac{p\pi}{c}\right)^2} \tag{10.13}$$

Where the values m, n and p are the eigenvalues of β_x β_y and β_z with the restriction if $m = p = 1$ the lowest order modes in the z direction.

$$\omega_y = \frac{1}{\sqrt{\mu_o\varepsilon_o}}\sqrt{\left(\frac{m\pi}{a}\right)^2 + \left(\frac{p\pi}{c}\right)^2} = c_o\sqrt{\left(\frac{m\pi}{a}\right)^2 + \left(\frac{p\pi}{c}\right)^2} \tag{10.14}$$

$$f_{101} = \frac{c_o}{2\pi}\sqrt{\left(\frac{m\pi}{a}\right)^2 + \left(\frac{p\pi}{c}\right)^2} \tag{10.15}$$

If for example, the lengths for a and b are 2 and 3 m respectively and this data is inserted into Equation 10.15, the result is that the cavity will resonate at 80 MHz (might be the resonant frequency in an elevator), as shown below.

$$f_{101} = \frac{3.1*10^8}{2*\pi}\sqrt{\frac{\pi}{2} + \frac{\pi}{3}} = 80\,\text{MHz}$$

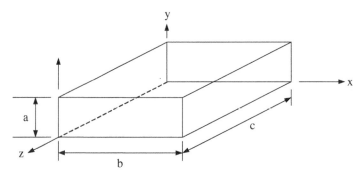

FIGURE 10.3   Cavity resonance example

Equation 10.16 is used throughout the calculations for electric and magnetic fields and Poynting vector.

$$\omega = \omega_y = 2\pi f_{101} \tag{10.16}$$

The electric field intensity E is calculated using equation 10.17 for x = 0.5 m and z = 1.0 m

$$\bar{E} = \hat{a}_y E_o \sin\left(\frac{\pi}{a}x\right)\sin\left(\frac{\pi}{c}z\right) \tag{10.17}$$

The result of inserting the values for the two variables x and z and constants a = 2 and c = 3 is as shown below.

$$\bar{E} = \hat{a}_y E_o \sin\left(\frac{\pi}{2}0.5\right)\sin\left(\frac{\pi}{3}1\right) = \hat{a}_y E_o(0.612)\text{V/m}$$

To calculate the H field the electric field is applied to Maxwell's equation; and the magnetic field intensity H field can be derived using Equations 10.18 and 10.19.

$$\bar{H} = \frac{1}{j\omega\mu_o}\nabla x\bar{E} = -\frac{1}{j\omega\mu_o}\left[-\hat{a}_x\frac{\partial E_y}{\partial z} + \hat{a}_z\frac{\partial E_y}{\partial x}\right] \tag{10.18}$$

$$\bar{H} = \frac{E_o}{j5.02*10^8*4\pi*10^{-7}}\left[\hat{a}_x 0.975 - \hat{a}_z\frac{\pi}{3}0.641\right] \tag{10.19}$$
$$= -jE_o\left(-\hat{a}_x 1.545 + \hat{a}_z 1.01\right)*10^{-3}\text{ A/m}$$

Inserting the same parameters as used for the calculation of the electric field intensity E and $\omega = 5.02*10^8$ rad/sec to the resulting H field is shown as follows:

$$\bar{H} = \frac{E_o}{j5.02*10^8*4\pi*10^{-7}}\left[\hat{a}_x 0.975 - \hat{a}_z\frac{\pi}{3}0.641\right]$$
$$= -jE_o\left(-\hat{a}_x 1.545 + \hat{a}_z 1.01\right)*10^{-3}\text{ A/m}$$

$$P_{av} = \frac{1}{2}\oiint_S(\bar{E}x\bar{H}^*)*dS = \frac{1}{2}\oiint_S(\bar{P}_{Poynt})*dS$$

$$\bar{P}_{Poynt} = \hat{a}_x\frac{E_o^2}{j2}\sin(\pi x)\cos(\pi x)\sin^2(\pi z) + \hat{a}_z\frac{1}{2}\frac{E_o^2}{j\omega\mu_o}\left(\frac{\pi}{c}\right)\sin^2\left(\frac{\pi}{a}x\right)\sin\left(\frac{\pi}{c}z\right)\cos\left(\frac{\pi}{c}z\right)$$

Inserting values for a, c, ω, x and z (at 30°, 45° and 60° respectively) and μ_o, the resulting equation is as shown below for the power density.

$$\bar{P}_{\text{Poynt}} = \bar{P}_{\text{Poynt}} = \hat{a}_x \frac{E_o^2}{j2} \sin(30°)\cos(30°)\sin^2(30°) + \hat{a}_z \frac{E_o^2}{j2} \sin^2(30°)\sin(30°)\cos(30°)$$

$$\bar{P}_{\text{Poynt}} = E_o^2 \left(0.054\hat{a}_x + 0.054\hat{a}_z\right) \text{at } 30°$$

$$\bar{P}_{\text{Poynt}} = \bar{P}_{\text{Poynt}} = \hat{a}_x \frac{E_o^2}{j2} \sin(45°)\cos(45°)\sin^2(45°) + \hat{a}_z \frac{E_o^2}{j2} \sin^2(45°)\sin(45°)\cos(45°)$$

$$\bar{P}_{\text{Poynt}} = E_o^2 \left(0.0625\hat{a}_x + 0.0625\hat{a}_z\right) \text{at } 45° \text{ W/m}^2$$

$$\bar{P}_{\text{Poynt}} = \bar{P}_{\text{Poynt}} = \hat{a}_x \frac{E_o^2}{j2} \sin(60°)\cos(60°)\sin^2(60°) + \hat{a}_z \frac{E_o^2}{j2} \sin^2(60°)\sin(60°)\cos(60°)$$

$$\bar{P}_{\text{Poynt}} = E_o^2 \left(0.162\hat{a}_x + 0.162\hat{a}_z\right) \text{at } 60° \text{ W/m}^2$$

The point of interest in this exercise is to find the radiation inside the cavity because it is of interest for personnel within the cavity itself, such as an elevator, telephone booth and so on where space is limited not for transmission but for the radiation itself within the confines of the cavity.

Example 10.4

A 5 A current flows through the cross-section with a radius a = 10^{-2} m of a copper wire, as shown in Figure 10.4. Consider the length infinite. The current density through the cross-section is non-linear. The objective here is to find the surface current ρ.

With the data provided, the attenuation factor is as shown below in Equation 10.20. The equation describing the current density is Equation 10.21 at 10 MHz.

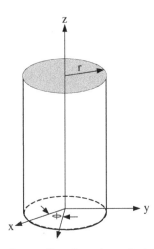

FIGURE 10.4 Current flow through a cylinderical copper wire

$$\alpha = \sqrt{\frac{\omega\mu\sigma}{2}} = 2\pi\sqrt{10*10^6 * 4*10^{-7} * 5.76*10^7} = 1.51*10^4 \qquad (10.20)$$

$$\bar{J} = \hat{a}_z J_o e^{-\alpha(a-\rho)} = \hat{a}_z J_o e^{-1.51*10^4(a-\rho)} \qquad J_o = \text{surface current} \qquad (10.21)$$

Then, the cross-section current is as shown in Equation 10.22.

$$I = \iint_S \bar{J} * \hat{a}_z d\bar{S} = \int_0^{2\pi}\int_0^a \hat{a}_z J_z * \hat{a}_z \rho d\rho d\phi = \int_0^{2\pi}\int_0^a J_o e^{-1.51*10^4(a-\rho)} \rho d\rho d\phi \quad (10.22)$$

The general integration form as taken from integral tables and used to integrate the variable ρ is Equation 10.23.

$$\int xe^{bx} dx = \frac{e^{bx}}{b}\left(1 - \frac{1}{b}\right) \qquad (10.23)$$

Then, inserting variables into the result of integration is shown below.

$$I = 2\pi J_o e^{-1.51*10^4 a}\left(\frac{e^{1.51*10^4\rho}}{1.51*10^4}\rho - \frac{1}{1.51*10^4}\right)\Bigg|_0^a$$

$$= 2\pi J_o e^{-1.51*10^4 a}\left(\frac{e^{1.51*10^4 a}}{1.51*10^4}a - \frac{1}{1.51*10^4} + \frac{1}{1.51*10^4}\right)$$

$$I = 2\pi J_o 6.62*10^{-5} a = 2\pi J_o 6.62*10^{-7} \text{ A for radius } a = 10^{-2} \text{m}.$$

$$0.368\left[4.159*10^{-5} J_o\right] = 2\pi J_o 6.62*10^{-7}\left[1.51*10^4\rho - 1\right]$$

$$153 = 1.51*10^4\rho - 1$$

$$\rho = 1.01\text{cm}$$

Example 10.5

A uniform plane wave inside the Earth (a perfect dielectric) has a relative permittivity $(\varepsilon_r = 9)$ at a frequency of 10 MHz. Find the velocity, wavelength, wave and intrinsic impedance. Equation 10.24 provides the phase velocity inside the Earth and Equation 10.25 is the wave and intrinsic impedance.

$$v = \frac{1}{\sqrt{\mu\varepsilon}} = \frac{1}{\sqrt{\mu_o\varepsilon_o\varepsilon_r}} = \frac{c_o}{\sqrt{\varepsilon_r}} = \frac{3.1*10^8}{\sqrt{9}} \approx 10^8 \text{ m/sec} \qquad (10.24)$$

$$Z_{wave} = \eta_o = \sqrt{\frac{\mu}{\varepsilon}} = \sqrt{\frac{\mu_o}{\varepsilon_o \varepsilon_r}} = \frac{377}{\sqrt{9}} = 125.67\,\Omega \tag{10.25}$$

Then

$$\lambda = \frac{v}{f} = \frac{10^8}{10^7} = 10\,\text{m}$$

Example 10.6

A plane wave this incident on an unbounded copper conducting surface the frequency and conductivity of f = 2 GHz and σ = 5.76*10⁷ S/m, respectively. Find the (wave and intrinsic) impedance and skin depth at penetration in meters.

$$\frac{\sigma}{\omega\varepsilon} = \frac{5.76*10^7}{2\pi(2*10^9)\left(\dfrac{10^{-9}}{36\pi}\right)} = 5.184*10^8 \gg 1 \qquad \text{Test for good conductor.}$$

Wave and intrinsic impedance:

$$\eta = \sqrt{\frac{\omega\mu}{2\sigma}}\lfloor 45° = \frac{2\pi(2*10^9)(4\pi*10^{-7})}{2*5.76*10^7}(1+j) = (0.137 + j0.137)\,\text{m}\Omega$$

Attenuation constant:

$$\alpha = \sqrt{\frac{\omega\mu\sigma}{2}} = \sqrt{\frac{(2\pi*2*10^9)(4\pi*10^{-7})(5.76*10^7)}{2}} = 2\pi\sqrt{2*10^9*5.76} = 6.744*10^5$$

When exponent is e⁻¹, that is reduction in magnitude is 33.3%.

$$n(0.333) = -1.1 = -6.744*10^5 * d$$

$$d = 1.631\,\mu\text{m} \qquad \text{Depth of penetration}$$

$$A_{att} = 20\log(0.333e^{-\alpha d}) = -9.55 - \alpha d * 20\log(e)$$

$$A_{att} = -85\,\text{dB} \qquad \text{Attenuation}$$

$$\beta = \sqrt{\frac{\omega\mu\sigma}{2}} = 6.744*10^5 \text{ rad/m} \qquad \text{Phase constant}$$

$$v = \frac{\omega}{\beta} = \frac{2\pi * 2 * 10^9}{6.744 * 10^5} = 1.863 * 10^4 \text{ m/s} \quad \text{Phase velocity}$$

$$\lambda = \frac{2\pi}{\beta} = \frac{2\pi}{6.744 * 10^5} = 9.32 \mu m \quad \text{Wavelength}$$

$$\delta = \frac{1}{\alpha} = 9.327 \mu m \quad \text{Skin depth}$$

Example 10.7

A radio station is operating in an area where the electric field intensity is 3 V/m and the operational frequency is 500 kHz. The soil around the area affected is considered moist earth with the following parameters $\sigma = 0.1$ S/m, $\varepsilon_r = 4$ and $\mu_r = 1$.

$$\frac{\sigma}{\omega\varepsilon} = \frac{.1}{2\pi (5*10^5) * 4 * \left(\frac{10^{-9}}{36\pi} \right)} = 90 >> 1 \quad \text{Test for good conduction}$$

$$\alpha = \sqrt{\frac{\omega\mu\sigma}{2}} = \sqrt{\frac{(2\pi * 5 * 10^5)(4\pi * 10^{-7}(0.1))}{2}}$$
$$= 2\pi\sqrt{5*10^{-3}} = 0.444 \quad \text{Attenuation constant}$$

$\delta = \frac{1}{\alpha} = \frac{1}{0.444} = 2.25 m$ Depth of penetration into the soil where the electric field intensity E is at 36.8%.

To reduce the magnitude of the E field to 1 V/m: what is the distance a 3 V/m wave must travel to reduce the wave magnitude to 1 V/m? The reduction ratio is 33.3%. The distance within the earth to result in this reduction is 2.477 m.

$$0.333 = e^{-\alpha d} \ln \quad (0.333) = -1.1 = -0.444 * d$$

$$d = 2.477 m \quad \text{Depth of penetration}$$

The resulting attenuation of the wave as it passes into the earth is 16.4 dB.

$$A_{att} = 20\log(0.333 e^{-\alpha d}) = -9.55 - \alpha d * 20\log(e)$$

$$A_{att} = -9.55 - 0.444 * 2.25 * 0.343 = 16.4 dB \quad \text{Attenuation}$$

To find the wave velocity, a calculation is made for the phase constant 0.25 rad/m using data from the calculations for phase constant, wavelength λ and phase velocity.

$\beta = \sqrt{f\mu_o\sigma} = \sqrt{5*10^5*4\pi*10^{-7}*0.1} = 0.25\mathrm{rad}/\mathrm{m}$. This calculation was done for conductive materials, as a previous test indicated.

$$v = \frac{\omega}{\beta} = \frac{2\pi*10^5}{0.25} = 2.513*10^6 \,\mathrm{m}/\mathrm{s} \qquad \text{Phase velocity}$$

$$\lambda = \frac{2\pi}{\beta} = 25.13\mathrm{m} \qquad \text{Wavelength}$$

The final calculation is for the intrinsic impedance.

$$\eta = \sqrt{\frac{\omega\mu}{2\sigma}}\big|45° = \sqrt{\frac{2\pi(5*10^5)(4\pi*10^{-7})}{2*0.1}}(1+j) = 2\pi\sqrt{\frac{5*10^{-2}}{0.1}} = 4.443\big|45° \,\Omega$$

This example allows the reader to look at all the possible calculations that can be made with a limited amount of data.

Example 10.8

A plane wave is incident on a lossless dielectric medium with $\varepsilon_r = 3.5$. The reflections can be eliminated if another dielectric material $\frac{1}{4}\lambda_a$ thick is inserted between the air and the dielectric with a permittivity of 4, as shown in Figure 10.5. The intrinsic impedance of the material to be inserted must be equal to $\varepsilon_r = 1.87$.

$$\eta_a = \sqrt{\eta_o\eta_2} = \left[\sqrt{\frac{\mu_o}{\varepsilon_o}}\frac{1}{1.87}\sqrt{\frac{\mu_o}{\varepsilon_o}}\right]^{1/2} = \frac{1}{1.87}\sqrt{\frac{\mu_o}{\varepsilon_o}} = \frac{1}{1.87}\eta_o$$

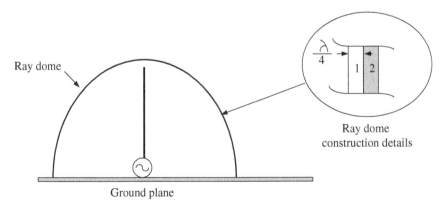

Ray dome

Ray dome construction details

Ground plane

FIGURE 10.5   Radiation through a ray dome cover

This analysis would be useful for designing a ray dome to cover antennas where the signal could go out without reflections internal to the ray dome. One such application is for an antenna cover to protect antenna farms on locomotives, buses, maintenance vehicles and other areas where the signal must pass through the cover that protects the equipment from the elements. In areas where there is mining of minerals such as iron, which is a good example, the dust in this particularly application is destructive to radiating signals from radio antennas. The dust generally consists of various forms of ferrites that can attenuate signals in either direction to/from the ray dome. Most of these ray domes can be made with a coating such as Teflon and various paints that will make the surface extremely slippery so that normal travel will keep them clear of dust particulates.

Example 10.9

If a Hertzian dipole (antenna length $L \ll \lambda$) is implemented in a radio system that is used for short range communication, that is a spread-spectrum 900 MHz radio, with an ERP of 680 mW and a wavelength of 0.1λ, the following equations are used to make the calculations for the various parameters. To find the radiation resistance use Equation 10.26.

$$R_R = \frac{2\eta\pi}{3}\left(\frac{L}{\lambda}\right)^2$$

Where $\lambda = 0.333$ m and $L = 0.1\lambda = 0.0333$ m and $\eta = 377\Omega$.
Then:

$$R_R = \frac{2*377\pi}{3}\left(\frac{0.0333\lambda}{\lambda}\right)^2 = 0.875\,\Omega \text{ and } I_{max} = \sqrt{\frac{0.680}{0.875}} = 881\,\text{mA} \quad (10.26)$$

To find electric intensity E field in spherical coordinates, Equation 10.27 is evaluated using all the known variables except the height of the antenna h.

$$E_\theta \simeq j\eta\frac{\beta I_o le^{-j\beta r}}{4\pi r}\sin\theta\left[2\cos(\beta h\cos\theta)\right]44\,\text{mV/m} \qquad z \geq 0 \qquad (10.27)$$

The resulting magnitude of the electric field intensity E is 44 mV/m. This is the far field where r = 1 m.

$$E_\theta \simeq j\eta\frac{\beta I_o le^{-j\beta r}}{4\pi r} = j377\frac{18.86*0.881*0.0333e^{-j0.033*1}}{4\pi*1} = 44\,\text{mV/m}$$

To plot the Eθ field intensity, Equation 10.28 will provide the elevation.

$$E_\theta \simeq 44\sin\theta\left[2\cos(\beta h\cos\theta)\right]\text{mV/m} \qquad (10.28)$$

$$E_\theta = 0 \qquad z < 0$$

The magnetic field intensity H_ϕ is the magnitude of the E field divided by η, with this result as Equation 10.29.

$$H_\varphi \approx j\frac{\beta I_o le^{-j\beta r}}{4\pi r}\sin\theta\left[2\cos(\beta h\cos\theta)\right] = 0.1167\sin\theta\left[2\cos(\beta h\cos\theta)\right]A/m$$
(10.29)

$$P_{average} = \eta\pi\left|\frac{I_o L}{\lambda}\right|^2 = 377*\pi\left|0.888*0.0333\right|^2 = 1.21\text{ W or }605\text{ mW}\quad(10.30)$$

$$P_{radiated} = P_{average}\qquad L<<\lambda$$

The RMS value for Equation 10.30 is 605 mW. This discrepancy from the original 680 mW is probably due to rounding errors made during the calculations. One of the reasons for an introduction to Hertzian dipole equations is that these may be extended and used for longer length antennas. An introduction of this technique using FDTD methods is discussed at www.wiley.com/go/electromagnetic compatibility, starting with my Hartzian approach and extending this to longer lengths of antennas.

Example 10.10

Equation 10.31 is the power intensity if h (distance from the bottom of the antenna to the ground plane) is at ground level, that is = 0, and θ is at 90°. Then U_{max} is as shown:

$$U = \frac{\eta}{2}\left|\frac{I_o L}{\lambda}\right|^2\sin^2\theta\cos^2(\beta h\cos\theta)\text{ W/steradians}$$
(10.31)

$$U_{max} = \frac{377}{2}\left(\frac{0.881*0.0333}{0.333}\right)^2 = 1.463$$

Then, using this result the D_0 directivity is as depicted below, using Equation 10.32.

$$D_0 = \frac{4\pi U_{max}}{P_{radiated}}$$
(10.32)

$$D_0 = \frac{4\pi*1.463}{1.21} = 15.2$$

Equation 10.33 is a depiction of D_0 at various distances above the ground plane, for example an antenna 2 m above the ground plane will have an effect on the directivity ($2\beta h = 75$ rad). Inserting this value into the second term of Equation 10.33 $D_o \approx 6$. The second and third term are negligible. The one-third term in the brackets is due to expansion of the Taylor series.

$$D_o = \frac{2}{\left\{(1/3) - \left[\cos(2\beta h)/(2\beta h)^2\right] + \left[\sin(2\beta h)/(2\beta h)^3\right]\right\}} = \frac{2}{(1/3)} = 6$$

(10.33)

$$\left[\frac{1}{3} - \frac{0.73}{(75)^2} - \frac{0.388}{(75)^3}\right] \approx \frac{1}{3}$$ Expansion of the denominator of Equation 10.33 is a series.

$$R_{\text{radiation}} = \frac{2P_{\text{radiation}}}{\left|I_o\right|^2}$$

(10.34)

Then, inserting values into Equation 10.34 for radiated power and antenna current, the radiation resistance is as follows for $h = 0$.

$$R_{\text{radiation}} = \frac{2 * 0.605}{0.881^2} = 1.5\Omega$$

$$R_R = 2\eta\pi\left(\frac{L}{\lambda}\right)^2\left[\frac{1}{3} - \frac{\cos(2\beta h)}{(2\beta h)^2} + \frac{\sin(2\beta h)}{(2\beta h)^3}\right]$$

(10.35)

If h is in meters and all of the values are inserted into Equation 10.35, the resulting radiation resistance R_R is as calculated below.

$$R_R = 2 * 377 * 0.0333^2 * \frac{1}{3} = 0.2787\Omega$$

These last two examples with Hertzian antennas represent the antennas for some of the wireless devices that may be found along the right of way, such as car monitoring and personnel with handheld spread-spectrum devices. As can be observed, the height of the radio above the ground plane can become critical and influence the radiation field. Generally in the far field for long antennas such as one-quarter wave monopole antennas radio transmission with an antenna definitely has a ground plane that is close to zero; the bracketed circular terms need not be considered.

Tables 10.1 and 10.2 are provided as a handy quick reference to some of the parameters used in the calculations for relative conductivity, permittivity and permeability. This is only a short list of the materials available for calculations and the various examples and calculations for review problems. These are some of the more important materials; however, other specialty items should be searched for online. It should be noted in Table 10.2 that the relative permeability for some of the exotic shielding materials is in the range of 20 000–100 000. These materials are especially

TABLE 10.1   Examples of Permittivity and Permeability Materials

Insulation/semiconductor	Rel σ_r	Rel ε_r
Glass	10^{-17}	2.32–2.4
Teflon	10^{-17}	2.1
Polyethylene	$2.47*10^{-3}$	2.26
Plexiglass	$5.1*10^{-3}$	3.4
Bakelite	10^{-9}	4.8
Leaded glass	10^{-9}	6.0
Ferrite	$1.5*10^{-3}$	12–16
Fresh water	10^{-2}	81
Salt water	4	81

TABLE 10.2   Metals with Shielding Properties for Magnetic Fields

Metal	Rel σ_r	Rel μ_r
Silver	1.064	1
Copper (reference metal = 1) $5.737*10^7$ Ω/m	1.000	1
Gold	0.700	1
Chromium	0.664	1
Aluminium (pure)	0.630	1
Aluminum (heat treated)	0.400	1
Aluminum (household foil)	0.530	1
Brass (66% Cu, 34% Zn)	0.350	1
Magnesium	0.380	1
Zinc	0.305	1
Tungsten	0.314	1
Beryllium	0.330	1
Cadmium	0.232	1
Platinum	0.170	1
Tin	0.151	1
Tantalum	0.120	1
Lead	0.079	1
Monel	0.041	1
Mercury	0.018	1
Supermalloy	0.023	100 000
78 Permalloy	0.108	8000
Iron (pure)	0.170	5000
Conetic AA	0.0310	20 000

(Continued)

TABLE 10.2 (Continued)

Metal	Rel σ_r	Rel μ_r
4-79 Permalloy	0.0314	20 000
Mumetal	0.0289	20 000
Permendur	0.247	800
Hypernick	0.0345	4500
45 Permalloy (annealed)	0.0384	2500
Hot rolled silicon steel	0.0384	1500
4% Silicon iron	0.0290	500
16 Alfenol	0.0113	4500
Hiperco	0.0690	650
Monimax	0.0216	2000
50% Nickel iron	0.0384	1000
Commercial iron 99.8% pure	0.170	200
Cold rolled steel	0.170	180
Nickel	0.230	100
Stainless steel	0.020	200
Rhometal	0.019	1000

Note: Relative permittivity and permeability are affected by frequency.

costly but they have extremely high shielding properties for magnetic fields. In Table 10.1 ferrite appears with an extremely high permittivity as a good insulator, but it can also have other properties when influenced by large magnetic fields. Most of the permittivity and permeability materials have imaginary components as alluded to previously at frequencies above 1 GHz, so all is not what it seems in these tables when operating well above 1 GHz.

REFERENCES

1. C.A. Balanis (1989) *Advanced Engineering Electromagnetics*, John Wiley & Sons Inc., New York.
2. C.A. Balanis (2005) *Antenna Theory, Analysis and Design, 3rd edn*, John Wiley & Sons Inc., New York.
3. K. Chamberlin (2015) Collected works, up to 2015; see www.wiley.com/go/electromag neticcompatibility. For example: (i) notes on ECE 993 *Electromagnetic Antenna Theory*; (ii) Measuring the impact of in-vehicle-generated EMI on VHF radio reception in an unshielded environment; (iii) a lecture on modeling FDTD. Other programs for digital modeling may be available upon request from Dr. Kent Chamberlin. He is considered an expert in electromagnetics and has taught many courses on the subject.
4. H.W. Ott (2009) *Electromagnetic Compatibility Engineering*, John Wiley & Sons Inc., New York. *Comments*: very simple analysis and easy to read; excellent for a novice to EMC.
5. D.R.J. White (2000) *EMC Handbook*, McGraw-Hill, New York.
6. J.D. Kraus, D.A. Fleisch (1999) *Electromagnetics with Applications, 5th edn*, McGraw-Hill, New York.
7. B.J. Kwaha, O.N. Inyang, P. Amalu (1985) The circular microstrip patch antenna – design and implementation, *D. Phys.*, 3, 405–406.
8. FCC Office of Engineering and Technology (1989) *Evaluation Compliance for Human Exposure to Radiofrequency Electromagnetic Fields*, Federal Communication Commission, Washington, D.C. *Comments:* with additional information for evaluation compliance of mobile and portable devices with FCC limits for human exposure to radiofrequency emissions.
9. US-DOT (1985) *Conductive Interference in Rapid Transit Signals Systems. Volume 1. Theory and Data*, US Department of Transportation, Washington, D.C., UMTA-MA-06-0153-85-5 or DOT-TSC-UMTA-85-21.
10. US-DOT (1986) *Conductive Interference in Rapid Transit Signaling Systems, Volume 2. Suggested Test Procedures*, US Department of Transportation, Washington, D.C., UMTA-MA-06-0153-86-7 or DOT-TSC-UMTA-86-7.

Electromagnetic Compatibility: Analysis and Case Studies in Transportation, First Edition.
Donald G. Baker.
© 2016 John Wiley & Sons, Inc. Published 2016 by John Wiley & Sons, Inc.
Companion website: www.wiley.com/go/electromagneticcompatibility

INDEX

Note: Page numbers in *italics* refer to Figures; those in **bold** to Tables.

access broadband over power line
(case study)
ARRL, 127–128
CCS *see* carrier current systems
certification technical report
requirements, 137–138
at distances, measurements, 130–131
extrapolated emission level, 129
inverse linear distance extrapolation
factor, 131, *131, 132*
measurement principles, 127, 132–133
on overhead power lines
extrapolated emission level, 138
slant-range distance, calculation of, 138
power line heights, 130, **130**
radiated emissions measurement
principles
on overhead line
installations, 134–135
in underground line installations,
135–136
responsibility of operator, 138
site-specific extrapolation factor,
131–132
slant range distance, *128*, 128–129
"smart grid," uses, 127
test environment
equipment under test, 133
in-situ testing, 133
transit systems, 127
AF track circuits *see* audio frequency
track circuits
AF Track Circuits Interference
Control Program
coupling paths, 316
EMC hardening flow diagram, 315, *316*
FCC CFR 47 part 15 B regulations, 316
radio intermodulation products, 315

Electromagnetic Compatibility: Analysis and Case Studies in Transportation, First Edition.
Donald G. Baker.
© 2016 John Wiley & Sons, Inc. Published 2016 by John Wiley & Sons, Inc.
Companion website: www.wiley.com/go/electromagneticcompatibility

aircraft industry, 9
American Public Transit Association, 284
American Railway Engineering and
        Maintenance of Way
        Association, 284, 291, 292
Ampere's law, 92, 185
amplifiers, CMRR
    balanced input, 68, *68*
    signal and ground lines, 67, *68*, 69
    specification, 69
antenna selection, OATS
    automatic switching, 19
    biconical, 19
    FCC certification, 20
    and ground plane, 14, **15,** 17
    height, 18
    3 m emission test, 16, 19–20, **20**
APTA *see* American Public Transit
        Association
AREMA *see* American Railway
        Engineering and Maintenance of
        Way Association
ATC *see* automatic train control
audio frequency track circuits
    blocks, 347–348
    bungalow/signal house, *347*, 347–348
    hot rail, magnetic coupling
        AC powers frequency, 349
        DC power, 348
        impedance data, 100 lb/yd running
            braille, 349, **350,** 351
        installation, 348, *348*
        power supply, 351
        running rail configurations, *348*, 349
        skin depth, 351
        traction motor, 349
    lightning stroke, 346
    man-made activities, 346
    maximum and minimum code rate, 345
    sensitivity, signals relay, 345, **345**
    signals equipment, 346
    signals testing, 346
    speed control method, 342
    step-down transformer, 346
    transmitter transformer, receiver and cab
        filters, 342, 344
    vehicle manufacturers, 346–347
automatic braking
    cabinets, 356–357

heat-related issues, 357–358
internal wiring, 357
sensor, filter and decoding equipment, 355
and speed control, 355–356, *356*
telecommunications equipment, 357
workstations, 356
wrong type and size of wiring, 357
automatic train control, 341, 342, 355

ballast
    definition, 354
    hot rail to running rail, crossover, 354–355
    signal indication, 355
    subway trains, 355
    three-block headway, 355
Bessel differential equation, 101
Bessel function equation, 101, 104, 105
Brewster angle, 109
broadband EMI, 308
broadband over power line (BPL)
        see access broadband over
        power line (case study)
bus and ferry system
    communication system
        bus driver operator console
            interfaces, 268
        communications bungalow, 266
        dead spots in radio
            communications, 269–270
        EDX and terrain analysis
            software, 265–266
        EMC/EMI issues, 267–269
        layout, *267*, 267–268
        microwave backhaul radio system,
            264–265, *265*
        multiple antenna installations, 266
        rack-mounted equipment, 269
        rooftop radio, 269
        self-contained inverter, 269
        tower detail, 264, *265*
    EMI/EMC issues, 264
    GPS, 263
    licensed radio antenna, 264
    microwave towers, communication
        system, 264
    onboard controls, 263
    reflections
        antenna characteristics, 274, **274**
        aperture equation calculation, 276

distance calculation, 273
electric field intensity, 275
EMS services, 278
free space loss calculation, 273
Fresnel zone, 272, *272*
Friis free space loss equation, 275, 276
patch antenna, bus roof, 273, *274*
power radiation, patch antenna, 275
Poynting vector, 275, 276
sea-going vehicles, 278
shipboard radiation, ground planes, 278
short-range mobile two-way
    radios, 278
steel rails, 278

Canadian Regulations, 14–15, 24
carrier current systems
    and in-house BPL, 136–137
    measurement principles, 132–133
Cartesian coordinate systems, 105
cavity resonance
    electric field intensity E, 367
    magnetic field intensity H, 367
    at 80 MHz, 366
    power density, 367–368
CCS *see* carrier current systems
central control system, 312
circuit theory
    automatic braking, 355–358
    ballast, 354–355
    geometric measurements, 354
common mode rejection ratio *see*
    amplifiers, CMRR
communication house (case study)
    aluminum alloys, 119–120
    electric field, 120
    halo installation, 118, *119*
    H fields, 118–119
    magnetic flux density, 120
    mass transit light rail system, 121
    mu-metal shielding, 120
    pulse width modulated drives, 121
    shielding effect, 121
    substation pole installation, 118, *119*
    transmission and reflections coefficient
        at each interface, 123, **123**, 124, **124**
        E and H fields, 122, 126
        at frequencies from traction
            motors, 124, **125**

halo in communication house, 124, 127
impedance of concrete and
    steel, 121–122, **122**
magnetic flux, magnitude of, 126
maintenance personnel, 125–126
resistance values, 121
signal/bungalow and track power
    houses, 125
communications equipment
    central control system, 312
    communications transmission
        system, 311–312
    controller, 311
    control panels, 311
    fire and intrusion alarms, 311
    LCD monitors, 310
    LRVs, 315
    PA system, 309–310
    radio subsystem, 312–313
    RTU, 312
    sensors, 311
    signals and propulsion
        equipment, 314–315
    switch machines, 315
    telephone, 313
    vehicle emissions, 313–314
    VMB, 310
    workstation LCD monitors, 312
communications network, 4
communications transmission
    system, 311–312
computer emissions, 9
conducted emission *see also* line impedance
    stabilization network
    DC/AC current measurement, 21–24, **22**
    single phase, 21, **22**
    three phases, 21, **22**
conductive paints, 239, 274, 320
conductor manufacturers, 320, **320**
control panels, 311, 323
copper wire, current density, 368–369
coupling EMI, 308
culprit, 2

Department of Transportation
    OATS ground plane and antennas,
        14, **15** *see also* open area test site
    radiation emission measurements,
        13–14, **14**

divergence theorem *see* Gauss
divergence theorem
DOT *see* Department of Transportation

electrical design
cable shielding, 327
capacitive coupling, *324*, 324–325
inductive coupling, 325–326, *326*
twisted pairs, 326, *327*
electromagnetic compatibility
aircraft industry, 9
conduction and radiation emission
sources, 9, **10**
control and test plan
communications contractor, 305–306
EMI general characteristics, 307
FCC Part 15 Regulation, 306
fire alarm subsystem, 306
definition, 2
equipment manufacturers, 16
fire and intrusion alarms, 7
Maxwell's equations, 8–9
medical facilities, 9
radiated emissions, 17
radiation exposure safety issues, 9
shielding, communication house, 9
techniques, 69, *70*
testing racks of equipment, 17
testing techniques, 8
wiring and coupling, 6
electromagnetic emission safety limits
B field flux density, 285–286, **286**
broadband harmonics, 282
burns, contact types, 288–289
capacitors, 289
current standards, 287
electric and magnetic fields, 282, **283**
electric traction power systems, 284–285
exposure limits *vs.* frequency, 283–284
high-voltage discharge, 289
IEEE and ICNIRP, 284
non-ionizing radiation, 281–282
occupational control exposure, 282, **283**
organizations, exposure limits, 282, **283**
personal wireless data and voice
devices, 285
planners and designers, 282
power distribution system, 285, **286**
standards, organizations, 287–288, **288**

transit systems, 284, 285
transmission and distribution lines, 286–287
electromagnetic field
Ampere's law, 92
attenuation factor, 97
Cartesian coordinate systems, 95
dell operator, 93
EMC field problems, **96**
Faraday's law, 92
Gauss divergence theorem, 93
Gauss' law, 92
Maxwell's equation differential
form, 91–92
Maxwell's equation integral form, 92–93
propagation constant calculations, 95
rectangular waveguide, 97, *97*
second order scalar differential wave
equations, 94
second order wave equations, 94
wave equations *see* wave equation
solutions
waves, transmission path, 97, **98**
electromagnetic interference *see also*
EMC/EMI design procedures
cable to cable coupling, 71
conducted emission measurement setup,
23, 23, 24
currents, 71–72
definition, 2, 307
E field, 71
H field, 71
licensed radiators, 71
measurement, units of, 308, **308**
prevention and control
DOT documents, 291
E and B fields, 290
frequency bands, 291
grounding methods, 291
heavy-duty vehicle, 290
measurements, 290
mitigation techniques, 290
organizations, 291–292
potential sources and hazards, 290
procedures, 290–291
transit systems, 291
receptors, 308
twisted pair, *72*, 72–73 *see also* shielded
twisted pair wiring
wiring and coupling, 16

electrostatic field shielding, 319–320
EMC *see* electromagnetic compatibility
EMC/EMI design procedures
  digital systems, 317
  electrical design, 324–327
  electromagnetic ambient, 317
  electrostatic field shielding, 319–320
  FCC Part 15 B regulations, 317
  grounding
    and bonding, 329, 331, *331*
    cable and rod size, 327, *329*
    communication houses, 327, 329, *329*
    operation control centers, 329, *330*
  hardening techniques, 317, 318
  magnetic H field shielding, 320–323
  mechanical design, 318
  non-ionizing radiation coupling and
    potential antennas, 331–332
  professional engineer's sealed
    documents, 332–333
  shielding materials
    copper and aluminum, 319
    magnetic and electric fields, 318,
      318, 319
    steel, 319
  subway cars, 332
EMI *see* electromagnetic interference
emission couplings
  audio line, telephone, 64
  capacitors, 61, *62*
  cell phone/two-way radio, 61
  E field, 61–62, *62*
  H field, *62*, 62–63
  inductance, 61, *62*
  length of wiring, 64
  shielded wire, 61, *62*
  types, 61, *62*
  wireless devices, 61
engineer designing system, 1
equipment under test, 133
  antenna height, 18
  attenuator/spectrum analyzer, 19
  circular ground planes, 18
  copper plate, LISN, 21, *23*
  pedestal, 20
  safety ground, 24
European Union Regulations, 24–25,
  26–56
EUT *see* equipment under test

extrapolated emission levels
  calculation formula, 141
  power line heights, comparisons,
    139–141
  slant-range method, overhead power
    lines, 138–139

Faraday's law, 92, 185, 194, 352
FCC 11-160 *see* access broadband over
    power line (case study)
FCC emission limit regulations
  classification, 12–13
  linear amplifiers, 11
  radiation measurement
    from 9 KHz to 30 MHz, **11**
    at 3 m, **11**
    at 10 m, **11, 12**
  radio signals, 12
  residential areas, 13
  two-way radios, 12
  wireless devices, frequency range, 11, 13
FCC Part 15 radiation measurements *see*
    open area test site
FCC regulations *see* Federal
    Communication Commission
    regulations
Federal Communication Commission
    regulations
  antenna farm, 301
  cordless phone standards and repeaters,
    298, **299**
  EIRP levels, 301
  European Union, 301
  occupational/controlled
    exposure, 297, **298**
  population/uncontrolled
    exposure, 298, **298**
  wireless emission standards, 299, **300**
ferry communication system *see also* bus
    and ferry system
  EMC/EMI issues, 271–272
  Fresnel zone, 272, *272*
  layout, 270, *270*
  onboard communications layout,
    271, *271*
  radio electric reflection off water,
    272, *272*
  shipboard radiation of ground
    planes, 278

filters
    active, digital and switched capacitor, 79
    compound, 80, *81*
    DC power supply, 80–81
    low pass, 79
    power line, 79–80
    products, 80
fire alarm subsystem, 306, 311
Fresnel zone, 272, *272*
Fresnel zone clearance, radio antenna
        obstruction, 333, *333*
    wave reflection, 334, *334*, 335
Friis free space loss equation, 240, 260,
        275–276, 365

Gauss divergence theorem, 93
Gauss' law, 92
ground planes, antenna farms
    circulation currents
        connectors and antenna
            mountings, 202
        electric field intensity, 201
        intrinsic wave impedance, 202
        permeability and permittivity,
            materials, 203
        reflections, antenna, 203
        skin depth, current, 202
        surface current density equation, 202
    edge effects, *229*
        distortion, 217, *218*
        elevation, power density and
            normalized field, 217
        incident and reflected diffracted
            waves, **231**
        incident and reflected waves, **231**
        Keller's equation, 230, 231
        P 25 radio antenna, 219
        reflected and diffracted electric
            intensity E field, 232–233
        reflected field calculation, 230
        subway cars and buses, 219
        total diffraction, 230
        tunnel operations, 219
OATS
    antenna selection, 19–20
    circular, 18
    conducted emission, 21–24
    EUT, 17–18
    measurement errors, 18–19

pre-compliance testing, 20–21
    reflections, 18
    wooden flooring, 18
reflection and ground waves at the edge
    attenuation constant, 238
    electric field impinging, *238*
    incident and reflected diffraction
        calculation, 236
    intrinsic impedance, concrete, 238
    radiation by diffraction, 234, *235*
    total electric field, surface, 238, *238*
    total incident and reflected electric
        intensity fields, 235

heat detectors, 311
Henkel function equation, 101, 105
Hertzian antennas, 375
Hertzian dipole, radiation
        resistance, 373–374

ICNIRP *see* International Committee for
        Non-Ionizing Radiation Protection
IEEE *see* Institute of Electrical and
        Electronic Engineers
immunity, 2 *see also* electromagnetic
        interference
IM products *see* intermodulation products
in-house BPL
    and CCS, measurement principles
        as computer peripheral, for testing,
            137
        overhead lines, *in-situ* testing, 137
        test environment and radiated
            emissions, *in-situ* testing, 136
    measurement principles, 133
Institute of Electrical and Electronic
        Engineers, 173, 284, 291, 292
intermodulation products, **65**
    antenna polarization, 65
    cell phones, aircraft
        carrier frequency, radios, 67
        harmonics, 66–67
        passengers, 66, *66*
    direct and non-direct hits, 65
    order of, 64–65
    radiators and radios, 64
    victim circuits and receptors, 67
International Committee for Non-Ionizing
        Radiation Protection, 284, 292

inter-system EMI, 308
intra-system EMI, 308
in-vehicle-generated EMI, VHF radio reception
  accuracy and repeatability, 216
  bypassing external radiation sources
    averaged signal over scan range, *214*
    bandwidth, function of sweep number, *214*
    convergence times, bands, 215
  computer-controlled radio, 212, *213*
  EMI data analysis
    averaged EMI, 215, *216*
    spectra, ambient noise, 215, *215*
  equipment, 212
  initial measurements, 212
  objectives, 212

Keller's equation, 229–231
Kharkov's voltage law, 108
Kirchhoff's current and voltage law, 107
knife edge diffraction loss calculation
  communication link, 337
  rounded obstruction edge, *336*, 336–337
  wavelengths, 335

LCD monitors, 310, 312
leaky radiating coaxial cable analysis
  construction details, 179, *179*
  copper, VHF and UHF frequencies, 178, **178**
  coupling loss, 178
  current stability criterion, 187
  distributed antenna system, 177
  impedance of coaxial line, 178
  lateral slots, 177
  length, aperture radiation, 180, *180*
  method of moments, 179
  radiating cable, 177
  simulation results
    cable characteristics, physical properties, 180–182, **183**
    cell, one-dimensional layout, 185, *186*
    E field simulation, 180, *181, 182*
    FDTD requirements, 185, 186
    Maxwell's equations, 185
    signal power, physical properties, 183, **184**
  for wireless communications, 178–179

Legendre functions, 105
lightning and transient protection
  communication and signal houses, 293–294
  copper lines, 293
  fire suppression, 293
  security phones, 294
  signs, railroad stations, 294
lightning rod ground EMC installation
  communications and station ground grid, 193
  electromotive force, determination, 194–195
  frequency of inverter, 198
  loop areas and impedance, 194, **194**
  loop noise voltages (mV), 197, **197**
  peak magnetic flux density, 195, **196**
  safety connection to cabinet, 192–193, *193*
  short-circuited version of loop, 197–198
  third rail (hot rail) current, 194
  traction motor current, 194, **195**
  wire resistance calculation, 192, **193**
lightning stroke, *82*
  AM radio, 81
  H (field), E (field) and V potential, 82–83
  protection devices, **83**
light rail transit vehicles
  antenna installation, radio mast case study *see* radio mast case study
  ground plane *see* ground planes, antenna farms
  in-vehicle-generated EMI, VHF radio reception, 212–216
  rooftop antenna application *see* rooftop antenna farm
line impedance stabilization network
  circuit design, 21, **23**
  instrumentation port and measurement device, 23–24
  power input and ground return, 21
  switching DC power supplies, 21
  120 V AC power, 24
LISN *see* line impedance stabilization network
LRT *see* light rail transit vehicles
LRVs, 315

magnetic H field shielding
  coaxial cable, 323
  gasket
    compression, 321, *321*
    pushbutton and lamp, 322–323
    types, compression, 322, **322**
  mechanical integrity enclosures, 320–321
  spring-loaded contacts, 321
Maxwell's equations, 8–9, 83, 102,
      110–112, 185, 194, 201, 318, 367
  differential form
    free space, 92
    with source charge, 91
  integral form
    free space, 92–93
    with source charge, 92
medical facilities, 9
method of moments, 179
MoM *see* method of moments

narrowband EMI, 308
National Electrical Code, 329
NEC *see* National Electrical Code

OATS *see* open air test systems; open
      area test site
OCC *see* operation control center
open air test systems, 12, 228
open area test site, 12, *15*, 17, 137, 228
  electromagnetic area, 228
  ground planes
    antenna selection, 19–20
    circular, 18
    conducted emission, 21–24
    EUT, 17–18
    measurement errors, 18–19
    pre-compliance testing, 20–21
    reflections, 18
    wooden flooring, 18
operation control center, 189
  AF track circuits *see* audio frequency
      track circuits
  automatic braking *see* automatic braking
  backup control room, 4
  elevators, 8
  primary and backup, 5, *5*
  project configuration, 2–3, *3*
  and RTUs *see* remote terminal units
  security, 8

temperature, 7
workstations, 3

PA systems *see* public address systems
patch antennas
  attenuation, signal level
    distance, **249**
    RADAR cross-section, **249**
  components, *251*
  construction details, 250
  data, rooftop antennae, 252, **252**
  low noise amplifiers, 248
  microstrip circular patch antenna, 250
  near and far field distances, 247, *248*
  physical material constants,
    substrates, 251
plane wave
  impedance and skin depth, 370–371
  ray dome, 372–373
PLC *see* programmable logic controller
power intensity
  ground plane, 374–375
  Hertzian antennas, 375
  metals, shielding properties, 375,
    376–377
  permittivity and permeability materials,
    375, **376**
  radiation resistance, 375
power line safety calculations
  duct banks, power transmission
    charge per linear foot, 296
    communication lines and signals, 296
    manhole layout, 295, *296*
    steel and fiberglass, 296
    voltage, surface of, 297
  parameters, 294–295
Poynting vector expression, 106, 107,
    189–190, 209, 222–223, 225, 228,
    233, 275–276, 366–367
pre-compliance testing, 20–21
professional engineer's sealed documents,
    332–333
programmable logic controller, 4, 118
public address systems
  AM detectors, 310
  band oscillations, audio amplifiers,
    309–310
  diodes, 310
  noise levels, 309

pulse width modulated drives, 80, 121, 188, 293, 341
PWM *see* pulse width modulated drives

radiation effects, electronic equipment, **60**
   conducted emissions, 59
   electric field, 59
radiation leakage
   communication cable coding, 190–191, **191**
   electromagnetic E fields, 191
   equipment, SCADA and signals, 190
   nine largest survey signals, 191, **192**
   SCADA clock, communication harmonics, 190–191, **191**
   test for radiation, 191
   2W two-way radios, 191–192
radio communications systems, 16
radio frequency interference, 2
radio mast case study
   antenna positions and wavelengths, relationship between, 204, **206**
   E field calculation, VHF array center, 208, 209
   VHF and UHF radio antennas
      antenna mast, *205, 207*
      construction, 204
      distances, **206**
      power, 209
      radiation pattern, *207, 208*
radio station, 165–166, 371–372
radio subsystem
   receivers, 313
   transmitter, 312
rails, shock hazard, 292–293
ray dome cover, 243–248, 252, *372,* 372–373
regulations
   Canadian standards, 14–15
   DOT *see* Department of Transportation
   European Union, 24–25, **26–56**
   United States FCC, 11–13
remote terminal units, 312
   definition, 4
   network functions, 5, **6–7**
   relay logic, 7–8
   server and switch, 4
RFI *see* radio frequency interference

rooftop antenna farm
   fade problem
      assumptions, 224
      calculations, 221–222
      diffraction loss, 224–225
      field electric field intensity, 225, *226*
      patch antennas, 227, *228*
      Poynting vector power, 228
      radiated power calculation, VHF, 222–223
      reflections off obstruction, 219, *220*
      transit systems, 226
      VHF 220 MHz radio, 220
   patch antennas *see* patch antennas
   with reflection
      assumptions, 240
      first scenario analysis, 245–247
      Friis equation, free space, 240
      ground plane size, 243
      installations, flat plate stainless steel, 240, *242*
      patch antennas, 240, 243
      Ray dome cover parameters, 245
      Ray dome details, 244, *244*
      RCS, obstructions, 240, **241**
RTUs *see* remote terminal units

SCADA system *see* supervisory control and data acquisition system
seawater surface
   relative permittivity and permeability, 362
   state trooper's radio antenna
      Fresnel zone, 365
      H field, 364–365
      160 MHz VHF radio, 363
      radiated power, 365
      reflection, 365
      transmission and reflection, 362, *363*
second order wave equations, 94
shielded twisted pair wiring
   common ground impedance, 77, *77*
   communication house
      Bessel functions, 75
      braided materials, 76
      carbon steel, 74
      circuit boards, 75
      far field, 75
      leaky enclosures, *74,* 74–75

shielded twisted pair wiring (*cont'd*)
  fiber optic circuits, 78
  isolation transformers, 78
  optical fiber, 78
  screen room, 73
  tin coating, 76
signal houses, 3–4, 8, 14, 25, 116–118, 165,
    175, 187, 293, 314, 344, 347
skin effect, wire
  copper wire, 84
  low current density J, 83
  plots and tables, 83
  power line current, 84–86, *85*
smoke detectors, 306, 311
SONET *see* synchronous optical
    networking
Stokes theorem, 93
strip line current density, 361–362, *362*
subway systems
  26 pair cable, rails
    operation control center, 189
    Poynting vector, 190
    pulse width modulated drives, 188
    steel conduit, shielding effect, 189
    subway cars, harmonic content, 189
    track and buried conduit layout,
      187, *188*
  types of equipment and bungalows/
      houses, **116**
    audio frequency track circuits, 117
    environmental factors, 116
    fire and intrusion alarm systems, 115
    interlockings, 117
    non-vital logic, 118
    PA and video equipment, 116
    radio communications, 115–116
    SCADA equipment, 115, 116
    signal house/bungalow, 116–117, **117**
    telephone and other functions, 118
subway tunnel, simulcast interference
  diffraction loss (knife edge) calculations
    Friis transmission evaluation,
      153–154, **154**
    knife edge loss calculations, 152, *153*
    radio distances from tunnel
      entrance, 152–153, **153**
    roadway tunnels, 155
    tunnel and entrance details,
      154–156, *155*

entrance of tunnel, survey of area
  calculations of shielding effects, 150
  signal reflection and knife edge
    effect, 147, *148*
  transmission obstructions, 150, *151*
  transmitter coverage and radio signal
    strength, 147, **148**
  transmitter coverage and UHF,
    147–149, **149**
  VHF and UHF worst-case power
    levels, 150, **151**
  VHF 160 MHz Exposition Blvd
    transmitter location, 147, **149**
leaky radiating cable attenuation, 151, **152**
  and signal power, 152, **152**
methodology, 146–147
shadow effect, subway cars, 163
subway radiating leaky cable and local
    radios, 145
tunnel and entrance layout, 145–146, *146*
wall reflections
  calculations, 157, 160
  electric and magnetic fields, 160
  leaky radiating cables, 165
  measurements, 164–165
  for 160 MHz, 156, *158*
  for 460 MHz, 156, *159*
  for 800 MHz, 156, *159*
  power reflections, 160, **161–162,**
    163, **164**
  reflection and transmission
    coefficients, 157, 160
  SCADA systems, 165
  tunnel entrances, 164, **164**
  for UHF 460/800 MHz radios at Blvd
    entrance, 156, **157**
  for VHF 160 MHz radios at Blvd
    entrance, 156, **156**
  worst-case reflections, tunnel
    entrances, 156, **158**
supervisory control and data acquisition
    system
  backup control room, 4
  monitoring, 7
  project configuration, OCC, 2–3
  radio signal strength, 7
  server and switch, RTUs, 4
  workstations, 3
susceptibility, 2, 16, 307

switch machines, 315
synchronous optical networking
  automatic switching, 5
  communications system, 5–7
  elevators, 8
  licensed emitters, 311
  non-licensed sources, 311
  primary OCC and backup OCC, 5
  radios, 311–312
  RTUs, 5, 7–8
  SCADA system, 7
  security, 8

TCRP *see* Transit Cooperative Research
    Program
telephone equipment, 313
TPS *see* Traction Power System
TPSS *see* Traction Power System
    Substation
track circuit relays, 314–315
track circuits and signals *see also* audio
    frequency track circuits
  ATC systems, 341, 342
  communications equipment, 358
  crossover (interlocking), 344
  Fortran 77 program, 358
  insulated joint, 342, *343*, 344
  loop calculations, hot rails
    circuit theory *see* circuit theory
    Faraday's law, 352
    impedance bonds, 353
    magnetic field density, 352
    peak voltage, 352–353
    subway cars, 353–354
    total rail resistance, 353
  signal house, 344
  step-down transformer, 342
  transmission line applications, 342, 358
tracks survey
  EMC track survey, 165–166
  equipment, set up, 166
  measurement methodology
    analyze radio communications, 170
    biconical antenna calibration
        sheet, 173, **174**
    correction factors, 173
    East horizontal polarization, 171, *172*
    East vertical polarization, 170–171,
        171, 172

equipment, set up, 169, *169*
frequency bands, 169
GPS, 169
log-periodic antenna calibration
    sheet, 173, **175**
and manufacturer's
    specifications, 173, 176
monopole antenna calibration
    sheet, 173, **176**
test across band (low accuracy),
    169–170, *170*
test cable loss *vs.* frequency, 173, **177**
West horizontal polarization, 171, *172*
West vertical polarization, 171, *171*,
    173, *173*
record sheet, 166, **167**
site data, 166
spread spectrum technology, 176
traction power system, 284, 290
traction power system substation, 16, 284,
    285, 307
transfer EMI, 308
Transit Cooperative Research
    Program, 284
Transportation Research Boards, 284
TRB *see* Transportation Research Boards
tunnel radiation, temporary antenna
    antennas, 143
  bidirectional communications, 145
  catwalk antenna radiation
    layout of antenna, 142, *142*
    maintenance personnel,
        overexposure, 144
    two-way radio operation, 144–145
  17 channel leaky coaxial cable
      installation, 145
  electric field and radiated power, 143
  EMC analysis, 144
  ERP, 143
  Friis transmission formula, 144
  measurement methodology, 168, *168*
  radio antennas, 143–144
  radio emanations, 144
  sample of test equipment, 168
  survey antenna position, 166–167, *168*
  tunnel construction, 142

United States FCC Regulations *see* FCC
    emission limit regulations

variable message board, 310
vehicle emissions
    audio frequency coded track circuits, 314
    loop detectors, 314
    propulsion to track circuit, 313–314
victim, 2
VMB *see* variable message board

wave equation solutions
    cylindrical coordinates
        angular components, 100
        Bessel differential equation, 101
        Bessel function equation, 101
        dell operator expression, 98, 99
        Henkel function equation, 101
        noise components, 102
        radial and angular components, 100
        scaler differential equation wave
            equations, 99

simple second order scalar differential
    equation, 101
spherical coordinate systems
    Brewster angle, 109
    critical phase angle, 112
    current, lossy strip, 108, *109*
    dell operator, 102
    EMC analysis, 106
    leakage capacitance, 107
    Legendre functions, 105
    reactive power, 106, 107
    scaler equations, all axes, 102, 104
    scaler solution, wave equation, 102, 103
    second order differential solutions,
        104, **104, 105**
    stray voltage, 108
    time-varying Poynting vector
        expression, 106, 107
workstation LCD monitors, 312